India's Journey Towards Sustainable Population

Bedprakas SyamRoy

India's Journey Towards Sustainable Population

 Springer

Bedprakas SyamRoy
Kolkata, West Bengal
India

ISBN 978-3-319-47493-9   ISBN 978-3-319-47494-6   (eBook)
DOI 10.1007/978-3-319-47494-6

Library of Congress Control Number: 2016953310

© Springer International Publishing Switzerland 2017
This work is subject to copyright. All rights are reserved by the Publisher, whether the whole or part of the material is concerned, specifically the rights of translation, reprinting, reuse of illustrations, recitation, broadcasting, reproduction on microfilms or in any other physical way, and transmission or information storage and retrieval, electronic adaptation, computer software, or by similar or dissimilar methodology now known or hereafter developed.
The use of general descriptive names, registered names, trademarks, service marks, etc. in this publication does not imply, even in the absence of a specific statement, that such names are exempt from the relevant protective laws and regulations and therefore free for general use.
The publisher, the authors and the editors are safe to assume that the advice and information in this book are believed to be true and accurate at the date of publication. Neither the publisher nor the authors or the editors give a warranty, express or implied, with respect to the material contained herein or for any errors or omissions that may have been made.

Cover illustration: Don Mammoser/Shutterstock.com

Printed on acid-free paper

This Springer imprint is published by Springer Nature
The registered company is Springer International Publishing AG
The registered company address is: Gewerbestrasse 11, 6330 Cham, Switzerland

*To my affectionate
Maniparna*

Preface

At 5 pm on 10 July 2015, the World Population Day, the population of India touched the magic figure of 12,7423,9769 constituting 17.25 % of the global population, as per the National Population Fund. With a rate of growth of 1.6 %, faster than China's rate of growth of population, India will have 1.63 billion people by 2050 and will surpass China. Population of India, as per the 2011 Census, is almost equal to the combined population of the USA, Indonesia, Brazil, Pakistan, Bangladesh and Japan. Unfortunately, for India, there is hardly any commensurate response to the enormity of this impending danger across the stakeholders. Population clock in India is, in fact, a mere national showpiece, a record keeper of number; it has not been of any use to generate or tick any alarming signal for the nation. This casual feeling has infected the administration of population control and family planning right from the policy-making areas down to its implementation at various layers of field functionaries. The absence of correct focus on an alarming situation has been the sad story of population control scenario in India. This is due to the fact that both the union government and the state governments have not been administering its constitutional mandate of population control and family planning, as enshrined at serial number 20A of the concurrent list of the Constitution of India. Instead, the country has taken up family welfare programme which lacks the robust vision of sustainable population in the country.

This laissez-faire approach in a runaway population scenario in our country has been compounded further by the demographic agenda of multiplying number by a particular religious group for improving its electoral voice through population ratio in the country. The porous international border together with a soft approach towards infiltration is responsible for continuous influx in the border states of India and in the consequent incremental population size. India has turned into a safe haven for illegal migrants. Since in our country vote bank politics is the order of the day, political parties of all descriptions are afraid to speak against illegal migrants. This is an irony of the role of political leaders in the population control programme in India.

The National Population Policy 2000 has since expired. The commitment and concern for the runaway population in India are still to be reflected in any policy of the day. The population control is still not an agenda of the work programme of the country; neither does it have any officially declared national norm of family size. It is hedged under the politically convenient concept of family welfare. Even the family welfare programme is now submerged under the National Health Mission. The big area of issue-based programme management is nowhere in sight and ironically, and the Department of Family Welfare has been wound up from the Ministry of Health and Family Welfare. The country has been left itself to an autogenerating process of demographic transition. The extent of carrying capacity of burgeoning population never bothers us.

India is an important signatory to the Declaration of Sustainable Development Goals (SDGs) at the Sustainable Development Summit 2015. The visionary and game-changing SDGs are now to be reached by the world in general and India in particular. The sustainable population is the prime requisite for achieving sustainable development and reaching SDGs.

This book intends to make a thorough investigation of the population problem issues in India from diverse angles- demographic, policy and programme etc. and attempts to capture the state of preparedness of our country to reach sustainable population.

In undertaking this study, I have made use of related publications of the Directorate of Census Operations, West Bengal and government of India. For the record, among others, I owe a great debt to the Websites of the Registrar General of Census, different ministries, government of India, Planning Commission, UN, UNDP, WHO, UNICEF, UNESCO, IIPS and also numerous others for making liberal use of relevant materials to prepare this book. I also express my thanks and gratitude to the library stuff of the Directorate of Census Operations, West Bengal, and also of the Bureau of Applied Economics and Statistics, government of West Bengal, for assistance. Moreover, I put on record the constant encouragement and support that I received from my daughter, Maniparna, to complete the study.

Finally, I deeply acknowledge the exemplary guidance and assistance that I received from Ms. Margaret Deignan, publishing editor, Springer Nature, and I am grateful to her and to the Springer Nature for agreeing to publish this book at the quickest time possible.

Salt lake Bedprakas SyamRoy
August 2016

About the Book

India faces the daunting challenge of alarming size of population. With 2.4 % geographical space, India houses 17.5 % of the global population and is soon to overtake China by 2030. The absence of correct focus on an alarming situation has been the sad story of population control scenario in India. Both the union government and the state governments have not been administering its constitutional mandate of 'population control and family planning', as enshrined at 20A of the concurrent list of the Constitution of India. Instead, the country took up family welfare programme which is now subsumed under the National Health Mission. The big area of issue-based programme management is nowhere in sight, and ironically, the Department of Family Welfare has been wound up. The country has been left itself to an autogenerating process of demographic transition. The extent of carrying capacity of burgeoning population never bothers us.

India is an important signatory to the Declaration of Sustainable Development Goals at the Sustainable Development Summit, 2015. The visionary and game-changing SDGs are now to be reached by the world in general and India in particular. The sustainable population is the prime requisite for achieving sustainable development and reaching SDGs.

This book intends to make a thorough investigation of the population problem issues in India from diverse angles- demographic, policy and programme, and attempts to capture the state of preparedness of our country to reach sustainable population.

Contents

Part I Theoretical Framework and Issues of Population Control and Family Planning

1	**Theoretical Framework on Studies of Population**............	3
2	**Issues Arising Out of Theories for Control of Population**........	9
	2.1 Population status..	9
	2.2 Food grains Production scenario........................	22
	2.3 Poverty status..	26
	2.4 Nutritional status......................................	34
	2.5 Unemployment scenario................................	41
	2.6 Human development profile............................	47
	2.7 Resources scenario....................................	52
	2.7.1 Forests Resources.............................	52
	2.7.2 Water Resources..............................	53
	2.8 Environmental and Climatic scenario....................	54
3	**Other Important Issues Relevant for Control of Population**......	57
	3.1 Demographic Issues....................................	58
	3.1.1 Population Projection..........................	59
	3.1.2 Crude Birth Rate..............................	64
	3.1.3 Total Fertility Rate (TFR)......................	65
	3.1.4 Other Instruments Available for Demographic Data..	66
	3.2 Social Issues..	68
	3.2.1 Age of Marriage..............................	68
	3.2.2 Dowry.......................................	70
	3.3 Cultural Issues..	71
	3.3.1 Girl Child....................................	71
	3.3.2 Sons Preference..............................	75
	3.4 Gender Issues...	75

		3.4.1	Male in Family Planning	76
		3.4.2	Women in Family Planning and Other Related Women Issues	79
	3.5	Socio-Religious Issues		82
		3.5.1	Minoritism	82
		3.5.2	Religious Issues	83
	3.6	Political Issues		89
	3.7	Contraceptive Issues		90
	3.8	Health Issues		99
		3.8.1	Pregnancy Care/Prenatal Care	99
		3.8.2	Post Partum Care/Post Natal Care	101
		3.8.3	Birth Spacing	102
		3.8.4	Breast Feeding	102
		3.8.5	Reproductive Tract Infection (RTI) & Sexually Transmitted Infection (STI)	103
		3.8.6	HIV and AIDS	104
		3.8.7	Adolescent Care	105
	3.9	Child Care Issues Including Child Nutrition		106
	3.10	Educational Issues		118
	3.11	Population Education		119
	3.12	Media Issues		121
	3.13	IEC Issues		122
	3.14	Legal Issues		131
		3.14.1	Constitutional Provisions	132
		3.14.2	Legal age of Marriage	132
		3.14.3	Registration of Marriage	133
		3.14.4	Dowry Prohibition Act, 1961	136
		3.14.5	The Preconception and Pre-natal Diagnostic Test Act	140
		3.14.6	The Medical Termination Of Pregnancy Act, 1971	146
		3.14.7	The Maharashtra Family Act, 1976	148
		3.14.8	Others	148
	3.15	Organisational Issues and Organisational Set-up		149
		3.15.1	Organisational Set-up	150
		3.15.2	Counselling Set-up and Counselling Issues	151
		3.15.3	Incentives Versus Disincentives	153
	3.16	Social Security Issues		154
4	**Issues Arising Out of the Health Policy and National Population Policy for Control of Population**			157
5	**Issues Emerging Out of the International Conference on Population and Development (ICPD) and Millennium Development Goals (MDGs)**			161

5.1	International Conference on Population and Development (ICPD)	161
	5.1.1 Programme of Actions	165
5.2	Millennium Development Goals (MDGs) and the Population Control Issues	167

6 Current Issues in the Post Census, 2011 171

Part II Requisites of the Population Control Programme

7 Conceptual Clarifications on Data Used in Population Control Area 177

8 Data Profiles Relevant for the Management of Population Control Mechanism 181
- 8.1 State Level Data 183
  - 8.1.1 Demographic Data 183
  - 8.1.2 Vital Statistics 184
  - 8.1.3 Family Planning/Welfare Data 184
- 8.2 District Level Data 184
  - 8.2.1 Demographic Data 184
  - 8.2.2 Vital Statistics 185
  - 8.2.3 Family Planning/Welfare Data 185
- 8.3 Sub-District Level Data 185
  - 8.3.1 Demographic Data 185
  - 8.3.2 Family Planning/Welfare Data 186

9 Family Welfare Structure in the Country and Issue-Based Management Support Structure 187
- 9.1 Apex Level 189
- 9.2 Directorate General Health Services 190

10 Decentralisation of Family Welfare Programmes in the Country 193

11 Financing the Population Control and Family Planning Programme in the Country 197

12 Monitoring Mechanism 209

Part III Sustainable Population in the Background of Sustainable Development Goals

13 Sustainable Population in the Background of Sustainable Development Goals 217
- 13.1 Background of Sustainable Development Goals—the End of Journey of Millennium Development Goals (MDGs) 223

13.2 Issues of Sustainable Population in the Background
of Sustainable Development Goals 243

Part IV Population Control Initiatives in India

14 Health Policy in India 247
   14.1 National Health Policy, 1983 248
   14.2 National Health Policy, 2002 250
       14.2.1 Objectives of the National Health Policy, 2002 250
   14.3 Draft National Health Policy 2015 252

15 Population Policy in India 255
   15.1 The National Population Policy, 2000 257
       15.1.1 Objectives of the Population Policy, 2000 259
       15.1.2 Legislation, Public Support and New Structures 268
   15.2 Outcome of the National Population Policy, 2000 271

16 Family Welfare Approaches in the Five Year Plans 273

17 Family Welfare Programmes in India 297

18 The Story of Achievements 313

**19 India's Perceived Challenges for Securing Sustainable Population
in the Context of Sustainable Development Goals** 329

Appendix A: FAQs on Sustainable Development Summit, 2015 339

**Appendix B: The Paris Agreement on Climatic Change Conference
at COP 21** 347

**Appendix C: China's Population Policy and Family Planning
Law—An Unofficial Version** 371

Appendix D: Quotations on Population Stabilisation 379

Index 381

About the Author

Dr. Bedprakas SyamRoy did his MA in economics in 1968 from the Calcutta University and Ph.D. from the North Bengal University in 1987. He started his career in the West Bengal Civil Service (1970) and was then selected in the IAS (1989). He is specialized in development administration, decentralized planning, and state planning. After retirement, he served as a member of the Third State Finance Commission, West Bengal. He authored nine books and several articles.

Acronyms

AIDS	Acquired immune deficiency syndrome or acquired immunodeficiency syndrome
AIR	All India Radio
ANC	Antinatal care
ANM	Auxiliary Nurse Midwifery
APY	Atal Pension Yojana
ARI	Acute Respiratory Infection
ASHA	Accredited Social Health Activists
ASRH	Adolescent Reproductive and Sexual Health Programme
AWC	Anganwadi Centre
AWW	Anganwadi Worker
BCC	Behaviour Change Communication
BMI	Body Mass Index
BPL	Below the Poverty Level
CBR	Crude Birth Rate
CDR	Crude Death Rate
CHCs	Community Health Centres
CMR	Child Mortality Rate
COCs	Combined oral contraceptive pills
CPR	Contraceptive Prevalence Rate
CRS	Civil Registration System
CSR	Child Sex Ratio
DHs	District Hospitals
EAG	Empowered Action Group
ECCR	Eligible Couple and Child Register
ECPR	Effective Couple Protection Rate
ECPs	Emergency Contraceptive Pills
EmOC	Emergency Obstetric Care
FP	Family Planning
GDP	Gross Domestic Product

GP	Gram Panchayat
HDI	Human Development Index
HDR	Human Development Report
HIV	Human Immunodeficiency Virus
ICDS	Integrated Child Development Services Scheme
ICPD	International Conference on Population and Development
IDGC	International Day of the Girl Child
IEC	Information, education and communication
IIPS	International Institute for Population Sciences
IMR	Infant Mortality Rate
IQ	Intelligent Quotient
ISM	Indian Systems of Medicine
ISMH	Indian System of Medicine & Homeopathy
IUCD	Intrauterine contraceptive devices
IUD	Intrauterine Methods
IVRS	Interactive Voice Response System
JSK	Jansankhya Sthirata Kosh
JSSK	Janani Shishu Suraksha Karyakarm
JSY	Janani Suraksha Yojana
KAP	Knowledge Aptitude and Practice
MCH	Mother and Child Health
MCTFC	Mother and Child Tracking Facilitation Centre
MCTS	Mother and Child Tracking System
MDGs	Millennium Development Goals
MIS	Management information systems
MMR	Maternal Mortality Rate
MMR	Maternal Mortality Ratio
MoHFW	Ministry of Health & Family Welfare
MTP	Medical Termination of Pregnancy
MUAC	Mid-upper arm circumference
NFBS	National Family Benefit Scheme
NFHS	National Family Health Survey
NGO	Non-governmental Organization
NHED	Nutrition, Health and Education
NHM	National Health Mission
NHP	National Health Policy
NIHFW	National Institute of Health and Family Welfare
NMBS	National Maternity Benefit Scheme
NMMUs	National Mobile Medical Units
NOAPS	National Old Age Pension Scheme
NPC	National Population Commission
NPP, 2000	National Population Policy, 2000
NRHM	National Rural Health Mission
NUHM	National Urban Health Mission
OCPs	Oral Contraceptive Pills

PCA	Primary Census Abstract
PCTE	Per capita Total Consumption Expenditure (PCTE)
PHCs	Primary Health Centres
PMJJBY	Pradhan Mantri Jeevan Jyoti Bima Yojana
PMSBY	Pradhan Mantri Suraksha Bima Yojana (PMSBY)
PNC	Postnatal care
PNDT Act	Preconception and Pre-natal Diagnostic Test Act
POPs	Progestin-only pills
PSE	(Non-formal)Pre-School Education
QA	Quality Assurance
RBSK	Rashtriya Bal Swasthya Karyakram
RCH	Reproductive and Child Health
RKS	Rogi Kalyan Samitis
RKSK	Rashtriya Kishor Swasthya Karyakram
RMNCH +A	Reproductive, Maternal, New born, Child and Adolescent Health Services
RTI	Reproductive Tract Infection
SAM	Severe acute malnutrition
SC	Scheduled Castes
SD	Standard Deviation
SDGs	Sustainable Development Goals
SDHs	Sub-District Hospitals
SMO	Social Marketing Organizations
SRH	Sexual and reproductive health
SRS	Sample Registration System
ST	Scheduled Tribes
STD	Sexually transmitted diseases (STD)
STI	Sexually Transmitted Infection
TFR	Total Fertility Rate
UHC	Universal Health Coverage
UIP	Universal Immunization Programme
UN	United Nations
UNCRC	United Nations and United Nations Convention on the Rights of the Child
UNDP	United Nations Development Programme
UNESCO	United Nations Educational Social Cultural Organization
UNICEF	United Nations International Children Emergency Fund
VAD	Vitamin A deficiency
VHSNC	Village Health Sanitation and Nutrition Committee
WHO	World Health Organization

Part I
Theoretical Framework and Issues of Population Control and Family Planning

Chapter 1
Theoretical Framework on Studies of Population

Abstract Population control and management is the most burning problem of the day and also a very challenging job in India. The programme managers and service providers of population control and management are tasked with high valued performance expectation to prevent runaway population growth in the states of India and restore population stabilisation in the country. Such functionaries are expected to be well conversant with the theoretical background of the areas of their functional domain. Generally speaking, population control and family planning does not belong to any particular of academic discipline. It happens to address a lot of cross cutting issues ranging from economics to medical science, from education to women empowerment, from child health to communication, from religious faith to legal provisions, and from human development to climatic change and sustainable development and so on. All these subject areas have their own big subject universe to significantly influence the reproductive behaviour of the citizens and impact the size of population. Since the knowledge of relevant subjects areas is crucial in the functioning areas of population control and family planning, it is only desirable that programme managers and service providers would need to scale up their skill and institutional capacity building to discharge their functional responsibilities. Economics happened to be the first discipline to take up seriously the problem of population in theoretical analysis. The broad outlines of theories of population, like Malthusian theory of population, the optimum theory of population and the theory of demographic transition have been touched, at the first instance, to familiarise with subject-environment before initiating programme intervention, its focused implementation and monitoring for the intended outcome.

Keywords Population control managers · Professional knowledge · Malthusian theory of population · The optimum theory of population · The theory of demographic transition

Population control and management is the most burning problem of the day and also a very challenging job in India. The programme managers and service providers of population control and management are tasked with high valued performance expectation to prevent runaway population growth in the states of India and restore population stabilisation in the country. Such functionaries are expected to be well conversant with the theoretical background of the areas of their functional domain. Generally speaking, population control and family planning does not belong to any particular of academic discipline. It happens to address a lot of cross cutting issues ranging from economics to medical science, from education to women empowerment, from child health to communication, from religious faith to legal provisions, and from human development to climatic change and sustainable development and so on. All these subject areas have their own big subject universe and some components of such big subjects do significantly influence the reproductive behaviour of the citizens and thereby impact the size of population in the given location. Since the knowledge of relevant subjects areas is crucial in the functioning areas of population control and family planning, it is only desirable that programme managers and service providers would need to equip themselves adequately to scale up their skill and institutional capacity building to discharge their functional responsibilities. Accordingly, an overview of all related subjects is taken up hereunder, in several chapters, to appreciate the theoretical background and the need to familiarise with subject-environment before initiating programme intervention, its focused implementation and monitoring for the intended outcome.

Economics happened to be the first discipline to take up seriously the problem of population in theoretical analysis. The broad outlines of such theories of population would be discussed hereunder. Historically, though Adam Smith and Benjamin Franklin happened to be pioneer in opening up population related issues at their time, it was Robert Tomas Malthus who, with his population doctrine, revolutionized the economic thinking by spelling out the linkage between population and food supply ever since the first publication of his Essay in 1798. After extensive tour and new findings, Malthus revised his first Essay in 1803 and finally summarized it in a short essay, A Summary View of the Principle of Population in 1830. The essence of Malthusian theory of population could be as follows:

Malthus predicted that food production would increase arithmetically (1, 2, 3, 4, 5, ...), but population would increase geometrically (1, 2, 4, 8, 16, ...). Eventually the demand for food by the increasing population would exceed the world's capacity to produce. The results would be war, famine, pestilence and strife.

Malthus was criticised mainly from two standpoints. The first was on the data used by him. The source of his data was based on America, a British colony and was provided by a close personal friend, Benjamin Franklin, prior to US independence. Population was growing rapidly there due to British migration. Food production did not need to grow as rapidly as the population since food was so abundant in the new world. As the population increased, more and more land was cultivated, and food production increased at the same rate as the population.

The second criticism centred on the assumption of constant production technology in agriculture. This was obvious. It was an era when the steam engine was the only source of power and its applicability was not developed for use in any form in agriculture. Production technology in agriculture has since come into being to improve increasing yields per acre.

The Optimum theory of population:

The optimum theory of population was propounded by Edwin Cannan in 1924 in his book Wealth and later popularised by Robbins, Dalton and Carr-Saunders. The proponents of this theory believes in the linkage between the size of population and the production of wealth as distinguished from Malthusian views between population and food supply.

The optimum population is the desired ideal number of the population that a country should have taking into consideration available resources at its command and also of means of production of the country. It will yield the maximum returns or income per head, considering its resources. Any deviation from this optimum-sized population will lead to a reduction in the per capita income. If the increase in population is followed by the increase in per capita income, the country is under-populated and it can afford to increase its population till it reaches the optimum level. On the contrary, if the increase in population leads to diminution in per capita income, the country is over-populated and needs a decline in population till the per capita income is maximised.

However, the optimum population level is not a fixed figure for any country for all time to come. Such a figure is variable with changes in the stock of natural resources and/or techniques of production. If there are improvements in the stock of natural resources or up-gradation in the methods and techniques of production, the output per head will rise and the optimum point will shift upward. Thus the optimum population is not anything static but a movable figure.

Let us now briefly touch upon the difference between the Malthusian Theory and the Optimum Theory:

(a) Malthusian theory dwelt on relationship between population growth and food production, whereas the optimum theory focuses on linkage between level of population and the resource base and its use as reflected in its per capita income level.
(b) Malthusian theory of population spoke of situation when a country exceeding a particular size of population would plunge in misery and distress. There is no rigidly fixed maximum population size in the optimum theory.
(c) For Malthus, famine, war and disease were the signs of over-population whereas the optimum theory maintains that even in the absence of such distressing phenomena, there can be over-population when its per capita income goes down.
(d) Malthus was haunted by the fear that population would outstrip food supply. The modern economists do not suffer from any such apprehensions.

Theory of Demographic transition:

The linkage between Population Growth and Economic Development is an eternal query for theoretical economists as it is linked with population stabilization and right kind of population size in a country. Admittedly, there is a mutual and close relation between the growth of population and the economic development of a country. One affects the other which, in turn, is affected by the other. This is brought out by the Theory of Demographic Transition:

(a) Effect of Growing Population on Economic Development:

Rapidly growing population impacts the economy in two ways: it stimulates economic growth on the one hand and affects it on the other. By enlarging effective demand, it releases economic force, motive and incentives to speed up initiatives and efforts to take up in a big way the challenges for economic development for the potential big market for goods and services and to make full use of the available resources by all possible ways. In the process, it accelerates economic growth.

On the other hand, hugely growing population appears to be a great obstacle for economic growth in a number of ways:

(i) The huge size of population very often compels such under developed country to import huge quantities of food grains on occasional crop failures and food shortages forcing diversion of essential foreign exchange for importing scarce capital goods so crucial (e.g., plant, machinery and equipment and accessories, and essential raw materials) for rapid economic growth.

(ii) Since the huge population contains a high percentage of unproductive population and a higher size of BPL, it forces the government to make provision for subsidy and other attendant social sector expenditures thereby compromising its own economic priority programme.

(iii) Similarly, a rapidly growing population aggravates the unemployment problems of all kinds—disguised or otherwise: under-employed or unemployed, both in rural and urban areas and forcing the government to divert country's resources to make provision for unemployment allowances and other related expenditures.

(iv) The huge size of population makes it difficult to improve quality of life and human development suffers.

(v) Finally, the most serious consequence of the fast growing population is that it affects the country's capacity to save and invest- a crucial factor for economic growth.

To sum up, a size of fast growing population acts as a drag on economic growth. Let us now see how economic growth impacts the population scenario:

(b) Effect of Economic Growth on Population Growth:

An underdeveloped country passes through a phase of high birth rates and high death rates. The birth rates are high because of the existence of universal and early marriages. The death rates are very high on ground of poor diet, bad sanitary conditions and absence of adequate preventive and curative medical facilities. The sociological features of such underdeveloped countries are characterised by high birth rates, high death rate, high IMR, inflexible social beliefs, rigid customs and non-negotiable religious attitudes and so on. The economic compulsion for supplementing family income also forces children to join in labour market and indirectly promotes for larger number of children in the family.

With the launching of economic development, the socio-economic features of the country undergoes gradual changes. Economic development brings in its trail improved agricultural practices, ensures increased food production and its availability. As a result per capita food intake increases and nutritional standard also improves. Along with the same, the medical facilities also go up and ensures better standard of living. All these improvements bring down the death rate. Initially, the birth rates also continue to rule high. As a result the population growth becomes still more rapid and the country faces a situation of 'population explosion'. However, with continued economic development, the birth rates of the country start falling. People realise the benefits of small families and the large number of children are no longer regarded as asset. The low IMR and small families become the accepted pattern of society. This, in short, is the Theory of Demographic Transition.

The demographic transition theory is a model that tries to explain the stages of population growth of a nation over a period of time. Economists could locate four stages in which such demographic transition usually takes place. The Stage one of the demographic transition model is associated with pre-industrial society where birth rate is high and death rate is also high resulting in the process a population of a relatively stable size and a slow growth rate. This stage was observable for most of human history. The Stage two of the demographic transition model takes place in a developing nation. In this second stage, although the birth rate remains high, the death rate drops, particularly of the category of infant mortality. As a result, a rapid population growth takes place there. This imbalance in the population size may however, be temporary. The Stage three of demographic transition can be seen in any newly developed nation when birth rates drop further while death rates remain constant. Rearing of children becomes more expensive and women also choose to have fewer children. The fourth stage of stable population occurs when both the birth rates and the death rates are low. Death rates are low because of improved quality of life in the form of improved medical facilities, required food and nutrition coverage. The birth rate is low because people find values of small family and have access to all kinds of contraception, and women are, by and large, gainfully employed or have other work opportunities for economic empowerment.

Chapter 2
Issues Arising Out of Theories for Control of Population

Abstract The theories of population threw up several issues governing population stabilisation of a country. Such issues look upon the population problem from its own domain perspective. Individually, such issues are important; collectively they explain genesis of the whole problem. The critical issues that have significant-bearing on sustainable population are the trend and the status of Population, Food grains Production scenario, Poverty status, Nutritional status, Unemployment scenario, Human Development Index, Resources scenario, Environmental and Climatic scenario. The trend and the current situation of such issues have been examined in the context of Indian situation and assessed the prospect of carrying further load of population.

Keywords Status of Population · Food grains production scenario · Poverty status · Nutritional status · Unemployment scenario · Human development profile · Resources scenario · Environmental and climatic scenario

The theories of population threw up several issues governing population stabilisation of a country. Such issues look upon the population problem from its own domain perspective. Individually, such issues are important; collectively they explain genesis of the whole problem. A select list of such issues are taken up here under in the context of our country.

2.1 Population status

Population is the most important asset of any country. Population is responsible in building up the economic edifice, cultural foundation and the standard of civilization of a country. On it also depends the quality of life of its citizens. However, the size of population is also the single most problem for any country when such size surpasses the optimum population and put a drag on economic prosperity and quality of life. With 2.4 % of the land territory of the world, India covers more than

Table 2.1 Decadal growth of the size of population of India from 1951–2011

Census Year	Size of population	Male population	Female population
1951	361,088,090	185,528,462	175,559,628
1961	439,234,771	226,293,201	212,941,570
1971	548,159,652	284,049,276	264,110,376
1981	683,329,097	353,374,460	329,954,637
1991	846,421,039	439,358,440	407,062,599
2001	1,020,193,422	531,277,078	495,738,169
2011	1,210,569,573	623,121,843	587,447,730

Sources Census publications of different years including PCA 2011

Table 2.2 Decadal incremental size of the population of India

Census year	Size of population	Incremental population	Decadal growth	Sex ratio
1951	361,088,090	–	13.31	946
1961	439,234,771	78,146,680	21.64	941
1971	548,159,652	108,924,881	24.80	930
1981	683,329,097	135,169,445	24.66	934
1991	846,421,039	163,091,942	23.86	927
2001	1,020,193,422	173,772,383	21.54	933
2011	1,210,569,573	190,376,151	17.7	943

Sources Census publications of different years including PCA 2011

Table 2.3 Growth of absolute number of population of India

Census Year	POPULATION OF INDIA	Times of population growth of India over the 1951 census	Incremental population over last census	Density of population
1951	361,088,090	–	–	117
1961	439,234,771	1.21	78,146,680	142
1971	548,159,652	1.51	108,924,881	177
1981	683,329,097	1.89	135,169,445	216
1991	846,421,039	2.34	163,091,942	267
2001	1,020,193,422	2.82	173,772,383	325
2011	1,210,569,573	3.35	190,376,151	382

Sources Census publications of different years including PCA 2011, India

17.5 % of the population of the world as per census 2011. The population size of India is unusually very big and the trend of growth of population is also very alarming. In a number of Tables from 2.1, 2.2, 2.3, 2.4, 2.5, 2.6, 2.7 and 2.8, the size of population, its decadal increment, decadal growth and density have been captured from censuses in India for the states of this country which are self-explanatory and indicate the critical population scenario of the country.

2.1 Population status

Table 2.4 Population size of the states of India

States	1951	1961	1971	1981	1991	2001	2011
Andhra Pradesh	31,115,259	35,983,447	43,502,708	53,551,026	66,508,008	76,210,007	84,580,777
Arunachal Pradesh	NA	336,558	467,511	631,839	864,558	1,097,968	1,383,727
Assam	8,028,858	10,837,329	14,625,152	18,041,248	22,414,322	26,655,528	31,205,576
Bihar	38,782,271	46,447,457	56,353,369	69,914,734	64,530,554	82,998,509	104,099,452
Chhattisgarh	–	–	–	–	17,614,928	20,833,803	25,545,198
NCT of Delhi#	1,744,072	2,658,612	4,065,698	6,220,406	9,420,644	13,850,507	16,787,941
Goa	596,059	626,667	857,771	1,086,730	1,169,793	1,347,668	1,458,545
Gujarat	16,262,657	20,633,350	26,697,475	34,085,799	41,309,582	50,671,017	60,439,692
Haryana	5,673,597	7,590,524	10,036,431	12,922,119	16,463,648	21,144,564	25,351,462
Himachal Pradesh	2,385,981	2,812,463	3,460,434	4,280,818	5,170,877	6,077,900	6,864,602
Jammu & Kashmir	3,253,852	3,560,976	21,843,911	4,616,632	5,987,389	10,143,700	12,541,302
Jharkhand	–	–	–	–	21,843,911	26,945,829	32,988,134
Karnataka	19,401,956	23,586,772	29,299,014	37,135,714	44,977,201	52,850,562	61,095,297
Kerala	13,549,118	16,903,715	21,347,375	25,453,680	29,098,518	31,841,374	33,406,061
Madhya Pradesh	26,071,637	32,372,408	41,654,119	48,566,242	52,178,844	60,348,023	72,626,809
Maharashtra	32,002,564	39,553,718	50,412,235	62,782,818	78,937,187	96,878,627	112,374,333
Manipur	577,635	780,037	1,072,753	1,420,953	1,837,149	2,166,788	2,570,390
Meghalaya	605,674	769,380	1,011,699	1,335,819	1,774,778	2,318,822	2,966,889
Mizoram	196,202	266,063	332,390	493,757	689,756	888,573	1,097,206
Nagaland	212,975	369,200	516,449	774,930	1,209,546	1,990,036	1,978,502
Odisha	14,645,946	17,548,846	21,944,615	26,370,274	31,659,736	36,804,660	41,974,218
Pondicherry#	317,253	369,079	471,707	604,471	807,785	974,345	1,247,953
Punjab	9,160,500	11,135,069	13,551,060	16,788,915	20,281,969	24,358,999	27,743,338
Rajasthan	15,970,774	20,155,602	25,765,806	34,261,862	44,005,990	56,507,188	68,548,437

(continued)

Table 2.4 (continued)

States	1951	1961	1971	1981	1991	2001	2011
Sikkim	137,725	162,189	209,843	316,385	406,457	540,851	610,577
Tamil Nadu	30,119,047	33,686,953	41,199,168	48,408,077	55,858,946	62,405,679	72,147,030
Tripura	639,029	1,142,005	1,556,342	2,053,058	2,757,205	3,199,203	3,673,917
Uttar Pradesh	63,219,672	73,754,573	88,341,521	110,862,512	132,067,653	166,197,921	199,812,341
Uttarakhand	–	–	–	–	7,050,634	8,489,349	10,086,292
West Bengal	26,299,980	34,926,279	44,312,011	54,580,647	68,077,965	80,176,197	91,276,115
Chadigarh#	24,261	119,881	257,251	451,610	642,015	900,635	1,055,450
Daman&Diu#					138,477	158,204	243,247
D&N Haveli#	41,532	57,963	74,170	103,676	138,477	220,490	343,709
Lakshadweep#	21,035	24,108	31,810	40,249	51,707	60,650	64,473
A&N Islands#	30,971	63,548	115,133	188,741	280,661	356,152	380,581
All India	361,088,090	439,234,771	548,159,652	6,833,290,971	846,421,039	1,028,610,328	1,210,569,573

Sources Census publications and PCA, 2011; # stands for Union Territories

2.1 Population status

Table 2.5 Decadal growth of population in the states of India

States	1951	1961	1971	1981	1991	2001	2011
Andhra Pradesh	14.02	15.65	20.90	23.10	24.20	14.59	11.0
Arunachal Pradesh	NA	NA	38.91	35.15	36.83	27.06	26.0
Assam	19.93	34.98	34.95	23.36	24.24	18.92	17.1
Bihar	10.58	19.79	20.91	23.38	24.16	28.62	25.4
Chhattisgarh	9.42	22.77	27.12	20.39	25.73	18.27	22.6
NCT of Delh#	90.00	52.44	52.93	53.00	51.45	47.02	21.2
Goa	1.21	7.77	34.77	26.75	16.08	15.21	8.2
Gujarat	18.69	26.88	29.39	27.67	21.19	22.66	19.3
Haryana	7.60	33.79	32.22	28.75	27.41	28.43	19.9
Himachal Pradesh	5.42	17.87	23.04	23.71	20.79	17.54	12.9
Jammu & Kashmir	10.42	9.44	29.65	29.69	30.34	29.43	23.6
Jharkhand	9.35	19.69	22.58	23.79	24.03	23.36	22.4
Karnataka	19.36	21.57	24.22	26.75	21.12	17.51	15.6
Kerala	22.82	24.76	26.29	19.24	14.32	9.43	4.9
Madhya Pradesh	8.38	24.73	29.28	27.16	27.24	24.26	20.3
Maharashtra	19.27	23.60	27.45	24.54	25.73	22.73	16.0
Manipur	12.80	35.04	37.53	32.46	24.29	24.86	18.6
Meghalaya	8.97	27.03	31.50	32.04	32.86	30.65	27.9
Mizoram	28.42	35.61	24.93	48.55	39.70	28.82	23.5
Nagaland	12.30	73.35	39.98	50.05	56.08	64.53	−0.6
Odisha	6.38	19.82	25.05	20.17	20.06	16.25	14.0
Pondicherry#	11.31	16.34	27.81	28.15	33.64	20.62	28.1
Punjab	−4.58	21.56	21.70	23.89	20.81	20.10	13.9
Rajasthan	15.20	26.20	27.83	33.97	28.44	28.41	21.4
Sikkim	13.34	17.76	29.38	50.77	28.47	33.06	12.9
Tamil Nadu	14.66	11.85	22.30	17.50	15.39	11.72	15.6
Tripura	24.56	27.03	31.50	32.04	32.86	16.03	14.8
Uttar Pradesh	11.78	16.38	19.54	25.39	25.55	25.85	20.2
Uttarakhand	12.67	22.57	24.42	27.45	24.23	20.41	18.8
West Bengal	13.22	32.80	26.87	23.17	24.73	17.77	13.8
Chadigarh#	7,47	394.13	114.59	75.55	42.16	40.28	17.2
Daman&Diu#	13.55	−24.56	70.85	26.07	28.62	55.73	53.8
D&N Haveli#	2.70	39.56	27.45	24.54	25.73	59.22	55.9
Lakshadweep#	14.60	14.61	31.95	26.53	28.47	17.30	6.3
A&N Islands#	−8.28	105.19	81.17	63.93	48.70	26.90	6.9
All India	13.31	21.64	24.80	24.66	23.86	21.54	17.7

Sources Census of India, paper 1 of 2011 and PCA, 2011; # stands for Union Territories

Table 2.6 Decadal rate of growth of some of the selected countries of the world

Countries	Reference date	Population in millions	Decadal change
China	01.11.2010	1341.0	5.43
India	01.03.2011	1210.5	17.64
USA	01.04.2010	308.7	7.26
Indonesia	31.05.2010	237.6	15.05
Brazil	01.08.2010	190.7	9.39
Pakistan	01.07.2010	184.8	24.78
Bangladesh	01.07.2010	164.4	16.76
Nigeria	01.07.2010	158.3	26.84
Russian Federation	01.07.2010	140.4	−4.29
Japan	01.10.2010	128.1	1.1
Other countries	01.07.2010	2844.7	15.43
World	01.07.2010	6908.7	12.97

Source Census of India, paper 1 of 2011

Table 2.7 Density of population in the states of India

States	1951	1961	1971	1981	1991	2001	2011
Andhra Pradesh	112	131	158	195	242	277	308
Arunachal Pradesh	–	4	6	8	10	13	17
Assam	102	138	186	230	286	340	398
Bihar	223	267	324	402	685	881	1106
Chhattisgarh	–	–	–	–	130	154	189
NCT of Delhi#	1176	1793	2742	4194	6352	9340	11,320
Goa	148	159	215	272	316	364	394
Gujarat	83	105	136	174	211	258	308
Haryana	128	172	227	292	327	478	573
Himachal Pradesh	43	51	62	77	93	109	123
Jammu & Kashmir	NA	NA	NA	59	77	100	124
Jharkhand	–	–	–	–	#276	338	414
Karnataka	101	123	153	194	235	276	319
Kerala	349	435	549	655	749	820	860
Madhya Pradesh	59	73	94	118	158	196	236
Maharashtra	104	129	164	204	257	315	365
Manipur	26	35	48	64	82	97	115
Meghalaya	27	34	45	60	79	103	132
Mizoram	9	13	16	23	33	42	52
Nagaland	13	22	31	47	73	120	119
Odisha	94	113	141	169	203	236	270
Pondicherry#	645	750	959	1229	1683	1989	2547
Punjab	182	221	269	333	403	484	551

(continued)

2.1 Population status

Table 2.7 (continued)

States	1951	1961	1971	1981	1991	2001	2011
Rajasthan	47	59	75	100	129	165	200
Sikkim	19	23	30	45	57	76	86
Tamil Nadu	232	259	317	372	429	480	555
Tripura	61	109	148	196	263	305	350
Uttar Pradesh	215	251	300	377	548	690	829
Uttarakhand	–	–	–	–	133	159	189
West Bengal	296	394	499	615	761	903	1028
Chadigarh#	213	1052	2257	3961	5632	7900	9258
Daman&Diu#	434	327	559	705	907	1425	2191
D&N Haveli#	85	118	151	221	282	449	700
Lakshadweep#	657	753	994	1258	1616	2022	2149
A&N Islands#	4	8	14	23	34	43	46
All India	117	142	177	216	267	325	382

Sources Census publications and PCA, 2011; # stands for Union Territories

Table 2.8 Relative density of population of some other populated countries of the world

Country	Population (in thousand)	Density of population
China	1,354,146	141
India	1,210,569	382
USA[a]	308	33
Indonesia	232,517	122
Brazil	195,425	23

[a]USA Census 2010
Source Census publication, 2011, India

The religion-wise growth of population of India is an issue of great demographic importance and is also of serious concern from population stabilisation point of view. While the Census of India shows the religious group under the Hindus, the Muslim, the Christain, the Sikh, the Buddhists, the Jains and others, the data analysis hereunder centers around on two principal religious groups, the Hindus and the Muslims, who share the major burden of population size in the country. The self-introductory Tables from 2.9, 2.10, 2.11, 2.12 and 2.13 reveal the relative contribution of the Hindus and the Muslims to the alarming population size and growth in the states of India.

To sum up, the total population of India at 0.00 hour of Ist March 2011 was 1210.6 million. Of this, the total rural population was 833.5 million and the urban population 377.1 million. In absolute numbers, out of the total increase of 182 million added to the last decade, the contribution of rural and urban areas is equal to 91.0 million each. Uttar Pradesh has the largest rural population of 155.3 million (18.6 % of the country's rural population) whereas Maharashtra has the highest urban population of 50.8 million (13.5 % of country's urban population) in the country.

Table 2.9 The muslim population in India and its growth rate

Census/Year	Total population	Muslim population	Proportion of muslim population to total population (%)	Incremental decadal size of muslim	Decadal growth rate of muslim population	Decadal growth rate of India
1951	361,088,090	35,856,047	9.93	NA	−16.5	13.31
1961	439,234,771	46,998,120	10.70	11,142,073	31.07	21.64
1971	548,159,652	61,448,696	11.21	14,459,576	30.74	24.8
1981	683,329,097	77,557,852	11.35	16,109,156	26.21	24.66
1991	846,421,039	102,586,957	12.12	25,029,105	32.27	23.86
2001	1,020,193,422	138,159,437	13.43	35,572,480	34.68	21.54
2011	1,210,569,573	172,245,158	14.88	34,085,721	24.67	17.7

Source Census of India in 2001 and 2011

Table 2.10 Decadal growth of the hindus in the states of India in 2011

States	Total population in 2011	Hindu population in 2011	Hindu population in 2001	Decadal growth of pop of the states in 2011	Decadal growth rate of hindus in 2011	Decadal growth rate of muslims in 2011
Andhra Pradesh	84,580,777	74,824,149	67,836,651	11.0	10.30	18.98
Arunachal Pradesh	1,383,727	401,876	379,935	26.0	5.77	30.81
Assam	31,205,576	19,180,759	17,296,455	17.1	10.89	29.59
Bihar	104,099,452	86,078,686	69,076,919	25.4	24.61	27.95
Chhattisgarh	25,545,198	23,819,789	707,978	22.6	20.73	25.73
NCT of Delhi[a]	16,787,941	13,712,106	11,358,049	21.2	20.72	32.96
Goa	1,458,545	963,877	886,551	8.2	8.72	31.83
Gujarat	60,439,692	53,533,988	45,143,074	19.3	18.58	27.30
Haryana	25,351,462	22,171,128	18,655,925	19.9	18.84	45.66
Himachal Pradesh	6,864,602	6,532,765	5,800,222	12.9	12.62	25.41
Jammu & Kashmir	12,541,302	3,566,674	3,005,349	23.6	18.67	26.12
Jharkhand	32,988,134	22,376,051	18,475,681	22.4	21.11	28.48
Karnataka	61,095,297	51,317,472	44,321,279	15.6	15.78	22.12
Kerala	33,406,061	18,282,492	17,883,449	4.9	2.23	12.84
Madhya Pradesh	72,626,809	66,007,121	55,004,675	20.3	20.00	24.29
Maharashtra	112,374,333	89,703,057	77,859,385	16.0	15.21	26.30
Manipur	2,570,390	1,181,876	996,894	18.6	18.55	25.61

(continued)

2.1 Population status

Table 2.10 (continued)

States	Total population in 2011	Hindu population in 2011	Hindu population in 2001	Decadal growth of pop of the states in 2011	Decadal growth rate of hindus in 2011	Decadal growth rate of muslims in 2011
Meghalaya	2,966,889	342,078	307,822	27.9	11.12	31.49
Mizoram	1,097,206	30,136	31,562	23.5	−4.51	46.87
Nagaland	1,978,502	173,054	153,162	−0.6	12.98	39.87
Odisha	41,974,218	39,300,341	34,726,129	14.0	13.17	19.64
Pondicherry[a]	1,247,953	1,089,409	845,449	28.1	28.85	27.29
Punjab	27,743,338	10,678,138	8,997,942	13.9	18.67	40.16
Rajasthan	68,548,437	60,657,103	50,151,452	21.4	20.94	29.81
Sikkim	610,577	352,662	329,548	12.9	7.01	28.26
Tamil Nadu	72,147,030	63,188,168	54,985,079	15.6	14.91	21.86
Tripura	3,673,917	3,063,903	2,739,310	14.8	11.84	24.21
Uttar Pradesh	199,812,341	159,312,654	133,979,263	20.2	18.90	25.19
Uttarakhand	10,086,292	8,368,636	7,212,260	18.8	16.03	38.99
West Bengal	91,276,115	64,385,546	58,104,835	13.8	10.80	21.81
Chadigarh[a]	1,055,450	852,574	707,978	17.2	20.42	44.73
Daman&Diu[a]	243,247	220,150	141,901	53.8	55.14	56.97
D&N Haveli[a]	343,709	322,857	206,203	55.9	56.57	98.07
Lakshadweep[a]	64,473	1788	2221	6.3	−19.49	7.54
A&N Islands[a]	380,581	264,296	246,589	6.9	7.18	9.83
All India	1,210,569,573	966,257,353	827,578,868	17.7	16.75	24.65

Source Census of India in 2001 and 2011
[a]Union territories

Table 2.11 Decadal growth of the hindus, muslims and the states of India in 2011

States	Total population in 2011	Hindu population in 2011	Muslim population in 2011	Decadal growth of pop of the state/India in 2011	Decadal growth rate of hindus in 2011	Decadal growth rate of muslims in 2011
Andhra Pradesh	84,580,777	74,824,149	8,082,412	11.0	10.30	18.98
Arunachal Pradesh	1,383,727	401,876	27,045	26.0	5.77	30.81
Assam	31,205,576	19,180,759	10,679,345	17.1	10.89	29.59
Bihar	104,099,452	86,078,686	17,557,809	25.4	24.61	27.95
Chhattisgarh	25,545,198	23,819,789	514,998	22.6	20.73	25.73
NCT of Delhi[a]	16,787,941	13,712,106	2,158,684	21.2	20.72	32.96
Goa	1,458,545	963,877	121,564	8.2	8.72	31.83

(continued)

Table 2.11 (continued)

States	Total population in 2011	Hindu population in 2011	Muslim population in 2011	Decadal growth of pop of the state/India in 2011	Decadal growth rate of hindus in 2011	Decadal growth rate of muslims in 2011
Gujarat	60,439,692	53,533,988	5,846,761	19.3	18.58	27.30
Haryana	25,351,462	22,171,128	1,781,342	19.9	18.84	45.66
Himachal Pradesh	6,864,602	6,532,765	149,881	12.9	12.62	25.41
Jammu & Kashmir	12,541,302	3,566,674	8,567,485	23.6	18.67	26.12
Jharkhand	32,988,134	22,376,051	4,793,994	22.4	21.11	28.48
Karnataka	61,095,297	51,317,472	7,893,065	15.6	15.78	22.12
Kerala	33,406,061	18,282,492	8,873,472	4.9	2.23	12.84
Madhya Pradesh	72,626,809	66,007,121	4,774,695	20.3	20.00	24.29
Maharashtra	112,374,333	89,703,057	12,971,152	16.0	15.21	26.30
Manipur	2,570,390	1,181,876	239,836	18.6	18.55	25.61
Meghalaya	2,966,889	342,078	130,399	27.9	11.12	31.49
Mizoram	1,097,206	30,136	14,832	23.5	−4.51	46.87
Nagaland	1,978,502	173,054	48,963	−0.6	12.98	39.87
Odisha	41,974,218	39,300,341	911,670	14.0	13.17	19.64
Pondicherry[a]	1,247,953	1,089,409	75,556	28.1	28.85	27.29
Punjab	27,743,338	10,678,138	535,489	13.9	18.67	40.16
Rajasthan	68,548,437	60,657,103	6,215,377	21.4	20.94	29.81
Sikkim	610,577	352,662	9867	12.9	7.01	28.26
Tamil Nadu	72,147,030	63,188,168	316,042	15.6	14.91	21.86
Tripura	3,673,917	3,063,903	316,042	14.8	11.84	24.21
Uttar Pradesh	199,812,341	159,312,654	38,483,967	20.2	18.90	25.19
Uttarakhand	10,086,292	8,368,636	1,406,825	18.8	16.03	38.99
West Bengal	91,276,115	64,385,546	24,654,825	13.8	10.80	21.81
Chandigarh[a]	1,055,450	852,574	51,447	17.2	20.42	44.73
Daman&Diu[a]	243,247	220,150	19,277	53.8	55.14	56.97
D&N Haveli[a]	343,709	322,857	12,922	55.9	56.57	98.07
Lakshadweep[a]	64,473	1788	62,268	6.3	−19.49	7.54
A&N Islands[a]	380,581	264,296	32,143	6.9	7.18	9.83
All India	1,210,569,573	966,257,353	172,245,158	17.7	16.75	24.65

Source Census of India in 2001 and 2011
[a]Union territories

2.1 Population status

Table 2.12 Trend of proportion of muslim population in the states of India 2001 and 2011

States	Total population in 2011	Muslim population in 2011	Muslim population in 2001	Incremental Muslim population in the decade, 2001–2011	Decadal growth of muslims in 2011	Proportion of muslims to total population in 2001	Proportion of muslims to Total population in 2011
Andhra Pradesh	84,580,777	8,082,412	6,793,240	1,289,172	18.98	9.2	9.56
Arunachal Pradesh	1,383,727	27,045	20,675	6370	30.81	1.9	1.95
Assam	31,205,576	10,679,345	8,240,611	2,438,734	29.59	30.9	34.22
Bihar	104,099,452	17,557,809	13,722,048	3,835,761	27.95	16.5	16.87
Chhattisgarh	25,545,198	514,998	409,615	105,383	25.73	2	2.02
NCT of Delhi[a]	16,787,941	2,158,684	1,623,520	535,164	32.96	11.7	12.86
Goa	1,458,545	121,564	92,210	29,354	31.83	6.8	8.33
Gujarat	60,439,692	5,846,761	4,592,854	1,253,907	27.3	9.1	9.67
Haryana	25,351,462	1,781,342	1,222,916	558,426	45.66	5.8	7.03
Himachal Pradesh	6,864,602	149,881	119,512	30,369	25.41	2	2.18
Jammu & Kashmir	12,541,302	8,567,485	6,793,240	1,774,245	26.12	67	68.31
Jharkhand	32,988,134	4,793,994	3,731,308	1,062,686	28.48	13.8	14.53
Karnataka	61,095,297	7,893,065	6,463,127	1,429,938	22.12	12.2	12.92
Kerala	33,406,061	8,873,472	7,863,842	1,009,630	12.84	24.7	26.56
Madhya Pradesh	72,626,809	4,774,695	3,841,449	933,246	24.29	6.4	6.57
Maharashtra	112,374,333	12,971,152	10,270,485	2,700,667	26.3	10.6	11.54
Manipur	2,570,390	239,836	190,939	48,897	25.61	8.8	9.33

(continued)

Table 2.12 (continued)

States	Total population in 2011	Muslim population in 2011	Muslim population in 2001	Incremental Muslim population in the decade, 2001–2011	Decadal growth of muslims in 2011	Proportion of muslims to total population in 2001	Proportion of muslims to Total population in 2011
Meghalaya	2,966,889	130,399	99,169	31,230	31.49	4.3	4.40
Mizoram	1,097,206	14,832	10,099	4733	46.87	1.1	1.35
Nagaland	1,978,502	48,963	35,005	13,858	39.87	1.8	2.47
Odisha	41,974,218	911,670	761,985	149,685	19.64	2.1	2.17
Pondicherry[a]	1,247,953	75,556	59,358	16,198	27.29	6.1	6.05
Punjab	27,743,338	535,489	382,045	153,444	40.16	1.6	1.93
Rajasthan	68,548,437	6,215,377	4,788,227	1,427,150	29.81	8.5	9.07
Sikkim	610,577	9867	7693	2174	28.26	1.4	1.62
Tamil Nadu	72,147,030	4,229,479	3,470,647	758,832	21.86	5.6	5.86
Tripura	3,673,917	316,042	254,442	61,600	24.21	8	8.60
Uttar Pradesh	199,812,341	38,483,967	30,740,158	7,743,809	25.19	18.5	19.26
Uttarakhand	10,086,292	1,406,825	1,012,141	394,684	38.99	11.9	13.95
West Bengal	91,276,115	24,654,825	20,240,543	4,414,282	21.81	25.2	27.01
Chadigarh[a]	1,055,450	51,447	35,548	15,899	44.73	3.9	4.87
Daman&Diu[a]	243,247	19,277	12,281	6996	56.97	7.8	7.92
D&N Haveli[a]	343,709	12,922	6524	6397	98.07	3	3.76
Lakshadweep[a]	64,473	62,268	57,903	4365	7.54	95	96.58
A&N Islands[a]	380,581	32,143	29,265	2878	9.83	8.2	8.45
All India	1,210,569,573	172,245,158	138,188,240	34,085,721	24.65	13.4	14.23

[a]Union territories

2.1 Population status

Table 2.13 Decadal growth rates of hindus and muslims since 1951

Census year	Hindu population in million	Decadal growth of Hindus	Muslim population in million	Decadal growth of Muslims	Muslim rate of growth more than Hindu rate of growth
1951	303.5	27.36	35.4	−16.5	–
1961	366.5	20.75	46.9	32.48	11.73
1971	453.3	23.68	61.4	30.92	7.24
1981	562.4	24.07	80.3	30.78	6.71
1991	690.1	22.71	106.7	32.87	10.16
2001	827.6	19.92	138.2	29.52	9.60
2011	966.2	16.75	172.2	24.60	7.85

Source IIPSIndia

The growth rate of population in India during 2001–2011 was 17.7 % (Rural-12.3 %, Urban-31.8 %). Meghalaya has recorded the highest growth rate in rural population and Daman&Diu (218.8) the highest decadal growth rate in urban population.

The population density in Census 2011 works out to be 382 showing an increase of 57 points from 2001. Delhi (11320) turns out to be the most densely inhabited followed by Chandigarh (9258) in all States/UTs, both in Census 2001 and 2011. Among the major States, Bihar occupies the first position with a density of 1106, surpassing West Bengal which occupied the first position during 2001. The minimum population density works out to be in Arunachal Pradesh (17) for both censuses.

The census 2011 has revealed that India is on way to overtake China, the most populous country of the World, that too, not even at a distant point of time. Unfortunately for India, there is hardly any commensurate response to the enormity of this impending danger across the stake holders. This casual feeling has its spread effects in the administration of population control and family planning right from the policy making areas down to its implementation at different layers of field functionaries in a hierarchical format. The absence of correct focus in an alarming situation has been the sad story of population control scenario in India. This is due to the fact that both the Union government and the State governments have not been administering its constitutional mandate of Population control and family planning, as enshrined at serial no 20A of the Concurrent List of the Constitution of India. Instead the country had taken up a water-down concept of Family Welfare which lacks the robust vision and kicking effect required for a sustainable population in the country. Incidentally, the population of India, as per the 2011 Census is almost equal to the combined population of USA, Indonesia, Brazil, Pakistan, Bangladesh and Japan. India will also overtake China, the most populous country in the world as per projection below:

Year	India	China	World
2009	1,160,813,000	1,338,612,968	6,786,743,939
2012	1,208,116,000	1,366,205,049	7,028,369,002
2015	1,254,019,000	1,393,417,233	7,269,526,256
2020	1,326,155,000	1,430,532,735	7,659,291,953
2025	1,388,994,000	1,453,123,817	8,027,490,191
2050	1,807,878,574	1,424,161,948	9,538,988,263

Sources National Commission on Population Govt. of India and U.S Census Bureau, International Database 29.05.2015

2.2 Food grains Production scenario

Connected with huge size of population is the need for adequate food grains to support required calories to such population to ensure a healthier people. It is true that just as the society evolves, food system also evolves. However, basic food grains for life support and food security remains almost the same. The position of food grains productions etc for the country is captured here to have an idea how far and to what extent the country is in a position to withstand burgeoning population pressure on the food grains front.

It would appear from above Table 2.14 that the area, production and yield under food grains has grown very slowly and even registered negative in a number of

Table 2.14 Food grains production scenario of India since 1951–1952 (unit—in million tons)

Year	Area	Production	Yield
1951–1952	76.5	47.0	65.8
1952–53	80.5	51.7	70.1
1953–1954	86.0	60.7	77.7
1954–1955	85.0	59.0	75.7
1955–1956	87.0	59.0	73.3
1956–1957	87.5	61.7	75.2
1957–1958	86.2	55.7	69.4
1958–1959	90.3	66.7	79.6
1959–1960	91.5	64.9	75.6
1960–1961	90.9	69.6	81.9
1961–1962	92.0	69.4	80.2
1962–1963	92.9	67.3	76.6
1963–1964	92.5	67.9	77.5
1964–1965	91.6	75.4	87.8
1965–1966	90.6	60.6	71.0
1966–1967	90.8	60.8	71.2
1967–1968	95.7	78.3	86.6
1968–1969	94.8	76.3	84.2
1969–1970	97.3	81.6	87.5
1970–1971	97.9	87.9	93.2

(continued)

2.2 Food grains Production scenario

Table 2.14 (continued)

Year	Area	Production	Yield
1971–1972	96.6	86.1	91.3
1972–1973	93.9	79.1	85.8
1973–1974	99.6	85.3	89.0
1974–1975	95.3	81.0	87.0
1975–1976	100.8	98.8	99.7
1976–1977	97.8	89.6	92.4
1977–1978	100.3	103.0	103.2
1978–1979	101.5	107.0	105.7
1979–1980	98.5	87.5	88.7
1980–1981	99.8	104.9	105.1
1981–1982	101.7	107.6	105.9
1982–1983	98.6	103.7	104.9
1983–1984	103.4	122.8	117.8
1984–1985	99.8	117.5	115.5
1985–1986	100.9	123.4	120.6
1986–1987	100.2	116.9	114.9
1987–1988	94.3	113.5	117.2
1988–1989	100.6	138.1	134.2
1989–1990	99.9	139.1	135.5
1990–1991	100.7	143.7	137.8
1991–1992	96.0	137.6	136.5
1992–1993	97.0	144.3	142.0
1993–1994	127.4	135.1	106.0
1994–1995	128.8	141.0	109.5
1995–1996	125.4	131.4	104.8
1996–1997	128.4	145.1	113.0
1997–1998	128.7	140.9	109.5
1998–1999	130.0	150.0	115.4
1999–2000	127.8	152.9	119.6
2000–2001	125.7	141.9	112.9
2001–2002	127.5	155.3	121.8
2002–2003	118.2	126.6	107.0
2003–2004	128.2	155.1	121.0
2004–2005	124.7	144.2	115.6
2005–2006	126.3	152.5	120.8
2006–2007	128.5	158.8	123.6
2007–2008	128.8	168.6	130.9
2008–2009	127.6	171.3	134.3
2009–2010	126.0	159.4	126.5
2010–2011	131.7	178.9	135.9
2011–2012	129.8	188.1	144.9

Source Ministry of Agriculture, Government of India

years. This is due to shifting of land use for non-agriculture purposes including industrialisation and urbanization and also due to adverse climatic factors. Besides, shrinkage of area under food grains was also due to transfer of food grains production area to non-food commercial crops area. Be that as it may, let us look at the annual growth rate of them over the years and then reposition them census year wise with decadal growth of population, as in Tables 2.15 and 2.16.

In the given scenario, it would be relevant to reorganize the per cent growth of areas, productions and yields vis-à-vis growth of population as per the census years:

The Provisional Census, 2011 Handbook published by the Registrar General of Census, Government of India also published a Table on census year wise population, GDP and Output of Food grains which is shown in Table 2.17.

The data, in brief, show a less than hopeful trend-scenario in the food-grains sector. The inelastic nature of availability of additional acreage of land for food-grains production together with not-commensurate growth of productivity is a real problem to withstand the burgeoning population size and its resultant

Table 2.15 Growth of Food grains Production scenario of India since 1951–1952 (unit—in million tons)

Year	Area	Growth (%)	Production	Growth (%)	Yield	Growth (%)
1951–1952	76.5		47		65.8	
1952–1953	80.5	5.23	51.7	10	70.1	6.53
1953–1954	86	6.83	60.7	17.41	77.7	10.84
1954–1955	85	−1.16	59	−2.80	75.7	−2.57
1955–1956	87	2.35	59	0.00	73.3	−3.17
1956–1957	87.5	0.57	61.7	4.58	75.2	2.59
1957–1958	86.2	−1.49	55.7	−9.72	69.4	−7.71
1958–1959	90.3	4.76	66.7	19.75	79.6	14.70
1959–1960	91.5	1.33	64.9	−2.70	75.6	−5.03
1960–1961	90.9	−0.66	69.6	7.24	81.9	8.33
1961–1962	92	1.21	69.4	−0.29	80.2	−2.08
1962–1963	92.9	0.98	67.3	−3.03	76.6	−4.49
1963–1964	92.5	−0.43	67.9	0.89	77.5	1.17
1964–1965	91.6	−0.97	75.4	11.05	87.8	13.29
1965–1966	90.6	−1.09	60.6	−19.63	71	−19.13
1966–1967	90.8	0.22	60.8	0.33	71.2	0.28
1967–1968	95.7	5.40	78.3	28.78	86.6	21.63
1968–1969	94.8	−0.94	76.3	−2.55	84.2	−2.77
1969–1970	97.3	2.64	81.6	6.95	87.5	3.92
1970–1971	97.9	0.62	87.9	7.72	93.2	6.51
1971–1972	96.6	−1.33	86.1	−2.05	91.3	−2.04
1972–1973	93.9	−2.80	79.1	−8.13	85.8	−6.02
1973–1974	99.6	6.07	85.3	7.84	89	3.73
1974–1975	95.3	−4.32	81	−5.04	87	−2.25

(continued)

2.2 Food grains Production scenario

Table 2.15 (continued)

Year	Area	Growth (%)	Production	Growth (%)	Yield	Growth (%)
1975–1976	100.8	5.77	98.8	21.98	99.7	14.60
1976–1977	97.8	−2.98	89.6	−9.31	92.4	−7.32
1977–1978	100.3	2.56	103	14.96	103.2	11.69
1978–1979	101.5	1.20	107	3.88	105.7	2.42
1979–1980	98.5	−2.96	87.5	−18.22	88.7	−16.08
1980–1981	99.8	1.32	104.9	19.89	105.1	18.49
1981–1982	101.7	1.90	107.6	2.57	105.9	0.76
1982–1983	98.6	−3.05	103.7	−3.62	104.9	−0.94
1983–1984	103.4	4.87	122.8	18.42	117.8	12.30
1984–1985	99.8	−3.48	117.5	−4.32	115.5	−1.95
1985–1986	100.9	1.10	123.4	5.02	120.6	4.42
1986–1987	100.2	−0.69	116.9	−5.27	114.9	−4.73
1987–1988	94.3	−5.89	113.5	−2.91	117.2	2.00
1988–89	100.6	6.68	138.1	21.67	134.2	14.51
1989–1990	99.9	−0.70	139.1	0.72	135.5	0.97
1990–1991	100.7	0.80	143.7	3.31	137.8	1.70
1991–1992	96	−4.67	137.6	−4.24	136.5	−0.94
1992–1993	97	1.04	144.3	4.87	142	4.03
1993–94	127.4	31.34	135.1	−6.38	106	−25.35
1994–1995	128.8	1.10	141	4.37	109.5	3.30
1995–1996	125.4	−2.64	131.4	−6.81	104.8	−4.29
1996–1997	128.4	2.39	145.1	10.43	113	7.82
1997–1998	128.7	0.23	140.9	−2.89	109.5	−3.10
1998–99	130	1.01	150	6.46	115.4	5.39
1999–2000	127.8	−1.69	152.9	1.93	119.6	3.64
2000–2001	125.7	−1.64	141.9	−7.19	112.9	−5.60
2001–2002	127.5	1.43	155.3	9.44	121.8	7.88
2002–2003	118.2	−7.29	126.6	−18.48	107	−12.15
2003–2004	128.2	8.46	155.1	22.51	121	13.08
2004–2005	124.7	−2.73	144.2	−7.03	115.6	−4.46
2005–2006	126.3	1.28	152.5	5.76	120.8	4.50
2006–2007	128.5	1.74	158.8	4.13	123.6	2.32
2007–2008	128.8	0.23	168.6	6.17	130.9	5.91
2008–2009	127.6	−0.93	171.3	1.60	134.3	2.60
2009–2010	126	−1.25	159.4	−6.95	126.5	−5.81
2010–2011	131.7	4.52	178.9	12.23	135.9	7.43
2011–2012	129.8	−1.44	188.1	5.14	144.9	6.62

Source Ministry of Agriculture, Government of India

Table 2.16 Decadal growth of food grains production scenario of India since 1951–1952

Year	Decadal growth of population	% of growth Area	% growth of food grains production	% growth of Yield
1951	13.31	NA	NA	NA
1961	21.64	1.21	−0.2.9	−0.2.08
1971	24.80	−1.33	−2.05	−2.04
1981	24.66	1.90	2.57	0.76
1991	23.86	−4.67	−4.24	−0.94
2001	21.54	1.43	9.44	7.88
2011	17.7	−1.44	5.14	6.62

Table 2.17 Census year wise population, GDP and output of food grains

Census Year	Population (in millions)	GDP (at constant prices in Rs. -crore)	Output of Food grains (million tons)
1950–1951	361	224,786	50.8
1960–1961	439	329,825	82.0
1970–1971	548	474,131	108.4
1980–1981	683	641,921	129.6
1990–1991	846	1,083,572	176.4
2000–2001	1028.7	1,864,300	196.8
2010–2011	1210.2	4,493,743	218.2

Source Census Handbook, 2011(provisional)

requirement for food-grains. The situation is bound to be more critical in years to come with the annual addition of around 19 million population per year together with possible demand of food-grains basket of energy requirement of 2400 kcal per person per day for the rural sector and 2100 kcal for the urban sector, a norm earlier set by the Planning Commission but not in place, is taken into consideration. Indeed, the hiatus of the growth of population and the growth of food-grains production is a pointer of the upcoming disaster in the food-grains scenario.

2.3 Poverty status

Poverty exists in all societies where a section or a proportion of its citizens fails to attain a level of wellbeing considered to be a reasonable minimum by the standard of the society. It varies from time to time and also from one country to another depending on the cost of living and the level of development. Historically, the first poverty line study in India, undertaken in 1962 by a Working Group set up by the Government of India, recorded a Per capita Total Consumption Expenditure (PCTE) of Rs. 20 per month in 1960–1961 prices. The Planning Commission defined the poverty level as below the average per capita daily intake of 2400 calories in rural

2.3 Poverty status

Table 2.20 Poverty estimates for 1993–1994 and 2003–2004 using Lakdawala Methodology

Sl No	Name of the state/union territory	1993–1994			2003–2004		
		Rural	Urban	Combined	Rural	Urban	Combined
1	Andhra Pradesh	15.9	38.3	22.2	11.2	28.0	15.8
2	Arunachal Pradesh	45.0	7.7	39.4	22.3	3.3	17.6
3	Assam	45.0	7.7	40.9	22.3	3.3	19.7
4	Bihar	58.2	34.5	55.0	42.1	34.6	41.4
5	Chhattisgarh	NA	NA	NA	40.8	41.2	40.9
6	Delhi	1.9	16.0	14.7	6.9	15.2	14.7
7	Goa	5.3	27.0	14.9	5.4	21.3	13.8
8	Gujarat	22.2	27.9	24.2	19.1	13.0	16.8
9	Haryana	28.0	16.4	25.1	13.6	15.1	14.0
10	Himachal Pradesh	30.3	9.2	28.4	10.7	3.4	10.0
11	Jammu & Kashmir	30.3	9.2	25.2	4.6	7.9	5.4
12	Jharkhand	NA	NA	NA	46.3	20.2	40.3
13	Karnataka	29.9	40.1	33.2	20.8	32.6	25.0
14	Kerala	25.8	24.6	25.4	13.2	20.2	15.0
15	Madhya Pradesh	40.6	48.4	42.5	36.9	42.1	38.3
16	Maharashtra	37.9	35.2	36.9	29.6	32.2	30.7
17	Manipur	45.0	7.7	33.8	22.3	3.3	17.3
18	Meghalaya	45.0	7.7	37.9	22.3	3.3	18.5
19	Mizoram	45.0	7.7	25.7	22.3	3.3	12.6
20	Nagaland	45.0	7.7	37.9	22.3	3.3	19.0
21	Orissa	49.7	41.6	48.6	46.8	44.3	46.4
22	Punjab	12.0	11.4	11.8	9.1	7.1	8.4
23	Rajasthan	26.5	30.5	27.4	18.7	32.9	22.1
24	Sikkim	45.0	7.7	41.4	22.3	3.3	20.1
25	Tamil Nadu	32.5	39.8	35.0	22.8	22.2	22.5
26	Tripura	45.0	7.7	39.0	22.3	3.3	18.9
27	Uttar Pradesh	42.3	35.4	40.9	33.4	30.6	32.8
28	Uttarakhand	NA	NA	NA	40.8	36.5	39.6
29	West Bengal	40.8	22.4	35.7	28.6	14.8	24.7
30	Pondicherry	32.5	39.8	37.4	22.9	22.2	22.4
Total	All India	37.3	32.4	36.0	28.3	25.7	27.5

Source Website of the Planning Commission

poor, i.e., reduced the number or the percentage of the population below the poverty line. Further, as in case of any price index, there is also a case for periodically raising the poverty line even beyond pure inflation adjustment in order to reflect growth of income in the economy.

Planning Commission had accordingly appointed the Tendulkar Committee in December 2005 to review alternate concepts of poverty and recommend changes in the existing procedures of official estimation of poverty. The Tendulkar Committee submitted its report in November 2009. The Committee reviewed various arguments advanced in favour of redefining the poverty line and came to the conclusion that some changes are necessary. However, it has not recommended a new basis for defining poverty in terms of calories, or any other minimum basic needs norm. Instead, it has decided (a) to locate the poverty line bundle of goods and services in the consumption pattern observed in the 2004–2005 NSS survey based on the mixed reference period; (b) and that the same bundle be made available to the rural population after correcting for the rural-urban price differential.

Needless to mention, when such revisions are made, the percentage of population below the poverty line is no longer comparable with the earlier estimates. The consequence of this procedure is that the rural poverty lines for 2004–2005 appeared to be too low compared to the corresponding urban poverty. Be that as it may, Poverty Lines and Poverty Head Count Ratio for 2004–2005, as per report of the Tendulkar Committee were as follows (Tables 2.21 and 2.22):

Table 2.21 Poverty lines and poverty head count ratio for 2004–2005 using Tendulkar methodology

No.	States	Poverty line (Rs)		Poverty Head Count Ratio (%)		
		Rural	Urban	Rural	Urban	Total
1	Andhra Pradesh	433.43	563.16	32.30	23.40	29.90
2	Arunachal Pradesh	547.14	618.45	33.60	23.50	31.10
3	Assam	478.00	600.03	36.40	21.80	34.40
4	Bihar	433.43	526.18	55.70	43.70	54.40
5	Chhattisgarh	398.92	513.70	55.10	28.40	49.40
6	Delhi	541.39	642.47	15.60	12.90	13.10
7	Goa	608.76	671.15	28.10	22.20	25.00
8	Gujarat	501.58	659.18	39.10	20.10	31.80
9	Haryana	529.42	626.41	24.80	22.40	24.10
10	Himachal Pradesh	520.40	605.74	25.00	4.60	22.90
11	Jammu & Kashmir	522.30	602.89	14.10	10.40	13.20
12	Jharkhand	404.79	531.35	51.60	23.80	45.30
13	Karnataka	417.84	588.06	37.50	25.90	33.40
14	Kerala	537.31	584.70	20.20	18.40	19.70
15	Madhya Pradesh	408.41	532.26	53.60	35.10	48.60
16	Maharashtra	484.89	631.85	47.90	25.60	38.10
17	Manipur	578.11	641.13	39.30	34.50	38.00
18	Meghalaya	503.32	745.73	14.00	24.70	16.10
19	Mizoram	639.27	699.75	23.00	7.90	15.30
20	Nagaland	687.30	782.93	10.00	4.30	9.00
21	Orissa	407.78	497.31	60.80	37.60	57.20
22	Punjab	543.51	642.51	22.10	18.70	20.90
23	Rajasthan	478.00	568.15	35.80	29.70	34.40

(continued)

2.3 Poverty status

Table 2.21 (continued)

No.	States	Poverty line (Rs)		Poverty Head Count Ratio (%)		
		Rural	Urban	Rural	Urban	Total
24	Sikkim	531.50	741.68	31.80	2.90	31.10
25	Tamil Nadu	441.69	559.77	37.50	19.70	28.90
26	Tripura	450.49	555.79	44.50	22.50	40.60
27	Uttar Pradesh	435.14	532.12	42.70	34.10	40.90
28	Uttarakhand	486.24	602.39	35.10	26.20	32.70
29	West Bengal	445.38	572.51	38.20	24.40	34.3
30	Pondicherry	385.45	506.17	22.90	9.90	14.10
Total	All India	446.68	578.80	41.80	25.70	37.20

Table 2.22 Number and percentage of population below poverty line by states—2011–2012 (Tendulkar methodology)

Sl No	States	Rural		Urban		Total	
		% of persons	No. of persons (Lakhs)	% of persons	No. of persons (Lakhs)	% of persons	No. of persons (Lakhs)
1	Andhra Pradesh	10.96	61.80	5.81	16.98	9.20	78.78
2	Arunachal Pradesh	38.93	4.25	20.33	0.66	34.67	4.91
3	Assam	33.89	92.06	20.49	9.21	31.98	101.27
4	Bihar	34.06	320.40	31.23	37.75	33.74	358.15
5	Chhattisgarh	44.61	88.90	24.75	15.22	39.93	104.1
6	Delhi	12.92	0.50	9.84	16.46	9.91	16.96
7	Goa	6.81	0.37	4.09	0.38	5.09	0.75
8	Gujarat	21.54	75.35	10.14	26.88	16.63	102.23
9	Haryana	11.64	19.42	10.28	9.41	11.16	28.83
10	Himachal Pradesh	8.48	5.29	4.33	0.30	8.06	5.59
11	Jammu & Kashmir	11.54	10.73	7.20	2.53	10.35	13.27
12	Jharkhand	40.84	104.09	24.83	20.24	36.96	124.33
13	Karnataka	24.53	92.80	15.25	36.96	20.91	129.76
14	Kerala	9.14	15.48	4.97	8.46	7.05	23.95
15	Madhya Pradesh	35.74	190.95	21.00	43.10	31.65	234.06
16	Maharashtra	24.22	150.56	9.12	47.36	17.35	197.92
17	Manipur	38.80	7.45	32.59	2.78	36.89	10.22
18	Meghalaya	12.53	3.04	9.26	0.57	11.87	3.61
19	Mizoram	35.43	1.91	6.36	0.37	20.40	2.27
20	Nagaland	19.93	2.76	16.48	1.00	18.88	3.76
21	Orissa	35.69	126.14	17.29	12.39	32.59	138.59
22	Punjab	7.66	13.35	9.24	9.82	8.26	23.18
23	Rajasthan	16.05	84.19	10.69	18.73	14.71	102.92
24	Sikkim	9.85	0.45	3.66	0.06	8.19	0.51

(continued)

Table 2.22 (continued)

Sl No	States	Rural		Urban		Total	
		% of persons	No. of persons (Lakhs)	% of persons	No. of persons (Lakhs)	% of persons	No. of persons (Lakhs)
25	Tamil Nadu	15.83	59.23	6.54	23.40	11.28	82.63
26	Tripura	16.53	4.49	7.42	0.75	14.05	5.24
27	Uttar Pradesh	30.40	479.35	26.06	118.84	29.43	598.19
28	Uttarakhand	11.62	8.25	10.48	3.35	11.26	11.60
29	West Bengal	22.52	141.14	14.66	43.83	19.98	184.98
30	Pondicherry	17.06	0.69	6.30	0.55	9.69	1.24
31	Andaman&Nicobar Islands	1.57	0.04	0.00	0.00	1.00	0.04
32	Chandigarh	1.64	0.004	22.31	2.34	21.81	2.35
33	Dadra Nagar	62.59	1.15	15.38	0.28	39.31	1.43
34	Daman&Diu	0.00	0.00	12.62	0.26	9.86	0.26
35	Lakshadweep	0.00	0.00	3.44	0.02	2.77	0.02
	All India	25.70	2166.58	13.70	531.25	21.92	2697.83

Notes
1. Population as on 1st March 2012 has been used for estimating number of persons below poverty line. (2011 Census population extrapolated)
2. Poverty line of Tamil Nadu has been used for Andaman and Nicobar Island
3. Urban Poverty Line of Punjab has been used for both rural and urban areas of Chandigarh
4. Poverty Line of Maharashtra has been used for Dadra & Nagar Haveli
5. Poverty line of Goa has been used for Daman&Diu
6. Poverty Line of Kerala has been used for Lakshadweep
Source website of the Planning Commission

A serious debate on Tendulkar methodology adopted for the poverty estimate began subsequent to publication of 2011–2012 poverty estimate and the Planning Commission, to revisit poverty estimates and related methodologies, set up an Expert Group to 'Review the Methodology for Measurement of Poverty 'under a Technical Group of eminent economists under the Chairmanship of Dr. C. Rangarajan. The Report, since published, mentionsthat three out of ten Indians are poor. The report also mentioned that poverty was at 38.2 % in 2009–2010 and then came down to 29.5 % in 2011–2012. The Rangarajan panel also recommended to raise the daily per capita expenditure to Rs. 32 from Rs. 27 for the rural poor.

Based on State-wise Poverty Line in Rural and Urban areas for 2011–2012, Poverty Ratio and Number of Poor in 2011–2012 for the States of India, as per the methodology of the Rangarajan Committee, is as follows Table 2.23:

The size of BPL population, worked out under whatever methodology, is significantly at a higher level in India. The size of BPL population across the States of India as revealed in the Rangarajan Committee Report, is massive even after more

2.3 Poverty status

Table 2.23 Poverty ratio and number of poor in 2011–2012

Sl. No	States	Rural % of persons	Rural No. of persons (in lakhs)	Urban % of persons	Urban No. of persons (in lakhs)	Total % of persons	Total No. of persons (in lakhs)
1	Andhra Pradesh	12.7	71.5	15.6	45.7	13.7	117.3
2	Arunachal Pradesh	39.3	4.3	30.9	1.0	37.4	5.3
3	Assam	42.0	114.1	34.2	15.4	40.9	129.5
4	Bihar	40.1	376.8	50.8	61.4	14.3	438.1
5	Chhattisgarh	49.2	97.9	43.7	26.9	47.9	124.8
6	Delhi	11.9	0.5	15.7	26.3	15.6	26.7
7	Goa	1.4	0.1	9.1	0.8	6.3	0.9
8	Gujarat	31.4	109.8	22.2	58.9	27.4	168.8
9	Haryana	11.0	18.4	15.3	14.0	12.5	32.4
10	Himachal Pradesh	11.1	6.9	8.8	0.6	10.9	7.5
11	Jammu & Kashmir	12.6	11.7	21.6	7.6	15.1	19.3
12	Jharkhand	45.9	117.0	31.3	25.5	42.4	142.5
13	Karnataka	19.8	74.8	25.1	60.9	21.9	136.7
14	Kerala	7.3	12.3	15.3	26.0	11.3	38.3
15	Madhya Pradesh	45.2	241.4	42.1	86.3	44.3	327.8
16	Maharashtra	22.5	139.9	17.0	88.4	20.0	228.3
17	Manipur	34.9	6.7	73.4	6.3	46.7	18.9
18	Meghalaya	26.3	6.4	16.7	1.0	24.4	7.4
19	Mizoram	33.7	1.8	21.5	1.2	27.4	3.1
20	Nagaland	6.1	0.8	32.1	1.9	14.0	2.8
21	Odisha	47.8	169.0	36.3	26.0	45.9	195.0
22	Punjab	7.4	12.9	17.6	18.7	11.3	31.6
23	Rajasthan	21.4	112.0	22.5	39.5	21.7	151.5
24	Sikkim	20.0	0.9	11.7	0.2	17.8	1.1
25	Tamil Nadu	24.3	91.1	20.3	72.8	22.4	163.9
26	Tripura	22.5	6.1	31.3	3.2	24.9	9.3
27	Uttarakhand	38.1	600.9	45.7	208.2	39.8	809.1
28	Uttar Pradesh	12.6	8.9	29.5	9.4	17.8	18.4
29	West Bengal	30.1	188.6	29.0	86.8	29.7	275.4

Source Rangarajan Committee Report

than four decades of planned development. In fact, the size of BPL population grows with the size of population, though the rate of its growth declines with time. The unsustainable size of population is the root cause of the existence of this huge size of parasitic BPL population. There is little scope to make use of this 'overpopulated' component of population size in India. The economic, social and political burden of the overpopulated population is one of the prime factors why India is yet to reach desired GDP, higher Per capita income or scale up its HDI.

2.4 Nutritional status

Nutrition is nourishment or energy that is obtained from food consumed or the process of consuming the proper amount of nourishment and energy. Good nutrition is a cornerstone of good health. Poor nutrition can lead to reduced immunity, increased susceptibility to disease, impaired physical and mental development, and reduced productivity. Malnutrition is directly related to inadequate dietary intake as well as disease, but indirectly to many factors, among others household food security, maternal and child care, health services and the environment.

Nutrition is usually having context centred meaning. Human nutrition seeks to obtain the essential nutrients necessary to support life and health. Economists on the other define nutrition in the context of poverty through a defined norm using a minimum dietary energy requirement norm. The Planning Commission has been following a nutritional norm of energy requirement of 2400 kcal per person per day for the rural sector and 2100 kcal for the urban sector while the Food and Agricultural Organisation of the United Nations (FAO) norm for India as a whole for 2003–2005 was 1770 kcal.

The nutritional status can be discussed from three angles:

- From poverty angle of average calorie intake
- From Child nutrition angle
- From Women nutrition angle

(a) From poverty angle of average calorie intake

As against the Planning Commission's norm of energy requirement of 2400 kcal per person per day for the rural sector and 2100 kcal for the sector, the trend of All-India picture of the proportion of per capita calorie intake (kcal) for the rural and urban area for the states of India for 2004–2005 and 2009–2010 have been captured at Table 2.24.

The per capita calorie intake per day during 2004–2005 to 2009–2010, reveals trend of decline in both in rural and urban areas. However, the extent of decline was more in urban areas. Among the major States, estimated per capita calorie intake (Kcal) per day in rural areas was highest in Punjab (2223 kcal) and lowest in Jharkhand (1900 kcal) in 2009–2010. In urban areas, the highest per capita calorie intake (Kcal) per day was reported in Odisha (2096 kcal) and lowest in West Bengal (1851 kcal) as against the national norm of 2100 kcal.

2.4 Nutritional status

Table 2.24 Estimated per capita calorie intake (kcal) per day in major states

States	Rural		Urban	
	2004–2005	2009–2010	2004–2005	2009–2010
Andhra Pradesh	1995	2047	2000	1975
Assam	2067	1974	2143	2003
Bihar	2049	1931	2190	2013
Chhattisgarh	1942	1926	2087	1949
Gujarat	1923	1982	1991	1983
Haryana	2226	2180	2033	1940
Jharkhand 1	1961	1900	2458	2046
Karnataka	1845	1903	1944	1987
Kerala	2014	1964	1996	1941
Madhya Pradesh	1929	1939	1954	1854
Maharashtra	1933	2051	1847	1901
Orissa	2023	2126	2139	2096
Punjab	2240	2223	2150	2062
Rajasthan	2180	2191	2116	2014
Tamil Nadu	1842	1925	1935	1963
Uttar Pradesh	2200	2064	2124	1923
Uttarakhand	2160	2179	2205	1984
West Bengal	2070	1927	2011	1851

Source NSS report 540, Nutritional intake in India

(b) From Child nutrition angle

Children usually face brunt of poverty and as a result malnourishment among children is very significant pointing to lack of food security and other essential child care facilities. The 'Prevalence of underweight children' in any country denotes the percentage of children under three years of age whose weight for age is less than minus two standard deviations from the median for the reference population aged 0–35 months.

At Table 2.25, the proportion of Underweight Children<3 years has been shown for the states of India. While the All India estimated proportion of Underweight Children<3 years in 1990 was estimated at 52 %, the position of improvement has not been satisfactory for a good number of states in 1992–1993(NFHS-1) , 1998–1999 (NFHS-2) and 2005–2006 (NFHS-3) as shown at Table 2.2. However, as per NFHS-3 survey results, 10 States namely Mizoram (14.2%), Sikkim (17.3%), Manipur (19.5%), Kerala (21.2%), Goa (21.3%), Punjab (23.6%), Nagaland (23.7%), Jammu & Kashmir (24%), Delhi (24.9%), and Tamil Nadu (25.9%) have already achieved the all India MDG target for prevalence of underweight children under three years of age and four more States i.e. Andhra Pradesh, Karnataka, Maharashtra and Uttarakhand are likely to achieve the target by 2015.

Children under age five years are classified as malnourished depending on three anthropometric indices of nutritional status: height for-age, weight-for-height, and

Table 2.25 Proportion of underweight children <3 years

Proportion of underweight children <3 years state	1990 estimated	NFHS-1 (1992–1993)	NFHS-2 (1998–1999)	NFHS-3 (2005–1906)	Likely achievement 2015	Target-2015
Andhra Pradesh	44.41	42.9	34.2	29.8	22.17	22.21
Arunachal Pradesh	28.62	32.1	21.9	29.7	25.50	14.31
Assam	43.48	44.1	35.3	35.8	29.48	21.74
Bihar	49.28	NA	52.2	54.9	59.00	24.64
Chhattisgarh	60.12	NA	53.2	47.8	41.02	30.06
Delhi	38.09	36.2	29.9	24.9	18.58	19.04
Goa	28.90	29.3	21.3	21.3	15.92	14.45
Gujarat	42.82	42.7	41.6	41.1	39.82	21.41
Haryana	28.60	31	29.9	38.2	43.29	14.30
Himachal Pradesh	40.35	38.4	36.5	31.1	26.78	20.17
Jammu & Kashmir	36.54	NA	29.2	24	18.14	18.27
Jharkhand	48.17		51.5	54.6	59.36	24.09
Karnataka	48.28	46.4	38.6	33.3	25.59	24.14
Kerala	22.25	22.1	21.7	21.2	20.54	11.12
Madhya Pradesh	43.75	NA	50.8	57.9	69.80	21.87
Maharashtra	52.24	47.3	44.8	32.7	25.39	26.12
Manipur	19.33	19.1	20.1	19.5	20.03	9.67
Meghalaya	32.02	6.9	28.6	42.9	44.17	16.01
Mizoram	19.27	17.2	19.8	14.2	13.03	9.63
Nagaland	17.36	18.7	18.8	23.7	27.66	8.68
Odisha	54.07	50	50.3	39.5	33.98	27.04
Punjab	39.66	39.9	24.7	23.6	14.79	19.83
Rajasthan	45.36	41.8	46.7	36.8	34.91	22.68
Sikkim	13.67	NA	15.5	17.3	20.24	6.84
Tamil Nadu	42.88	0.7	31.5	25.9	18.06	21.44
Tripura	42.67	42.1	37.3	35.2	30.36	21.34
Uttar Pradesh	56.78	NA	48.1	41.6	33.81	28.39
Uttarakhand	42.38	NA	36.3	31.7	26.12	21.19
West Bengal	56.11	53.2	45.3	37.6	28.79	28.05
All India	52.00	51.5	42.7	40.4	32.85	26.00

Source NFHS, M/o HFW

weight-for-age. The height-for-age index is an indicator of growth retardation and cumulative growth deficits. Children whose height-for-age Z-score in the states below minus two standard deviations (-2 SD) are considered short for their age (stunted) and are chronically malnourished. Children below minus three standard deviations (−3 SD) are considered to be severely stunted. Stunting reflects the long-term effects of malnutrition in a population. The weight-for-height index measures body mass in relation to body length and describes current nutritional status. Children whose Z-score is below minus two standard deviations (-SD) are considered thin (wasted) for their height and are acutely malnourished. Wasting represents inadequate food intake or a recent episode of illness causing loss of weight and the onset of malnutrition. Children whose weight-for-height is below minus three standard deviations (−3 SD) are considered to be severely wasted. Weight-for-age is a composite index of height-for-age and weight-for-height. It takes into account both acute and chronic malnutrition. Children whose weight-for-age is below minus two standard deviations are classified as underweight. Children whose weight-for-age is below minus three standard deviations (−3 SD) are considered to be severely underweight. The percentage of such children for of India in 2005–2006 is shown at Table 2.26.

It would appear from Table 2.26 that inadequate nutrition is a problem throughout India, but there are variations within the states in India. The above table shows that under nutrition is most pronounced in Madhya Pradesh, Bihar, and Jharkhand.

(c) From Women nutrition angle

The height and weight measurements in NFHS-3 are used to calculate the BMI. The BMI is defined as weight in kilograms divided by height in metres squared (kg/m2). A cut-off point of 18.5 is used to define thinness or acute under nutrition and a BMI of 25 or above indicates overweight or obesity. A woman's nutritional status has important implications for her health as well as the health of her children. A woman with poor nutritional status, as indicated by a low body mass index (BMI), short stature, anaemia, or other micronutrient deficiencies, has a greater risk of obstructed labour, having a baby with a low birth weight, having adverse pregnancy outcomes, producing lower quality breast milk, death due to postpartum haemorrhage and illness for herself and her baby. Table 2.27 captures Low Body Mass Index (BMI) and Anaemia in Women (%) in the states of India.

At Table 2.27, the nutritional status of women for the states of India has been presented by means of two indicators—BMI and Anaemia. It would appear there from that the position of West Bengal, Odisha, Jharkhand, Chattisgarh and Bihar is very alarming in respect of both the counts and much above the national average. A good number of states posted better status like Andhra Pradesh, Arunachal Pradesh, Delhi, Goa, Gujarat, Haryana, Himachal Pradesh, Jammu & Kashmir and all North Eastern and Southern States. In respect of Anaemia in ever married Women (15–49) too, the percentage for Andhra Pradesh, Assam, Bihar, Jharkhand, Odisha, Sikkim, Tripura and West Bengal is alarmingly high and much above the

Table 2.26 Nutritional status of children by states

State	Height for Age			Weight for Height				Weight for Age			
	Percentage below −3 SD	Percentage below −2 SD	Mean Z score (SD)	Percentage below −3 SD	Percentage below −2 SD	Percentage above +2 SD	Mean Z score SD	Percentage below −3 SD	Percentage below −2 SD	Percentage above +2 SD	Mean Z score (SD)
India	23.7	48.0	−1.9	6.4	19.8	1.5	−.0	15.8	42	50.4	−1.8
North											
Delhi	20.4	42.2	−1.6	7.0	15.4	4.0	−0.5	8.7	26.1	1.0	−1.3
Haryana	19.4	45.7	−1.8	5.0	19.1	1.4	−1.0	14.2	39.6	0.2	−1.7
Himachal Pradesh	16.0	38.6	−1.5	5.5	19.3	1.1	−1.0	11.4	36.5	0.5	−1.6
Jammu & Kashmir	14.9	35.0	−1.3	4.4	14.8	2.3	−0.7	8.2	25.6	0.5	−1.3
Punjab	17.3	36.7	−1.5	2.1	9.2	1.5	−0.5	8.0	24.9	0.5	−1.2
Rajasthan	22.7	43.7	−1.7	7.3	20.4	1.6	−1.1	15.3	39.9	0.4	−1.7
Uttaranchal	23.1	44.4	−1.8	5.3	18.8	2.3	−0.9	15.7	38.0	0.3	−1.7
Central											
Chhattisgarh	24.8	52.9	−2.0	5.6	19.5	1.3	−1.1	16.4	47.1	0.0	−1.9
Madhya Pradesh	26.3	50.0	−2.0	12.6	35.0	1.0	−1.6	27.3	60.0	0.1	−2.3
Uttar Pradesh	32.4	56.8	−2.2	5.1	14.8	1.2	−0.8	16.4	42.4	0.1	−1.8
East											
Bihar	29.1	55.6	−2.1	8.3	27.1	0.3	−1.4	24.1	55.9	0.1	−2.2
Jharkhand	26.8	49.8	−1.9	11.8	32.3	0.6	−1.5	26.1	56.5	0.2	−2.2
Orissa	19.6	45.0	−1.7	5.2	19.5	1.7	−1.0	13.4	40.7	0.5	−1.7
West Bengal	17.8	44.6	−1.7	4.5	16.9	1.9	−0.9	11.1	38.7	0.5	−1.6

(continued)

2.4 Nutritional status

Table 2.26 (continued)

State	Height for Age			Weight for Height				Weight for Age			
	Percentage below −3 SD	Percentage below −2 SD	Mean Z score (SD)	Percentage below −3 SD	Percentage below −2 SD	Percentage above +2 SD	Mean Z score SD	Percentage below −3 SD	Percentage below −2 SD	Percentage above +2 SD	Mean Z score (SD)
Northeast											
Arunachal Pradesh	21.7	43.3	−1.6	6.1	15.3	3.4	−0.7	11.1	32.5	0.6	−1.4
Assam	20.9	46.5	−1.8	4.0	13.7	1.1	−0.8	11.4	36.4	0.3	−1.6
Manipur	13.1	35.6	−1.4	2.1	9.0	2.2	−0.6	4.7	22.1	0.5	−1.2
Meghalaya	29.8	55.1	−2.0	19.9	30.7	2.6	−1.2	27.7	48.8	0.2	−2.0
Mizoram	17.7	39.8	−1.6	3.5	9.0	4.3	−0.3	5.4	19.9	1.2	−1.1
Nagaland	19.3	38.8	−1.4	5.2	13.3	4.7	−0.5	7.1	25.2	0.8	−1.2
Sikkim	17.9	38.3	−1.4	3.3	9.7	8.3	−0.1	14.9	19.7	1.3	−0.9
Tripura	14.7	35.7	−1.5	8.6	24.6	2.2	−1.2	15.7	39.6	0.1	−1.7
West											
Goa	10.2	25.6	−1.1	5.6	14.1	4.3	−0.7	6.7	25.0	1.9	−1.1
Gujarat	25.5	51.7	−2.0	5.8	18.7	1.2	−1.0	16.3	44.6	0.1	−1.8
Maharashtra	19.1	46.3	1.8	5.2	16.5	2.8	−0.9	11.9	37.0	0.9	−1.6
South											
Andhra Pradesh	18.7	42.7	−1.7	3.5	12.2	2.2	0.7	9.9	32.5	0.6	−1.5
Karnataka	20.5	43.7	1.7	5.9	17.6	2.6	−1.0	12.8	37.6	0.5	−1.6
Kerala	6.5	24.5	−1.1	4.1	15.9	1.2	−0.9	4.7	22.9	0.4	−1.2
Tamil Nadu	10.9	30.9	−1.1	8.9	22.2	3.6	−1.0	6.4	29.8	1.9	−1.3

Source NFHS-3 Chapter-10

Table 2.27 Low body mass index (BMI) and Anaemia in women (%)

State	Women with BMI below normal		Anaemia in ever married women (15–49)	
	NFHS-2 (1998–1999)	NFHS-3 (2005–2006)	NFHS-2 (1998–1999)	NFHS-3 (2005–2006)
Andhra Pradesh	37.4	33.5	49.8	62.9
Arunachal Pradesh	10.7	16.4	62.5	50.6
Assam	27.1	36.5	69.7	69.5
Bihar	39.3	45.1	63.4	67.4
Chhattisgarh	48.1	43.4	68.7	57.5
Delhi	12.0	14.8	40.5	44.3
Goa	27.1	27.9	36.4	38.0
Gujarat	37.0	36.3	46.3	55.3
Haryana	25.9	31.3	47.0	56.1
Himachal Pradesh	29.7	29.9	40.5	43.3
Jammu & Kashmir	26.4	24.6	58.7	52.1
Jharkhand	41.1	43.0	72.9	69.5
Karnataka	38.8	35.5	42.4	51.5
Kerala	18.7	18.0	22.7	32.8
Madhya Pradesh	38.2	41.7	54.3	56.0
Maharashtra	39.7	36.2	48.5	48.4
Manipur	18.8	14.8	28.9	35.7
Meghalaya	25.8	14.6	63.3	47.2
Mizoram	22.6	14.4	48.0	38.6
Nagaland	18.4	17.4	38.4	NA
Odisha	48.0	41.4	63.0	61.2
Pondicherry				
Punjab	16.9	18.9	41.4	38.0
Rajasthan	36.1	36.7	48.5	53.1
Sikkim	11.2	11.2	61.1	60.0
Tamil Nadu	29.0	28.4	56.5	53.2
Tripura	35.2	36.9	59.0	65.1
Uttar Pradesh	35.8	36.0	48.7	49.9
Uttarakhand	32.4	30.0	45.6	55.2
West Bengal	43.7	39.1	62.7	63.2
All India	35.8	35.6	51.8	55.3

Source NHFS-2 and NHFS-3

national average of 55.3 %. The position of ArunachalPradesh, Delhi, Goa, Gujarat, Haryana, Himachal Pradesh, Jammu & Kashmir and all Southern States is better.

To sum up, the nutritional status of population is linked with quality of life of its citizens. Better is the nutritional status of its citizens, better is the chance of its citizens to have improved quality of life. It is also linked with human development of its citizens as well. Human development is up-scaled by the quality of its nutritious citizens. Given such premises, it reveals from Tables, as above, that there is strong correlation between the size of population and the corresponding nutritional standard. The nutritional status in the states of India is poorer where the size of its population is also on higher side. It is akin to primary level arithmetic of dividing a cake by the number of children in the family. Nutritional status of the country in a way reflects the syndrome of its over populated size. However, this simple fact is seldom focused in any discourse on nutritional status of children or of others.

2.5 Unemployment scenario

From human development angle, access to gainful employment is an essential condition to earn livelihood and economic wellbeing. This is equivalent to exercising economic rights in a market economy. The nature of livelihood reflected through its employment is now the determining factor in shaping the pattern of population growth linked as it is with marriage and procreation. The size of unemployed population is linked with another fundamental aspect of the economy in that whether the country is in a position to make economic use of its population and sustain them. The declining trend of unemployment gives a signal that the country has reached somewhat at a level of its potential strength and it cannot absorb the current size of incremental unemployed population. The growing trend of unemployment status is decisively an indicator of un-sustained nature of population and a signal over population. It is also an emphatic signal that oft-quoted population dividend in India is confusing, misleading and inappropriate for any country struggling to reach full employment situation.

With the above premise in view, it would be worthwhile to examine the unemployment status of India in general and the states in particular through a series of self-spoken Tables. It starts with a Working paper of the Planning Commission.

The Working paper on Status of Employment made use of by the Planning Commission for formulating the Draft Twelfth Five Year Plan gives a state wise Employment status of our country for the year 2004–2005 and 2009–2010 as shown at Table 2.28.

As per the Report compiled from the NSS (66th round), list of Major States of India ranked according to unemployment published by the Ministry of Statistics and Programme Implementation, Government of India for the year 2009–2010 is also shown at Table 2.29. As per the said list, Kerala has the highest unemployment rates and ranks worst, while Rajasthan and Gujarat has the least unemployment rate among major States of India. Incidentally, a higher rank

Table 2.28 State wise Absolute Employees (in millions) by Major Sectors and Share of Employment (2004–2005 & 2009–2010)

States	Absolute employees in (millions) by major sectors 2004–2005 & 2009–2010									
	2004–2005					2009–2010				
	Agriculture	Manufacturing	Non-manufacturing	Services	Total of Sectors	Agriculture	Manufacturing	Non-manufacturing	Services	Total of Sectors
Andhra Pradesh	20.5	4.6	2.7	10.9	38.8	20.4	4.4	5.4	9.7	39.9
Assam	7.7	0.4	0.3	2.4	10.8	6.9	0.5	0.6	3.0	10.9
Bihar	21.3	1.4	0.9	4.3	27.8	17.2	1.4	2.9	5.5	26.9
Chhattisgarh	8.6	0.4	0.5	1.3	10.8	6.3	0.5	1.7	1.5	10.0
Delhi	0.1	1.3	0.3	3.5	5.2	0.0	1.6	0.3	4.0	5.9
Gujarat	15.7	3.2	1.3	5.1	25.3	12.9	3.4	1.8	6.6	24.7
Haryana	5.0	1.1	0.7	2.3	9.2	4.3	1.5	1.1	2.7	9.6
Himachal Pradesh	2.1	0.2	0.3	0.7	3.3	2.2	0.1	0.5	0.6	3.4
Jammu & Kashmir	2.8	0.4	0.3	0.7	4.3	2.9	0.4	0.4	1.0	4.7
Jharkhand	7.7	0.9	1.3	1.8	11.7	4.9	0.7	2.3	2.2	10.1
Karnataka	17.6	2.6	1.2	6.0	27.4	15.3	2.7	2.1	6.7	26.8
Kerala	5.1	1.7	1.5	4.4	12.7	4.2	1.6	2.1	5.1	12.9
Madhya Pradesh	18.0	2.5	1.5	6.3	28.2	18.4	1.8	4.0	4.4	28.5
Maharashtra	22.0	7.1	3.0	16.5	48.1	26.0	5.3	3.2	14.6	49.1
Odisha	11.2	1.5	1.1	2.9	16.7	10.1	1.4	2.0	2.8	16.2
Punjab	3.6	1.7	1.4	4.1	10.7	4.7	1.3	1.4	3.0	10.4
Rajasthan	17.4	2.2	2.5	4.4	26.5	13.0	1.6	7.4	5.2	27.2
Tamil Nadu	14.5	6.1	2.2	8.5	31.3	12.6	5.2	4.2	8.1	30.0

(continued)

2.5 Unemployment scenario

Table 2.28 (continued)

States	Absolute employees in (millions) by major sectors 2004–2005 & 2009–2010									
	2004–2005					2009–2010				
	Agriculture	Manufacturing	Non-manufacturing	Services	Total of Sectors	Agriculture	Manufacturing	Non-manufacturing	Services	Total of Sectors
Uttar Pradesh	43.3	7.2	3.0	11.7	65.2	39.8	6.4	7.2	12.6	65.9
Uttarakhand	2.7	0.2	0.3	0.8	4.0	2.4	0.3	0.5	0.9	4.0
West Bengal	15.5	5.3	1.6	9.3	31.7	14.8	6.3	2.7	10.4	34.2
Total across states	262.5	52.7	27.9	107.7	449.6	236.1	48.1	56.8	110.4	451.4

States	Share of employees across sectors 2004–2005 & 2009–10							
	2004–2005				2009–2010			
	Agriculture	Manufacturing	Non-manufacturing	Services	Agriculture	Manufacturing	Non-manufacturing	Services
Andhra Pradesh	52.8	11.9	7.0	28.1	51.2	11.0	13.5	24.3
Assam	71.3	3.6	2.8	22.2	62.9	4.1	5.8	27.2
Bihar	76.5	5.0	3.1	15.4	63.8	5.1	10.7	20.4
Chhattisgarh	79.3	4–1	4.8	11.7	63.2	5.0	17.2	14.6
Delhi	1.0	24.8	6.2	67.9	0.2	27.4	4.9	67.5
Gujarat	62.1	12.6	5.1	20.2	52.2	13.7	7.3	26.8
Haryana	54.8	12.2	7.8	25.2	44.8	15.4	11.9	27.9
Himachal Pradesh	63.6	6.1	9.1	21.2	64.2	3.9	15.2	16.7
Jammu & Kashmir	66.1	9.6	7.1	17.2	61.6	7.6	8.9	21.9
Jharkhand	65.4	7.9	11.1	15.6	49.1	6.7	22.5	21.7
Karnataka	64.4	9.4	4.4	27.7	57.3	9.9	7.7	25.1
Kerala	40.1	13.5	11.6	34.8	32.1	12.4	16.3	39.2

(continued)

Table 2.28 (continued)

States	Share of employees across sectors 2004–2005 & 2009–10							
	2004–2005				2009–2010			
	Agriculture	Manufacturing	Non-manufacturing	Services	Agriculture	Manufacturing	Non-manufacturing	Services
Madhya Pradesh	63.8	8.9	5.3	22.3	64.4	6.3	14.0	15.3
Maharashtra	45.7	14.8	6.2	34.3	52.9	10.8	6.3	29.8
Odisha	67.1	8.9	6.7	17.3	62.2	8.3	12.1	17.4
Punjab	33.6	15.9	13.1	38.3	45.0	12.7	13.2	29.1
Rajasthan	65.8	8.3	9.4	16.5	47.7	5.9	27.3	19.1
Tamil Nadu	46.4	19.6	7.0	27.1	41.8	17.2	14.0	27.0
Uttar Pradesh	66.4	11.1	4.6	17.9	60.4	9.6	10.9	19.1
Uttarakhand	68.6	4.3	7.5	19.6	60.5	6.3	11.8	21.4
West Bengal	49.0	16.7	50	29.2	43.4	18.4	7.9	30.3
Total across states								

Source Working Group on Twelfth Plan: Employment Planning Policy

2.5 Unemployment scenario

Table 2.29 Unemployment rates-2009–2010

		Unemployment Rates-2009–2010 (per 1000)		
Rank	State	Rural	Urban	Total
14	Kerala	75	73	148
13	Bihar	20	73	93
12	Assam	39	52	91
11	Punjab	26	48	74
10	Odisha	30	42	72
9	Himachal Pradesh	16	49	65
8	West Bengal	19	40	59
**	All India	16	34	50
7	Tamil Nadu	15	32	47
6	Andhra Pradesh	12	31	43
6	Haryana	18	25	43
5	Uttar Pradesh	10	29	39
4	Maharashtra	6	32	38
3	Madhya Pradesh	7	29	36
2	Karnataka	5	27	32
1	Rajasthan	4	22	26
1	Gujarat	8	18	26

Source Ministry of Statistics and Programme Implementation, Government of India

represents higher unemployment among the population. National average stands at 50. This is shown at Table 2.29.

The Labour Bureau, Ministry of Labour & Employment, Government of India publishes state-specific data based on employment and unemployment survey conducted by it from time to time. The second employment and unemployment survey conducted in 2011–2012 reveals startling facts on employment and unemployment states of India. Data of a select States of India for Unemployment Rate (Per 1000) for Persons above 15 years & above according to usual participatory approach (ps) have been shown in Table 2.30.

The survey reveals that the lowest unemployment rate was in Gujarat while the largest numbers of unemployed persons were found in Kerala and West Bengal. The All-India unemployment rate was estimated at 3.8 % while the All India female unemployment rate was of 6.9 %. The report also revealed the fact that Gujarat has got very low female unemployment rate while Sikkim, Tripura and West Bengal had highest unemployment rates.

The third Annual Employment and Unemployment Survey for 2012–2013, pegged the all-India unemployment rate at 4.7 % in 2012–2013 with urban unemployment at 5.7 % and rural employment at 4.4 %. The unemployment rate amongst workers between 15 and 29 years was, however, pegged at 13.3 %. The survey also revealed that unemployment rate per 1000 persons aged more than

Table 2.30 Unemployment Rate (Per 1000) for Persons above 15 years & above according to usual participatory approach (ps) for each state

Sl. no	States	Rural			Urban			Rural + Urban		
		Male	Female	Person	Male	Female	Person	Male	Female	Person
1	Andhra Pradesh	21	24	22	46	111	61	27	35	30
2	Arunachal Pradesh	40	86	56	99	252	142	47	101	65
3	Assam	48	128	62	44	206	73	47	138	63
4	Bihar	59	205	85	45	181	64	58	203	83
5	Chhattisgarh	11	6	9	27	59	35	13	11	12
6	Delhi	25	175	45	32	149	49	31	153	48
7	Goa	94	549	231	62	290	109	80	462	179
8	Gujarat	5	13	7	12	42	15	8	8	10
9	Haryana	26	71	30	24	120	36	25	88	32
10	Himachal Pradesh	41	12	28	38	130	63	40	17	31
11	Jammu & Kashmir	33	157	50	44	260	71	36	182	56
12	Jharkhand	37	89	47	50	120	59	39	93	48
13	Karnataka	24	21	23	27	35	29	25	24	25
14	Kerala	32	214	82	40	375	145	34	262	99
15	Madhya Pradesh	18	28	20	44	96	51	24	37	27
16	Maharashtra	19	26	21	23	107	42	20	47	28
17	Manipur	25	23	24	87	65	80	39	34	37
18	Meghalaya	24	40	31	42	100	64	27	49	36
19	Mizoram	10	9	10	8	119	43	10	32	19
20	Nagaland	65	48	59	63	72	65	65	52	60
21	Odisha	28	29	28	36	107	43	29	34	30
22	Punjab	13	66	17	13	86	21	13	74	18
23	Rajasthan	12	30	16	16	64	20	13	33	17
24	Sikkim	82	163	113	105	511	229	85	194	126
25	Tamil Nadu	19	24	21	19	41	25	19	29	22
26	Tripura	47	310	115	139	446	236	66	344	141
27	Uttarakhand	35	108	55	21	142	40	32	113	52
28	Uttar Pradesh	22	33	23	32	71	35	24	40	25
29	West Bengal	43	159	61	89	402	139	53	212	78
	All India	27	56	34	34	125	50	29	69	38

Source Report on Employment and Un-employment Survey-2011–2012

15 years was highest in Sikkim at 136, followed by Arunachal Pradesh at 130, Tripura at 126, Goa at 107 and Kerala at 104. In contrast, Chhattisgarh had the lowest unemployment rate of 14, followed by Karnataka at 20, Madhya Pradesh at 22, Andhra Pradesh at 25 and Gujarat at 27 (Table 2.31).

To sum it up, in brief, the above Survey Tables indicate that the country is unable to make full use of its growing labour force in productive and gainful employment. The unsustainable size of population and its overtime growth has been reflected in all Annual Employment and Unemployment Surveys indicating therein the limits of absorbing potential of the economy to sustain them. Truly, the population of India is too large to secure full employment.

2.6 Human development profile

The concept of human development is a paradigm shift to development assessment. The objective of human development is to create an enabling environment for people to enjoy long, healthy and creative lives. An index, called the Human Development Index (HDI) has come into being as an alternative to the common practice of evaluating a country's progress in development based on per capita Gross Domestic Product. HDI is a summary measure of human development and is worked through a composite statistics of life expectancy, education and income indices. Human Development Report captures the HDI of the country and its overtime reports reveal the trend of such human development. The Human Development Report in India was first published in 2001, a summary position of which is shown in Table 2.32.

The HDI for the country as whole has increased to 0.470 in 2001 from 0.302 in 1981. For the States, it varies between 0.638 for Kerala to 0.365 in case of Bihar. Among better of States, Punjab, Tamil Nadu and Maharashtra had a HDI value of above 0.52. At the other hand, the States like Uttar Pradesh, Assam and Madhya Pradesh had value less than 0.400. By and large States maintained their relative position between 1981 and 2001.

The India Human Development Report 2011, prepared by Institute of Applied Manpower Research, was published by the Planning Commission. It captured state wise Human Development scenario from 1999–2000 to 2007–2008 as could be seen in Table 2.33.

India Human Development Report 2011, as shown at Table 2.33 estimated the HDI for the beginning of the decade, and for the latest year for which data were available for preparing the Report. The top five ranks in both the years are captured by the states of Kerala, Delhi, Himachal Pradesh, Goa and Punjab. It further appeared that States that perform better on health and educational outcomes are also the states with higher HDI and of higher per capita income. West Bengal's position is far from worth mentioning. It continues to remain at rank 13 in the all India position while even the seven north eastern states (excluding Assam) 1 have done

Table 2.31 Unemployment Rate (Per 1000) for Persons above 15 years & above according to usual participatory approach (ps) for each state

Sl. no	States	Rural			Urban			Rural +Urban		
		Male	Female	Person	Male	Female	Person	Male	Female	Person
1	Andhra Pradesh	21	15	19	31	88	44	24	27	25
2	Arunachal Pradesh	120	148	130	88	262	128	114	159	130
3	Assam	46	142	61	63	244	97	48	154	65
4	Bihar	54	106	60	54	145	58	54	107	60
5	Chhattisgarh	8	11	9	23	106	40	11	21	14
6	Delhi	73	409	115	40	128	51	43	154	57
7	Goa	39	243	93	55	289	123	46	265	107
8	Gujarat	13	72	24	15	166	32	14	94	27
9	Haryana	44	67	46	36	149	52	41	96	48
10	Himachal Pradesh	49	83	63	34	153	60	47	87	63
11	Jammu & Kashmir	67	210	83	67	282	101	67	234	88
12	Jharkhand	77	131	87	77	184	88	77	136	87
13	Karnataka	16	20	17	13	69	27	15	34	20
14	Kerala	46	242	102	44	274	110	45	251	104
15	Madhya Pradesh	18	14	17	29	86	39	21	25	22
16	Maharashtra	29	26	28	41	96	52	33	43	36
17	Manipur	29	44	33	25	52	32	28	46	32
18	Meghalaya	19	34	25	102	150	121	37	60	46
19	Mizoram	35	35	35	28	13	22	32	24	29
20	Nagaland	54	102	70	71	116	85	58	105	73
21	Odisha	48	62	51	70	243	93	51	76	56
22	Punjab	22	166	40	46	153	61	32	160	48
23	Rajasthan	27	29	27	40	103	48	30	41	32
24	Sikkim	113	184	134	97	295	147	110	204	136
25	Tamil Nadu	30	43	34	31	83	44	30	55	38
26	Tripura	78	244	111	133	341	194	87	268	126
27	Uttarakhand	50	46	49	35	340	81	46	106	57
28	Uttar Pradesh	52	97	56	68	232	82	55	125	61
29	West Bengal	63	122	73	58	180	75	62	133	74
	All India	40	58	44	42	128	57	40	72	47

Source The third Annual Employment and Unemployment Survey for 2012–2013

2.6 Human development profile

Table 2.32 Human Development Index for India-combined

Human development index for india-combined

States	1981 value	1981 rank	1991 value	1991 rank	2001 value	2001 rank
Andhra Pradesh	0.298	9	0.377	9	0.416	10
Assam	0.272	10	0.348	10	0.386	14
Bihar	0.237	15	0.308	15	0.367	15
Gujarat	0.360	4	0.431	6	0.479	6
Haryana	0.360	5	0.443	5	0.509	5
Karnataka	0.346	6	0.412	7	0.478	7
Kerala	0.500	1	0.591	1	0.638	1
Madhya Pradesh	0.245	14	0.328	13	0.394	12
Maharashtra	0.363	3	0.452	4	0.523	4
Orissa	0.267	11	0.345	12	0.404	11
Punjab	0.411	2	0.475	2	0.537	2
Rajasthan	0.256	12	0.347	11	0.424	9
Tamil Nadu	0.343	9	0.466	3	0.531	3
Uttar Pradesh	0.255	13	0.314	14	0.388	13
West Bengal	0.305	8	0.404	8	0.472	18
All India	0.302		0.381		0.472	

Source Human Development Report, India, 2001
Note The HDI for 2001 has been estimated only for a few selected States for which some data, including the Census 2001 was available

Table 2.33 Human development scenario of India from 1999–2000 to 2007–2008

State	HDI (1999–2000)	HDI (2007–2008)	Rank (1999–2000)	Rank (2007–2008)
Kerala	0.677	0.790	2	1
Delhi	0.783	0.750	1	2
Himachal Pradesh	0.581	0.652	4	3
Goa	0.595	0.617	3	4
Punjab	0.543	0.605	5	5
North Eastern States (excluding Assam)	0.473	0.573	9	6
Maharashtra	0.501	0.572	6	7
Tamil Nadu	0.480	0.570	8	8
Haryana	0.501	0.552	7	9
Jammu & Kashmir	0.465	0.529	11	10
Gujarat	0.466	0.527	10	11
Karnataka	0.432	0.519	12	12
West Bengal	0.422	0.492	13	13

Source HDR of India, 2011

Table 2.34 Forest cover in states in India (in sq.km)

Sl. No.	States	Geographical area	Forest coverage areas			
			Very dense forest	Moderately dense forest	Open forest	Total forest area	Percentage of G.A
1	Andhra Pradesh	275,069	850	26,242	19,297	46,389	16.86
2	Arunachal Pradesh	83,743	20,868	31,519	15,023	67,410	80.5
3	Assam	78,438	1444	11,404	14,825	27,673	35.28
4	Bihar	94,163	231	3280	3334	6845	7.27
5	Chhattisgarh	135,191	4163	34,911	16,600	55,674	41.18
6	Delhi	1483	6.76	49.48	119.96	176.2	11.88
7	Goa	3702	543	585	1091	2219	59.94
8	Gujarat	196,022	376	5231	9012	14,619	7.46
9	Haryana	44,212	27	457	1124	1608	3.64
10	Himachal Pradesh	55,673	3224	6381	5074	14,679	26.27
11	Jammu & Kashmir	222,236	4140	8760	9639	22,539	10.14
12	Jharkhand	79,714	2590	9917	10,470	22,977	28.82
13	Karnataka	191,791	1777	20,179	14,238	36,194	18.87
14	Kerala	38,863	1442	9394	6464	17,300	44.52
15	Madhya Pradesh	308,245	6640	34,986	36,074	77,700	25.21
16	Maharashtra	307,713	8736	20,815	21,095	50,646	16.46
17	Manipur	22,327	730	6151	10,209	17,090	76.54
18	Meghalaya	22,429	433	9775	7067	17,275	77.02
19	Mizoram	21,081	134	6086	12,897	19,117	90.68
20	Nagaland	16,579	1293	4931	7094	13,318	80.33
21	Odisha	155,707	7060	21,366	20,477	48,903	31.41
22	Pondicherry	480	0	35.37	14.69	50.06	10.43
23	Punjab	50,362	0	736	1028	1764	3.5
24	Rajasthan	342,239	72	4448	11,567	16,087	4.7
25	Sikkim	7096	500	2161	698	3359	47.34
26	Tamil Nadu	130,058	2948	10,321	10,356	23,625	18.16
27	Tripura	10,486	109	4686	3182	7977	76.07
28	Uttar Pradesh	240,928	1626	4559	8153	14,338	5.95
29	Uttarakhand	53,483	4762	14,167	5567	24,496	45.8
30	West Bengal	88,752	2984	4646	5365	12,995	14.64
	All India	3,287,263	83,471	320,736	287,820	692,027	21.05

Source India State of Forest Report 2011

2.6 Human development profile

Table 2.35 State wise percentage of forest to total geographical area (1995–2011)

Sl. No	States	1995	1997	1999	2001	2003	2005	2007	2011[a]
1	Andhra Pradesh	23.17	23.2	23.2	23.2	23.2	23.2	23.2	16.86
2	Arunachal Pradesh	61.55	61.55	61.55	61.55	61.55	61.55	61.55	80.5
3	Assam	39.15	39.15	39.15	34.45	34.45	34.21	34.21	35.28
4	Bihar	16.81	16.81	16.81	6.45	6.87	6.87	6.87	7.27
5	Chhattisgarh				43.85	44.21	44.21	44.21	41.18
6	Delhi	2.83	2.83	5.73	5.73	5.73	5.73	5.73	11.88
7	Goa	32.93	37.34	37.34	33.07	33.06	33.06	33.06	59.94
8	Gujarat	9.89	9.89	9.89	9.69	9.75	9.67	9.66	7.46
9	Haryana	3.82	3.78	3.78	3.51	3.52	3.53	3.53	3.64
10	Himachal Pradesh	67.52	63.6	63.6	66.52	66.52	66.52	66.52	26.37
11	Jammu & Kashmir	9.08	9.08	9.08	9.1	9.1	9.1	9.1	10.14
12	Jharkhand				29.61	29.61	29.61	29.61	28.82
13	Karnataka	20.15	20.19	20.19	20.19	22.46	19.96	19.96	18.87
14	Kerala	28.88	28.87	28.87	28.87	28.99	28.99	28.99	44.52
15	Madhya Pradesh	35.07	34.84	34.84	30.89	30.89	30.72	30.72	25.21
16	Maharashtra	20.75	20.8	20.8	20.13	20.13	20.13	20.13	16.46
17	Manipur	67.87	67.87	67.87	78.01	78.01	78.01	78.01	76.54
18	Meghalaya	42.34	42.34	42.34	42.34	42.34	42.34	42.34	77.02
19	Mizoram	75.59	75.59	75.59	75.59	79.3	79.3	79.3	90.68
20	Nagaland	52.02	52.05	52.05	52.05	50.05	52.05	55.62	80.34
21	Odisha	36.73	36.73	36.73	37.34	37.34	37.34	37.34	31.41
22	Punjab	5.64	5.76	5.76	6.07	6.12	6.12	6.07	3.5
23	Rajasthan	9.22	9,26	9.26	9.49	9.49	9.49	9.54	4.7
24	Sikkim	37.34	37.34	37.34	81.24	82.31	82.31	82.31	47.34
25	Tamil Nadu	17.45	17.4	17.4	17.59	17.59	17.59	17.59	18.11
26	Tripura	60	60.01	60.01	60.01	60.01	60.02	60.02	76.07
27	Uttar Pradesh	17.49	17.55	17.55	6.98	6.98	6.97	6.88	5.95
28	Uttarakhand				64.81	64.81	64.79	64.79	45.8
29	West Bengal	13.38	13.38	13.38	13.38	13.38	13.38	13.38	14.64
	All India	23.36	23.28	23.28	23.38	23.57	23.41	23.41	21.05

Source Compendium of Environment Statistics, 2011, India State of Forest Report, 2011
[a]2011 figures corresponds to forest cover

remarkably well in human development outcomes to climb up three rungs from 1999–2000 to 2007–2008.

The moot point for bringing human development in the discourse is that huge load of population of India stands in the way of achieving desired scale of human development in meeting essential requisites for people to enjoy long, healthy and

Table 2.36 Water resources of India at a glance

Estimated annual precipitation (including snowfall)	4000 km^3
Run-off received from upper riparian countries	(Say) 500 km^3
Average annual natural flow in rivers and aquifers.	1869 km^3
Estimated utilisable water (i) Surface (ii) Ground	1123 km^3 (i) 690 km^3 (ii) 433 km^3
Water demand ≈utilization (for year 2000) (i) Domestic (ii) Irrigation (iii) Industry, energy and others	634 km^3 (i) 42 km^3 (ii) 541 km^3 (iii) 51 km^3

Source Comprehensive Mission Document of National Water Mission, India

creative lives. As a result, the country continues to have relatively low HDI, and the prospect of better HDI score for such huge mass of population has become rather impossible. Huge size of population is a permanent drag on any developmental efforts, including Human Development, in India.

2.7 Resources scenario

Natural resources are provided by the Mother Earth to make the earth system liveable for all its creatures. Such natural resources can be classified under biotic and abiotic. Biotic resources originate from living things or organic materials such as plants, animals, and fossil fuels whereas abiotic resources include nonliving and inorganic materials such as air, sunlight, and water. Minerals (gold, copper, iron, diamonds etc.) are also considered abiotic. Renewable resources are those natural resources that are liable to be replenished after its use. The sun and wind are replenished by the Earth system as a part of its natural order. On the other hand, plants and water can be replenished by judicious use of such assets, by its conservation and replanting, and making optimum use of water resources and allowing replenishable time of ground level water- recharging. In this section, among the biotic and non-biotic resources, only two items are taken up hereunder, namely Forests Resources and Water Resources for understanding the current status in our country.

2.7.1 Forests Resources

The Proportion of land area covered by forest is a strong indicator on environmental sustainability. On it also depends how efficiently the effects of climate change can be mitigated. Additionally, it helps to improve water security, safeguards rich biodiversity and provides livelihood security for large number of people. As per

2011 assessment, the Country has a forest cover of 692,027 km^2, which is 21.05 % of the Country's geographical area. Incidentally, the forest cover (revised) estimate for 2009 shows total forest cover of 692,394 km^2 which indicates a decline of 367 km^2 in 2011. The quality of forest coverage area depends on dense forest, moderately dense forests and open forests and extent of such coverage. At Table 2.34, the forests coverage and per cent of forest coverage over the years respectively in the States of India have been shown:

Table 2.35 reveals the declining trend of forest coverage in the States of India being factored, among other things, by the encroachment of increasing human habitation to accommodate surging human population and also due to non-availability of 'spare able' land for spurting new forest growth and coverage. Thus, the prospect of growth elasticity of forestry in the States of India is shockingly negative and with annual addition of human number the status of forest cover would change from bad to worse.

2.7.2 Water Resources

'The main water resources of India consists of the precipitation on the Indian territory which is estimated to be around 4000 km^3/year, and trans-boundary flows which it receives in its rivers and aquifers from the upper riparian countries. Out of the total precipitation, including snowfall, the availability from surface water and replenishable groundwater is estimated as 1869 km^3. Due to various constraints of topography, uneven distribution of resource over space and time, it has been estimated that only about 1123 km^3 including 690 km^3 from surface water and 433 km^3 from groundwater resources can be put to beneficial use.

Precipitation over a large part of India is concentrated in the monsoon season during June to September/October. Precipitation varies from 100 mm in the western parts of Rajasthan to over 11,000 mm at Cherrapunji in Meghalaya (Table 2.36).

There takes place extreme natural occurrences in the country—floods are followed by droughts and in some cases they exist simultaneously in different locations of the country. Due to excess rainwater, floods occur in certain parts. It has been estimated by RashtriyaBarhAyog (RBA) that 40 mha of area is flood-prone and this constitute 12 % of total geographical area of the country. Droughts are also experienced due to deficient rainfall. It has been found that 51 mha area is drought prone and this constitute 16 % of total geographical area.

This availability of water resources is impacted by the growing population of the country. Accordingly, the per capita availability of water for the country as a whole has decreased from 5177 m^3/year in 1951–1654 m^3/year in 2007. Due to spatial variation of rainfall, the per capita water availability also varies from basin to basin. The distribution of water resources potential in the country shows that the average per capita water availability in Brahmaputra & Barak basin was about 14,057 m^3/year whereas it was 308 m^3/year in Sabarmati basin in year 2000. Meanwhile, with

ongoing climatic change and the increasing size of population, the prospect of water resources h
as gone down substantially Further, in the context of projected population size of India being 1.39 billion in 2025 and 1.80 billion in 2050 along with emerging high intensity impact of climatic change, the prospect for water availability would be fearsome for multiple needs for development and per capita human needs for domestic use.

2.8 Environmental and Climatic scenario

Human population growth is a major contributor to global warming. The use of fossil and its increasing use shape up the lifestyles of growing number of population which is indeed alarming. The large number of population means more demand for oil, gas, coal and other fuels from below the Earth's surface. When those are mined, drilled and then burned, increasing volume of carbon dioxide (CO_2) is spewed into the atmosphere to trap warm air inside like a greenhouse.

As per the estimate of the United Nations Population Fund, with 6.1 billion people during the course of the 20th century. Emissions of CO_2, the leading greenhouse gas, grew 12-fold and with worldwide population expected to surpass nine billion over the next 50 years, environmentalists and others are worried about the ability of the planet to withstand the added load of greenhouse gases entering the atmosphere and wreaking havoc on ecosystems down below.

Though developed countries consume the lion's share of fossil fuels right at the present moment, fast-growing developing countries will contribute significant proportion of global CO_2 emissions by 2050. It is widely believed that, just like China, India with its massive population size and significant growth rates will also become very active contributors in magnifying the impacts of global warming. Incidentally as per estimate of National Population Fund, India will have 1.80 billion people by 2050 and will surpass the estimated population size of China of 1.42 billion in 2050. The increasing size of population along with significant growth rate will alter the status of contribution of CO2 emissions of India. There is imperative need to minimize the damage making potential of CO_2 emission by India and the country needs to undertake harsh measures to check the rush of population numbers to assume a responsible role in the management of global Environmental and Climatic change scenario.

The Ministry of Environments and Forests, Government of India released its greenhouse gas inventory of 2007 emissions recently, making it the first developing country to publish such emissions data. The released emissions data was based on 1994 figures. Since then, India's emissions have grown at an average annual rate of 3.3 per cent, increasing from 1.25 billion tons in 1994 to 1.9 billion tons in 2007. The report analysed emissions from electricity use, transportation, agriculture, and land use change. Land use serves as a net carbon sink in India, in contrast to many developing countries where deforestation is a major source of emissions. India is

2.8 Environmental and Climatic scenario

now the world's fifth-largest emitter of greenhouse gases, ranking behind China, the United States, the European Union, and Russia. When releasing the data, the former Environment Minister, however, had pointed out that India's emissions are still one-quarter of those of the top emitters, the United States and China. He had further highlighted that in the same period, from 1994 to 2007, India reduced the emissions intensity of its economy by 30 per cent. Be that as it may, unless India reduces emissions intensity by a further 20–25 % between 2005 and 2020, the position indeed would be very bleak. It thus sends a clear message that India has to make a very serious effort on population stabilization front and arrest its runaway population size and growth in order to minimise the demand pull factor to increased use of oil, gas, coal and other fuels from below the Earth's surface.

Chapter 3
Other Important Issues Relevant for Control of Population

Abstract The Population Control and Family Planning (or Family Welfare, as in India) is a multi-disciplinary subject with a host of issues relevant to each discipline. Such issues are functionally very important and needs appropriate address at the time of programme implementation. The most crucial task is to identify the related issues that do not get escaped from the domain of programme management. Adequate time and application of mind have, therefore, to be invested to locate the issues relevant for the nationally important population control and family planning. Earlier, at Chap. 2, the issues arising from theories of economics, which border on macro level management area, have been discussed. Since economic development is also impacted in a very significant way by the population size and its load, the crucial micro issues assume equal importance in containing the growth of population. Such micro-issues happen to be very important from programme management angle. Issue-based micro-management can only ensure intended outcome. Such issues include, Demographic Issue, Social Issue, Cultural Issue, Gender Issue, Socio-religious Issue, Religious Issue, Political Issue, Contraceptive Issue, Health Issue, Child Care Issue, Educational Issue, Population Education issue, Media Issue, Information, Education and Communication Issue, Legal Issue. Organisational Issues and Social Security Issue In this chapter, attempts have been made to analyse the contributing power and strength of each of these issues to make big impact to check in a significant manner the runaway population growth of our country.

Keywords Demographic issue · Gender issue · Socio-religious issue · Political issue · Contraceptive issue · Population education issue

The Population Control and Family Planning (or Family Welfare, as in India) is a multi-disciplinary subject with a host of issues relevant to each discipline. Such issues are functionally very important and needs appropriate address at the time of programme implementation. The most crucial task is to identify the related issues that do not get escaped from the domain of programme management. Adequate time and application of mind have, therefore, to be invested to locate the issues relevant for the nationally important population control and family planning. At Chap. 2, the issues arising from theories of economics, which border on macro level management area, have been discussed. Since economic development is also impacted in a very significant way by the population size and its load, other micro issues assume equal importance in containing the growth of population. Such issues happen to be very crucial from programme management angle. Issue-based management can only ensure intended outcome. In this chapter, attempts have been made to enlist the discipline which are germane to population control and family planning and then have focussed on discipline-specific issues that have bearing in a significant manner in controlling the population problem. They are as follows:

3.1 Demographic Issues

Demography is the study of statistics such as births, deaths, income and other incidences which illustrate the changing structure and the composition of a human population in a particular geographical location. Thus, from population control point of view, demographic issues relate to the kind of population size over a period of time, the trend of its growth rate, its birth rate including crude birth rate, death rate, infant mortality rate including child death rate, Sex ratio, density of population etc. Studied data on such components reveal the nature of prevailing demographic scenario. If the data reveal any area of concern, it demands adequate notice and of its critical appraisal vis-a-vis the current policy and its linked programme, if any. The size of the population and its incremental addition is an indicator whether the concerned location is in a position to carry its load from economic and environmental angle. The growth rate is also a pointer whether the fertility behaviour and or other immigration/infiltration factors is a problematic zone and requires resolution. Similarly, declining sex ratio, adverse neonatal mortality rate, infant mortality rate or child death rates are areas that have wide implications on population issues and call for dedicated actions. Since the demographic data, in good measure, give an overview of the characters of the given population, unless such data of any given location are objectively studied before launching any programme of action, the population control programme and or family planning programme of such location would cease to be meaningless and void of any objective reality as obtained in the field. Unfortunately, demographic issues seldom get its due importance in population control and family planning programme at the state, district and sub-district levels in our country explaining why such programme, where ever it exists, does not yield intended result. Understanding the demographic scenario of any given location is the most important job for population control and family planning.

The national census provides database at every 10 year interval and is the appropriate source to understand the latest demographic status of a given location. It covers the size of population, its decadal growth rate, sex ratio, rural and urban population growth and its ratio, age-specific population including those in the 0–6 age group, marital status data, educational data, religion data and the like. Earlier in

3.1 Demographic Issues 59

Chap. 2, some important demographic data from Censuses conducted in India since 1951–2011 have been captured. In this chapter other demographic data, as relevant, for our country are presented.

3.1.1 Population Projection

The National Commission on Population appointed a Technical Group on Population Projections to have a fair idea about the population scenario for a period spanning from 2001 to 2026. The said Technical Group on Population submitted its Report in May 2006. The Report projected quite an alarming trend of the size of population of India and for some of the states in particular as mentioned at Table 3.1

The projected population of states, which appear to be very cautious and conservative, opens up the enormity of serious demographic challenges which the political executives at different levels in India needs to ponder. The Technical Group on Population Projections observed as follows:

"The State, which is expected to have least growth in the quarter century (2001–2026) is Tamil Nadu (15 %), followed by Kerala (17 %). In contrast, Delhi will have the highest projected growth of 102 % during 2001–2026. States, which will have projected growths in the range of 20–30 %, are Himachal Pradesh, Punjab, West Bengal, Orissa, Andhra Pradesh and Karnataka. The population in the states of Haryana, Rajasthan, Uttar Pradesh and Madhya Pradesh is projected to increase by 40–50 % during 2001–2026, which is above the national average of 36 %. The population of Uttar Pradesh is expected to be highest among all the states of the country at almost 249 million in 2026.

Of the projected increase in population of 371 million in India during 2001–26, 187 million is likely to occur in the seven States of Bihar, Chhattisgarh, Jharkhand, Madhya Pradesh, Rajasthan, Uttar Pradesh and Uttaranchal (termed as BIMARU states, since it was so before division). Thus nearly 50 % of India's demographic growth during this period of twenty five years, is projected to take place in these seven states. Twenty two per cent of the total population increase in India of 371 million during 2001–26 is anticipated to occur in Uttar Pradesh alone. The population in these seven states together is expected to grow at 1.5 % per annum during 2001–26. In contrast, the contribution of the four southern states, namely Andhra Pradesh, Karnataka, Kerala and Tamil Nadu, to the total increase in population size of the country during 2001–2026 is expected to be 47 million-thirteen per cent of total Demographic growth of the country. The population in these four states together is expected to grow at 0.8 % per annum during 2001–26. Proportion of contribution of some of the other states to the total increase in population size during 2001–26 is Maharashtra (10 %), West Bengal and Gujarat (5 % each) and Delhi (4 %). These four states together thus contribute 24 % of the total increase in population size during 2001–2026. The contribution of the remaining states and union territories to the total increase in population size during 2001–2026 is 13 %."

Table 3.1 Population projections of India and States, 2001–26

Year	Projected population					
	India	Jammu & Kashmir	Himachal Pradesh	Punjab	Chandigarh	Uttaranchal
2001	1,028,610	10,144	6078	24,359	901	8489
2002	1,045,547	10,301	6151	24,699	922	8634
2003	1,062,388	10,461	6234	25,041	957	8780
2004	1,079,117	10,622	6309	25,384	1000	8927
2005	1,095,722	10,783	6383	25,724	1050	9073
2006	1,112,186	10,941	6455	26,059	1103	9219
2007	1,128,521	11,099	6526	26,391	1161	9365
2008	1,144,734	11,257	6595	26,722	1227	9511
2009	1,160,813	11,414	6662	27,048	1297	9656
2010	1,176,742	11,568	6728	27,368	1368	9800
2011	1,192,506	11,718	6793	27,678	1438	9943
2012	1,208,116	11,865	6856	27,981	1508	10,084
2013	1,223,581	12,010	6918	28,279	1580	10,224
2014	1,238,887	12,152	6978	28,568	1651	10,362
2015	1,254,019	12,289	7037	28,846	1719	10,499
2016	1,268,961	12,419	7095	29,112	1780	10,632
2017	1,283,600	12,545	7151	29,372	1859	10,761
2018	1,298,041	12,665	7206	29,625	1941	10,887
2019	1,312,240	12,780	7259	29,868	2028	1101p
2020	1,326,155	12,888	7311	30,101	2122	11,129
2021	1,339,741	12,987	7361	30,323	2226	11,241
2022	1,352,695	13,086	7408	30,542	2301	11,351
2023	1,365,302	13,180	7453	30,753	2374	11,457
2024	1,377,442	13,269	7497	30,956	2438	11,558
2025	1,388,994	13,353	7537	31,154	2488	11,655
2026	1,399,838	13,434	7575	31,345	2518	11,746
Year	Projected population					
	Haryana	Delhi	Rajasthan	Uttar Pradesh	Bihar	Sikkim
2001	21,145	13,851	56,507	166,198	82,999	541
2002	21,579	14,273	57,664	169,547	84,612	548
2003	22,015	14,698	58,825	172,944	86,194	555
2004	22,400	15,129	59,984	176,374	87,745	562
2005	22,883	15,569	61,136	179,824	89,264	569
2006	23,314	16,021	62,276	183,282	90,752	576
2007	23,743	16,484	63,408	186,755	92,208	583
2008	24,171	16,955	64,534	190,254	93,633	591
2009	24,597	17,457	65,650	193,763	95,026	598
2010	25,020	17,935	66,750	197,271	96,389	605
2011	25,439	18,451	67,830	200,764	97,720	612

(continued)

3.1 Demographic Issues

Table 3.1 (continued)

Year	Projected population					
	Haryana	Delhi	Rajasthan	Uttar Pradesh	Bihar	Sikkim
2012	25,854	18,983	68,892	204,250	99,020	619
2013	26,266	19,529	69,940	207,739	100,289	626
2014	26,675	20,092	70,969	211,217	101,526	633
2015	27,079	20,676	71,973	214,671	102,732	640
2016	27,477	21,285	72,948	218,088	103,908	647
2017	27,868	21,896	73,924	221,469	105,064	653
2018	28,253	22,523	74,884	224,829	106,192	660
2019	28,631	23,164	75,828	228,152	107,293	667
2020	29,002	23,818	76,759	231,425	108,372	673
2021	29,362	24,485	77,676	234,631	109,431	679
2022	29,720	25,162	78,521	237,676	110,410	686
2023	30,071	25,852	79,339	240,651	111,352	692
2024	30,416	26,553	80,116	243,517	112,245	698
2025	30,755	27,263	80,841	246,234	113,081	704
2026	31,087	27,982	81,501	248,763	113,847	709
Year	Projected population					
	Arunachal Pradesh	Nagaland	Manipur	Mizoram	Tripura	Meghalaya
2001	1098	1990	2167	889	3199	2319
2002	1112	2016	2195	900	3241	2349
2003	1127	2042	2223	912	3283	2379
2004	1141	2068	2251	923	3324	2409
2005	1155	2094	2280	935	3366	2440
2006	1169	2119	2308	946	3407	2470
2007	1184	2145	2336	958	3449	2500
2008	1198	2171	2364	970	3491	2530
2009	1212	2197	2393	981	3532	2560
2010	1227	2223	2421	993	3574	2591
2011	1241	2249	2449	1004	3616	2621
2012	1255	2275	2478	1016	3658	2651
2013	1270	2301	2506	1026	3700	2682
2014	1284	2327	2534	1039	3742	2712
2015	1299	2354	2563	1051	3784	2743
2016	1313	2380	2592	1063	3826	2773
2017	1327	2405	2619	1074	3867	2803
2018	1341	2430	2646	1085	3906	2832
2019	1354	2454	2673	1096	3946	2860
2020	1367	2477	2698	1106	3983	2887
2021	1379	2500	2723	1116	4019	2914
2022	1392	2522	2747	1126	4056	2940

(continued)

Table 3.1 (continued)

Year	Projected population					
	Arunachal Pradesh	Nagaland	Manipur	Mizoram	Tripura	Meghalaya
2023	1404	2544	2771	1136	4091	2965
2024	1415	2566	2794	1146	4125	2990
2025	1427	2586	2817	1155	4159	3014
2026	1438	2606	2839	1164	4191	3038

Year	Projected population					
	Assam	West Bengal	Jharkhand	Orissa	Chhattisgarh	Madhya Pradesh
2001	26,656	80,176	26,946	36,805	20,834	60,348
2002	27,071	81,278	27,443	37,241	21,197	61,581
2003	27,478	82,320	27,922	37,670	21,553	62,799
2004	27,878	83,316	28,388	38,085	21,904	64,006
2005	28,273	84,277	28,846	38,490	22,251	65,202
2006	28,665	85,216	29,299	38,887	22,594	66,390
2007	29,053	86,125	29,745	39,276	22,934	67,569
2008	29,435	86,995	30,181	39,655	23,269	68,737
2009	29,814	87,839	30,611	40,025	23,600	69,897
2010	30,191	88,669	31,040	40,389	23,929	71,050
2011	30,568	89,499	31,472	40,750	24,258	72,200
2012	30,945	90,320	31,904	41,105	24,585	73,344
2013	31,319	91,122	32,334	41,453	24,909	74,482
2014	31,693	91,920	32,766	41,797	25,232	75,614
2015	32,069	92,725	33,203	42,138	25,555	76,745
2016	32,449	93,550	33,652	42,479	25,879	77,875
2017	32,810	94,334	34,069	42,808	26,186	78,964
2018	33,166	95,109	34,483	43,132	26,488	80,042
2019	33,516	95,875	34,887	43,450	26,782	81,101
2020	33,856	96,633	35,278	43,762	27,066	82,134
2021	34,183	97,383	35,652	44,068	27,337	83,135
2022	34,495	98,075	36,018	44,349	27,605	84,111
2023	34,796	98,747	36,375	44,620	27,865	85,064
2024	35,084	99,383	36,718	44,876	28,117	85,989
2025	35,354	99,988	37,046	45,112	28,359	86,879
2026	35,602	100,534	37,356	45,324	28,591	87,729

Year	Projected population					
	Gujarat	Daman & Due	Dadra & Nagar Haveli	Maharashtra	Andhra Pradesh	Karnataka
2001	18,930	57	50	41,101	20,809	17,962
2002	19,412	60	54	42,137	21,098	18,369
2003	19,895	63	61	43,195	21,382	18,778
2004	20,379	65	70	44,270	21,661	19,188

(continued)

3.1 Demographic Issues

Table 3.1 (continued)

Year	Projected population					
	Gujarat	Daman & Due	Dadra & Nagar Haveli	Maharashtra	Andhra Pradesh	Karnataka
2005	20,864	66	81	45,359	21,935	19,599
2006	21,351	68	93	46,456	22,205	20,012
2007	21,839	69	106	47,565	22,470	20,426
2008	22,328	70	121	48,688	22,730	20,842
2009	22,818	70	137	49,821	22,986	21,259
2010	23,310	71	154	50,960	23,238	21,677
2011	23,803	71	172	52,100	23,487	22,098
2012	24,298	72	190	53,244	23,732	22,520
2013	24,794	72	208	54,395	23,973	22,944
2014	25,291	73	227	55,549	24,210	23,369
2015	25,790	73	246	56,701	24,445	23,796
2016	26,290	74	264	57,846	24,676	24,225
2017	26,779	74	285	59,006	24,902	24,643
2018	27,266	75	307	60,170	25,123	25,059
2019	27,749	75	330	61,337	25,339	25,471
2020	28,225	76	354	62,507	25,551	25,877
2021	28,690	76	379	63,680	25,758	26,272
2022	29,756	79	396	66,078	26,510	27,166
2023	30,956	83	411	68,762	27,393	28,170
2024	32,427	88	424	72,012	285,044	29,400
2025	34,308	94	431	76,112	30,001	30,968
2026	36,737	103	431	81,341	31,999	32,990

Year	Projected population					
	Goa	Lakshadweep	Kerala	Tamil Nadu	Pondicherry	Andaman & Nicobar Island
2001	671	27	8267	27,484	649	116
2002	684	28	8329	28,388	650	122
2003	707	28	8390	29,300	664	128
2004	738	28	8456	30,218	686	135
2005	772	28	8508	31,140	714	142
2006	808	28	8565	32,063	745	150
2007	848	28	8621	32,988	780	158
2008	893	27	8675	33,918	822	167
2009	941	26	8727	34,850	869	176
2010	989	26	8778	35,780	915	185
2011	1034	25	8826	36,708	960	193
2012	1079	25	8872	37,634	1005	202
2013	1124	24	8917	38,560	1051	211
2014	1168	23	8959	39,484	1097	219

(continued)

Table 3.1 (continued)

Year	Projected population					
	Goa	Lakshadweep	Kerala	Tamil Nadu	Pondicherry	Andaman & Nicobar Island
2015	1208	23	8998	40,403	1138	227
2016	1241	23	9035	41,314	1172	235
2017	1286	22	9070	42,208	1221	244
2018	1333	22	9102	43,096	1271	253
2019	1381	21	9132	43,972	1325	263
2020	1434	21	9160	44,831	1383	273
2021	1491	20	9284	45,667	1447	285
2022	1545	20	9415	46,920	1517	297
2023	1599	21	9691	48,258	1591	311
2024	1654	21	10,060	49,776	1671	326
2025	1710	23	10,572	51,570	1758	342
2026	1766	24	11,272	53,734	1854	360

Report of the technical group on population projections constituted by the national commission on population, May 2006

3.1.2 Crude Birth Rate

From population control point of view, the Crude birth rate is the primary indicator of the fertility behaviour of the population. Crude Birth Rate (CBR) is defined as a ratio of the total number of births during a given year and on a given geographical area to the average (or mid-year) population ever lived in that year and geographical area. It is the simplest method used to measure fertility in a given area in a given period. There is, therefore, close correspondence between crude birth rate and the size of population in a given state. Any programme of population stabilization has to address this critical area. The Technical Group on Population Projections addressed the crucial area and located the states which need to give more attention on this area from country's population stabilization programme point of view. The projected crude birth rates of the states of India are indicated at Table 3.2

As against the above projections, censusindia.gov.in, based on SRS reports, records in 2012 that at the national level the CBR is 21.6 which varies from 23.1 in rural areas and 17.4 in urban areas. Andhra Pradesh, Delhi, Himachal Pradesh, Jammu & Kashmir, Kerala, Maharashtra, Punjab, Tamil Nadu are the major states having birth rate below 20, both in urban and rural areas. On the other hand, Madhya Pradesh has the highest birth rate in rural areas (28.3) and Uttar Pradesh has the highest in urban areas (23.5) followed by Rajasthan. The lowest CBR recorded in rural areas (15.1) was in Kerala and in urban areas (11.6) it was Himachal Pradesh.

3.1 Demographic Issues

Table 3.2 Crude birth rate of India and States of India from 2001–2025

India and States	2001–2005	2006–2010	2011–2015	2016–2020	2021–2025
India	23.2	21.3	19.6	18.0	16.0
Jammu & Kashmir	23.4	21.7	19.3	16.4	14.3
Punjab	19.0	17.8	15.9	14.0	12.6
Uttaranchal	25.2	23.5	21.4	18.9	16.5
Haryana	22.5	20.4	18.2	16.0	14.1
Delhi	16.0	15.4	15.9	15.5	14.3
Rajasthan	27.1	24.4	21.7	19.7	16.7
Uttar Pradesh	30.2	28.4	26.1	23.8	20.5
Bihar	27.5	24.2	21.6	19.7	17.4
Assam	24.0	22.0	20.8	19.0	16.6
West Bengal	18.8	16.6	15.9	15.3	14.1
Jharkhand	24.7	22.0	21.0	19.2	17.1
Orissa	21.0	18.9	17.5	16.2	14.4
Chhattisgarh	26.2	23.7	22.0	19.7	17.6
Madhya Pradesh	28.1	25.4	23.2	20.7	18.0
Gujarat	21.5	19.1	17.3	15.1	14.3
Maharashtra	19.8	18.6	16.9	15.5	13.8
Andhra Pradesh	19.2	17.4	16.2	15.1	13.7
Karnataka	19.3	17.8	16.6	14.9	13.8
Kerala	16.3	15.4	14.2	13.1	12.3
Tamil Nadu	17.3	15.7	14.6	13.4	12.5
North Eastern States excluding Assam	18.1	17.5	17.2	15.9	14.8

Source Report of the technical group on population projections constituted by the national commission on population, May 2006

3.1.3 Total Fertility Rate (TFR)

From population stabilisation angle, the status of TFR is another important indicator. The TFR means the number of children which a woman would bear during her life time if she were to bear children throughout her life. The TFR of level 2.1 children per woman in a population is generally taken as replacement level fertility. It is usually considered to be a precondition for population stabilization. However, if a population attains TFR of level 2.1, it does not mean the size of population get stabilized immediately. This takes some time to take effect because acceleration in population growth persists due to larger number of couples already existing in the reproductive ages in that population.

The Technical Group on Population Projections also worked on the status of the TFRs of the states in India and pointed out the possible journey of population stabilization of such states. In a way, it is implied that the programme of population

Table 3.3 Year by which projected TFR will be 2.1

India and major states	Year by which projected TFR will be 2.1
1 Andhra Pradesh	2002
2 Assam	2019
3 Bihar	2021
4 Chhattisgarh	2022
5 Delhi	Achieved in 2001
6 Gujarat	2012
7 Haryana	2012
8 Himachal Pradesh	Achieved in 2002
9 Jammu & Kashmir n.a.	n.a.
10 Jharkhand	2018
11 Karnataka	2005
12 Kerala	Achieved in 1988
13 Madhya Pradesh	2025
14 Maharashtra	2009
15 Orissa	2010
16 Punjab	2006
17 Rajasthan	2021
18 Uttar Pradesh	2027
19 Uttaranchal	2022
20 Tamil Nadu	Achieved in 2000
21 West Bengal	2003
22 North-East (Excl. Assam)	2005
India (*Weighted*)	*2021*
n.a.	n.a.

n.a. not available
Report of the technical group on population projections constituted by the national commission on population, May 2006

control need to be redesigned in such states on the basis of its TFR status in order to arrest the runaway population size of the country. Table 3.3 as below, gives us the position of TFRs:

3.1.4 Other Instruments Available for Demographic Data

- Sample Registration System and
- Civil Registration System

3.1.4.1 Sample Registration System

The data on fertility and mortality are available only after 10 years. Demographers feel it difficult to forecast the fertility behaviour of the country with the census data. Accordingly, the Government of India established the National Sample Registration

System based on Dual Recording System in the late 1960s. The field investigation under Sample Registration System consists of continuous enumeration of births and deaths in a sample of villages/urban blocks by a resident part time enumerator, and an independent six monthly retrospective survey by a full time supervisor. The data obtained through these two sources are matched. The unmatched and partially matched events are re-verified in the field to get an unduplicated count of correct events. The advantage of this procedure, in addition to elimination of errors of duplication, is that it leads to a quantitative assessment of the sources of distortion in the two sets of records making it a self-evaluating technique.

The main objective of SRS is to provide reliable estimates of birth rate, death rate and infant mortality rate at the natural division level for the rural areas and at the state level for the urban areas. Natural divisions are National Sample Survey (NSS) classified group of contiguous administrative districts with distinct geographical and other natural characteristics. It also provides data for other measures of fertility and mortality including total fertility, infant and child mortality rate at higher geographical levels. Such SRS data is a compulsory read for population control programme functionaries of the related state.

3.1.4.2 Civil Registration System (CRS)

The Registration of Births and Death Act, 1969 was enacted by the Parliament to enforce uniform civil Registration throughout the country. The relevant Rules under the Act are passed by the concerned State. It provides a forum for registration of births and deaths. Unlike census; it captures births and deaths on a day to day basis and is very useful for validation of Eligible Couple List in the local areas. Such CRS data has also to be used compulsorily by population control programme workers at the related local area.

Immigration and emigration:

There are four variables which govern changes in population size.

- births
- deaths
- immigration
- emigration.

A population gains individuals by birth and immigration, and loses individuals by death and emigration. While the SRS data gives us births rate and death rate, migration data are available in census publications.

3.2 Social Issues

Individual opinion per se is important for any decision concerning the welfare of an individual. However, man is a social animal and is usually subject to be influenced by the opinion prevailing in the given society. It often happens that man surrenders his own judgement and allows him to be guided by the common understanding of society in the larger belief that it would be in the best interest for him/her as well as for the society to obey them. Society is in fact the repository of all values and practices concerning the pattern of life and such invisible and unwritten social code are binding on all. Such values and practices transcend over generations and have become a part of hereditary structure of the society with its almost universal acceptance. Population control and family planning is associated with social mind-set. As a matter of fact, the Population control programme is more of a social issue management programme than a legal one. The area of social mind-set, social awareness and social ownership ultimately decides the societal approved population size and number. A number of social issues individually and collectively play their role in forming social opinion which impacts on the marriage and related fertility behaviour of the eligible couples. It would, therefore, be desirable to enlist them, discuss each of them and then find a way out to address them formally. They are as follows:

3.2.1 Age of Marriage

Generally speaking, the beginning of a family corresponds with the date of marriage. It is the turning point of life of a man and a woman to commence conjugal life of their own and take decision on such reproductive matter as they deem proper. However, for taking reproductive decision they need to be matured enough to take right decision which would be good to them, as husband or wife. Decision taking power depends on earning maturity both in terms of age and other factors. In the case of marriage, it also depends on right kind of physical and emotional build-up which is an essential requisite for leading a healthy sexual life. In such context, the concept of age of marriage has come into being. Age of marriage or Marriageable age (or marriage age) is the age at which a person is perceived to have attained certain level of maturity to lead sexual life and is allowed by law in the given society to marry either as a right or as an act subject to parental or other forms of consent. Age and other prerequisites to qualify for marriage vary between countries; but at a general rule, it is set at eighteen. All marriages, below 18 years of age, are usually called Child marriages which deny the right of women to marry at her majority and very often devastate the lives of such under-aged girls, their families and their communities as they are soon burdened with too many pregnancies. There are other dimensions as well. Such under-aged girl suffer health risks associated with early sexual activity and childbearing, leading to high rates of maternal and child mortality as well as sexually transmitted infections, including HIV. Very often they happen to be victims of domestic violence, sexual abuse and social isolation. Additionally, such child brides are pulled out of school as a part of societal practices depriving them of education and scope for meaningful work later. Consequently, it perpetuates the cycle of poverty in the life of such under aged girls. The deprivation of education and denial of work opportunities stand as a stumbling block of women empowerment and nip in the bud any scope to improve the quality of their human development. This harmful practice is in place usually in rural areas and more particularly among the relatively poor communities. However, this practice is also prevalent in several urban pockets among urban poor as well. Connected with this practice is the plea of a section of population to carry on the practice of under aged marriage of girls on the pretext of Personal Laws of their religious faith where community leaders seldom come forward to mobilise public opinion against it. Incidentally, all such under aged marriages in the country, irrespective of any religious faith and belief, are prohibited under 'The Prohibition Of Child Marriage Act, 2006(6 of 2007)'. The Prohibition of Child Marriage Act states that a girl in India can't marry before the of 18, and a boy before 21.

Over the years, the family planning workers and civil society partners have been trying to educate and mobilize public opinion against the under-aged marriages but not with very significant success. According to UNICEF study, 47% of girls are married by 18 years of age and 18 % of the girls by 15 years of age in India. Child marriage rate estimates vary significantly between sources and size of local survey samples. Table 3.4 below provides some of child marriage estimates of India along with the nature of data collection.

The methodology followed in small sample surveys explain different results on the extent of child marriages in India. For example, NFHS-3 data for 2005, as mentioned above, used a survey of the women in the age group 20–24, where they were asked if they were married before they were 18. The NFHS-3 also surveyed older women, up to the age of 49, asking the same question. The survey found that more number of the 40–49 women were married before they turned 18, than the 20–24 age women.

According to a Registrar General of India report in 2009, the states with highest observed marriage rates for under-18 girls, were Jharkhand (14.1 %), West Bengal

Table 3.4 Child marriage estimates of India

Source	% females married (<18)	Data Year	Sampling method
ICRW	47	1998	Small sample survey
UN	30	2005	Small sample survey
NFHS-3	44.5	1998–2002	Small sample survey
Census of India	43.4	1981	Nationwide census
Census of India	35.3	1991	Nationwide census
Census of India	14.4	2001	Nationwide census
Census of India	3.7	2011	Nationwide census

Source Wikipedia

(13.6 %), Bihar (9.3 %), Uttar Pradesh (8.9 %) and Assam (8.8 %). According to this report, despite sharp reductions in child marriage rates since 1991, still 7 % of women passing the age of 18 in India were married as of 2009. According to 2011 nationwide census of India, the average age of marriage for women in India is 21.2 and the extent of under-aged marriage is 3.7 %. The Annual Report of the Ministry of Health and Family Welfare, however, mentions that age of marriage of 22.1 % of girls is below 18 years of age.

The upshot of the discussion on the age of marriage is that it has one to one correspondence with fertility rate, population control and family planning linked as it is with reproductive space and reproductive behaviour. It is also linked with quality of children of the country. Children born out of under-aged girls are potentially malnourished and chance of being healthy citizen is very suspect.

3.2.2 Dowry

Dowry is an ancient custom in marriage system and was conceived as an affectionate parting gift to a daughter at the time of her marriage. It was often considered as transfer of parental property to a daughter at the time of her marriage rather than after the owner's death. This was also justified on the ground that dowry enables a daughter to have start-up fund to begin its new phase of life and may as well provide an element of financial security in times of crisis and also in the event of any tragic incident in the family. This solemn intention has now taken a backstage in a repackaged form to include durable goods, cash, jewellery, electrical appliances, furniture, bedding, crockery, utensils and other household items and real or movable property that the bride's family gives or bound to give to the bridegroom, his parents, or his relatives as a condition of the marriage. Even after the passing of the Hindu Succession Act, when the daughter enjoys equal right like the sons of the original family, the menace of dowry system has been on the rise. Apart from putting great financial burden on the bride's family with scaled-up demands of dowry, it often leads to crime against women, ranging from emotional abuse, injury to even deaths. As per legal position in the country, the payment of dowry, in any form, is now prohibited under the Dowry Prohibition Act, 1961 and punishable under Sections 304B and 498A of the Indian Penal Code. Even then, there is no let-down of the practice even in the educated families or also even when the bride is employed and does not depend on husband for financial support for maintenance. The scourge of dowry system has to end in the interest of women of the country. It is outright against dignity of women and a negative force altogether to equality of men and women.

The relevance of abolition of Dowry system is linked with the possible voice of the women to have a say on reproductive decision as equal partner. Since in the rural areas it is linked with the age of marriage, it tends to promote under-aged marriage to claim discount on dowry on count of younger age of the bride. Thus, indirectly the dowry system abets more children in the family.

3.3 Cultural Issues

Culture in popular sense signals a pattern of human behaviour, either explicit or implicit, in a social context of a given location. It has many dimensions and has been looked upon by different people in different ways in different context. The essential characteristics of culture consist of traditional ideas and especially their attached values. While a cultural system may be considered as a product of action, it is at the same a conditioning element for further action. Culture is acquired and shared by members of the society and is transmitted from one generation to another in a very silent mode. It shapes the pattern of human behaviour and its core feature is the value orientation embodied in the beliefs of the people.

Culture and the role of culture has been a key issue in the sociological discourse during the early stages of India's economic development when people in general were found to take for granted the century old belief-system and guided their mind-set and social responses to any development issue. Incidentally, it was an era of low level of literacy and educational attainment and the imprint of traditional religious faith was very prominent. Consequently, macro-level questioning mind was not in place. With the spread of literacy and education, knowledge and value orientation began to change for the better. The hard shell of resistance of traditional belief and faith began to melt down. From its static situation of stagnancy it started to move on to a new level of buoyancy and the social mind-sets have gone a transformational change, to a great extent, to respond, assess and decide on new ideas if those are properly posed and packaged.

However, it is a fact of life that culture dies hard. More so of issues concerning his/her personal way of looking at family building and all issues connected therewith. Two important cultural issues, namely Girl child and son's preference impact significantly in the reproductive behaviour of a traditional society.

3.3.1 Girl Child

A child is the best gift of nature. It is gender neutral. A girl child is as precious as a boy child. A family is formed by a man and a woman as joint partner. There is, therefore, no reason for a girl child to have a different status than that of a boy. However, in practice, a girl is usually *discriminated* both inside and outside the family. Inside the family, she often carries a transitional existence and burdened image and never treated alike like a boy in relation to food, clothing and other material comforts. As a result, her nutritional status remains relatively poor. Moreover, the priority in good number of families is not her education but in her marriage though education has been receiving attention of late, ever since the enactment of the Right to education Act. Moreover, the absence of school education closes her option to pursue higher education and a different career, and altogether a new life, and denies as well the privilege of a requisite forum to obtain knowledge and skills to qualify for any self-employment or other job. Needless to mention, education enables a child to realize her full potential, to think, question and judge independently; to be a wise decision-maker, develop civic sense and learn to respect, love her fellow human beings and to be a good citizen.

The issue of girl child is often discussed at the international level from the premise of human development rights of the girl child. On the Human Rights of the girl child, it states that "Human Rights are universal, and civil, political, economic, social and cultural rights belong to all human beings, including children and young people. Children and youth also enjoy certain human rights specifically linked to their status as minors and to their need for special care and protection. Girl-children are particularly vulnerable to certain human rights violations, and therefore require additional protections".

The human rights of children and the girl-child are explicitly set out in the Convention on the Rights of the Child, which includes the following indivisible, interdependent and interrelated human rights:

- The human right to freedom from discrimination based on gender, age, race, colour, language, religion, ethnicity, or any other status, or on the status of the child's parents.
- The human right to a standard of living adequate for a child's intellectual, physical, moral, and spiritual development.
- The human right to a healthy and safe environment.
- The human right to the highest possible standard of health and to equal access to health care.
- The human right to equal access to food and nutrition.
- The human right to life and to freedom from prenatal sex selection.
- The human right to freedom from cultural practices, customs and traditions harmful to the child, including female genital mutilation.
- The human right to education—to free and compulsory elementary education, to equal access to readily available forms of secondary and higher education and to freedom from all types of discrimination at all levels of education.
- The human right to information about health, sexuality and reproduction.
- The human right to protection from all physical or mental abuse.
- The human right to protection from economic and sexual exploitation, prostitution, and trafficking. The human right to freedom from forced or early marriage.
- The human right to equal rights to inheritance.
- The human right to express an opinion about plans or decisions affecting the child's life.'

Against this wider background, three issues on girl child have become most crucial in relation to population control and family planning in India. These may be listed as below:

3.3 Cultural Issues 73

- Declining child sex ratio

The demographic balance of any healthy society depends on the prevalence of a natural sex ratio. Sex ratio means the ratio of females to males in a given population. Such sex ratio may be looked upon from three angles: primary, secondary and tertiary. The primary sex ratio is the ratio at the time of conception; secondary sex ratio is the ratio at time of birth while tertiary sex ratio is the ratio of mature organism. The secondary sex ratio is commonly assumed, in human system, to be 105 boys to 100 girls and is often referred as "a ratio of 105".

The child sex ratio of our country has been declining as a whole and it has gone down from 927 in 2001 to 919 in 2011. The Child Sex Ratio means number of female children per 1000 male children in the age group 0–6 years of age. It indicates the gap between the natural Child Sex Ratio vis-à-vis the actual sex ratio at the early years of age group. It reveals societal attitude and outlook towards girl child. It is also a powerful indicator of the trend of social response on female children. Such data is also very important connected as it is with vital events in the society in the areas of marriage, labour force, age structure, migration etc. Deficit in Child Sex Ratio leads to structural imbalance in human system with spin off adverse consequences on the society and pose a grave concern. At Table 3.5, the sex ratio and child sex ratio of India since 1951 has been shown.

Incidentally, the Beti Bachao, Beti Padhao was introduced as a pilot programme in October, 2014 to address the issue of declining Child Sex Ratio (CSR) in 100 selected districts with low in CSR. The programme has been extended nationwide by the Prime minister on 22 January 2015 to address the dipping child sex ratio and empower girl child in the country.

- Child-marriage

The culture of child marriage is a deep rooted social evil. It deprived the child to grow till she passes her adulthood, loosened her bondage with the original family and then almost got uprooted in her family ties after being denied her lawful share at the time of inheritance though the Hindu girls have been protected after the

Table 3.5 Sex ratio and child sex ratio of India in India from 1951–2011

Census Year	Sex ratio	Child sex ratio
Census 1951	946	983
Census 1961	941	976
Census 1971	930	964
Census 1981	934	962
Census 1991	929	945
Census 2001	933	927
Census 2011	943	919

Source Census publications including PCA India, 2011

coming into being of the Hindu Succession Act, 1956. The child marriage practice is widely and strongly in force in Muslim families in both in rural and urban areas and the relevant Personal Law does not provide for any corresponding legal right in inheriting the property rights to such married girls in the original family. The position is really very sad.

The subject of child marriage has been separately discussed earlier. As mentioned therein, there are wide spread prevalence in child marriage in almost all the states in India. According to a Registrar General of India report in 2009, the states with highest observed marriage rates for under-18 girls, were Jharkhand (14.1 %), West Bengal (13.6 %), Bihar (9.3 %), Uttar Pradesh (8.9 %) and Assam (8.8 %). According to this report, despite sharp reductions in child marriage rates since 1991, still 7 % of women in India before passing the age of 18 were married off as of the year 2009. According to 2011 nationwide census of India, the average age of marriage for women in India is 21.2 and the extent of under-aged marriage is 3.7 %.

The upshot of the discussion on the age of marriage is that under-age marriage has one to one correspondence with population control and family planning linked as it is with reproductive space and reproductive behaviour. It is also linked with quality of children of the country. Children born out of under-aged girls are potentially malnourished and chance of being healthy citizen is very suspect.

The Prohibition Of Child Marriage Act, 2006(6 of 2007) came into being precisely to prevent child marriage for being one of the prime reasons for number of child births, poor health of the Children and the mother. Child marriage is also important causal factor for high IMR and MMR in our country. The Act raised the age limit of the male at 21 years of age and female at 18 years of age for the purpose of marriage. The Act did not, however, stipulate such Child marriage as void. The said Act extends to the whole of India and it applies also to all citizens of India within and beyond India.

Incidentally, in 2011, a United Nations resolution established 11 October as the International Day of the Girl Child (IDGC), a day designated for promoting the rights of girls and addressing the unique challenges they face. The inaugural day in 2012 focused on the issue of ending child marriage.

- Female foeticide:

Amniocentesis started in India in 1974 to detect fetal abnormalities. These tests were used to detect gender for the first time in 1979 in Amritsar, Punjab. Later the test was stopped by the Indian Council of Medical Research. However, with the coming into being of the ultra sound techniques, the detection of unborn girl child and its killing takes place in substantial number in many parts of the country, Haryana and Punjab earning special mention in this regard. A UN figure states that about 750,000 girls are aborted every year in India.

3.3.2 Sons Preference

The traditional preference for sons is deeply rooted in the structure of a feudal society, and more often in a agriculture based rural economy where men folk share a major portion of hard labour. Besides, the religious text of important faiths enjoins only men to perform the traditional ancestor cult. Since girls leave the original family after marriage into another family, sons stay in place to take care of the parents in old age in an undivided joint family. It is for the same reason that sons are perceived to carry on family name and lineage. Another factor that goes in favour of preference for son is due to the burden of the cost of the marriage of daughters, especially the ever increasing dowry associated with marriage. All these factors coalesces sons preference over a period of time and has become a part of the cultural belief system.

With the break-down of the joint family system and coming into being of the nucleus family, the traditional values and customs are changing fast. The historical dependence on sons for old age security is giving place to institutional old age related security scheme. Moreover, the spread of girl education and the consequent economic empowerment enables such girl children to take care of their parents as part of their duty even after their marriage. In that aspect, daughters are no different from sons. Till then, age old preference for son remains very strong and universal though many mothers want a balance of son and daughter.

The importance of son's preference in population control and family planning is immense. It is the driving force to have any number of children till a son is born in the family. Thus the quest for a son compels the wife to bear rounds of pregnancy till a son is born affecting in the process the health of the mother and existing children in the family, and adding in the process a number of additional children which would have been avoided otherwise. Additional births on this count calls for vigorous social education and awareness programme. Such a mental set-up is also the root cause of prenatal sex determination and subsequent female feticide.

3.4 Gender Issues

Family planning is the joint responsibility of any eligible couple-the male and the female. The role of the male or the female has not been uniformly fair in our country in sharing the family planning responsibility and has thus given rise to gender issue in the area of family planning. It is addressed now:

3.4.1 Male in Family Planning

Family planning is a programme meant for the family unit-the husband and the wife. As partner, they are normally supposed to take decision together jointly on composition of family unit and also its timing. However, this ideal situation is not the real story in the field and in a tradition bound society, the role of the husband and the wife is hardly equal. The reality obtaining in the social system is that the man often has a major say in decision on child bearing and family planning. The wife is normally expected to abide by the decision. Since the women bear the burden of pregnancy, the focus on family planning has usually been centred on women. Additionally, since family welfare programme is serviced primarily by ANM in rural areas, they interact with women only with the women-centric contraceptive methods to prevent further pregnancies. The ANMs, being female, do not approach, by tradition, to the men folk to take up male specific contraceptives. As a result, the male folk remained largely ignorant about male-specific contraception and of their role. Such an approach has, by default, led the men out of the direct contract on family planning services and enabled husband to give an excuse to shift his contraceptive responsibility on to his wife giving rise to the general impression that husbands do not show much interest in responsible parenthood. This macro mind-set, though changing, needs to be further activated to enable male to go in for male contraceptives; at least in the following areas male participation is actively required:

Temporary method:

(a) Periodic abstinence—Periodic abstinence is the traditional mode of contraception where the male partner has to abstain from physical union on days when his partner continues to remain in her fertile period.
(b) Withdrawal—Withdrawal is another form of contraception where participation of male partner is essential. The effectiveness of this method depends on the ability of male partner to withdraw before ejaculation. The failure to withdraw in time leads to unwanted pregnancies. With cautious practice the retention period can be increased and safe withdrawal ensured. This is a very cost effective method like periodical abstinence.
(c) Condoms—Condom is a non-intrusive and safe method providing dual protection against diseases as well as pregnancy. It is the most popular method for contraception by male partner and is also most effective if used with perfect consistency. It also protects against any virus of HIV.

Permanent Method:

(d) Vasectomy—Vasectomy is a safe and permanent method of male sterilization. It is completely effective. Now a days, the process of permanent sterilization has been made very simplified under non-scalpel vasectomy. Non-Scalpel vasectomy is convenient and easy option for male sterilization.

To sum up, in a traditional society like ours, men are usually by nature very conservative and do not of their own seek information on family planning methods from others except from female partner. The lack of quality communication on family planning method between wife and husband is one reason why husbands, with relatively less knowledge, is unable to make right choice among the available contraception and appears to be less supportive. There is, therefore, a strong case for knowledge empowerment of male on family planning by way of male participation in the clinic and also outside the clinic.

Further, the family planning is a unique case of shared responsibility. There are male methods of contraception just as there are female methods. All the methods have come into being out of sheer demand of acceptors. The age of the couple, their reproductive behaviour, physical condition of the couple at a given time, access to family planning services including supplies, the scope of consultation with medical and other health professional, the number of children already born in the family etc. determine the kind of contraception needed for good of the family. There is no scope for thrusting any method, either temporary or permanent, on to the women only. The male has to take appropriate role to limit the number of children in the family by way of accepting temporary or permanent method for health of the wife and the children. Additionally, in cases where the husband has multiple number of wives or also where divorce and remarriages are very common and frequent, the case for male contraception is all the more strong. In the interest of family planning and also in the interest of male involvement in family planning, the role of male motivator for males in family has increased tremendously. The purpose of family planning would be defeated if the men do not take appropriate contraception.

The Family Planning Professionals have been working on ways how the male participation can be improved and the hard-shell of resistance or wilful indifference of participation can be broken. They recommended the following for improvement of male involvement in Family Planning:

- Offer FP services at times that are convenient for men, including late evenings and some Sundays and holidays. Staff could be given compensatory leave in lieu of additional pay for the extra hours worked.
- Train service providers how to recognize and address the ways in which masculine identities impinge upon gender equality and sexual and reproductive health (SRH).
- Ensure that men's perspectives and motivations are integral in the design of programme activities.
- Broaden the concept of "male involvement" to "male responsibility" but understand that doing so demands changes in educational campaigns and motivational efforts. Men and women need to be educated about how contraception relates to gender equality and to their reproductive rights and responsibilities, and not only how FP helps achieve demographic goals.
- Service delivery systems and community-based service organizations must find coordinated and formal mechanisms for fully involving men in order to ensure their voluntary participation as change agents.

- Health professionals should be trained on how to sensitively address issues of sexuality. Discussion about sexuality is one of the key entry points for engaging boys and men on the topics of gender equality, gender-based violence, and sexual and reproductive health and rights.
- Experiences with male engagement in programmes on HIV prevention and gender-based violence have demonstrated that the "bigger" issues of sexuality and masculinity are at the center of men's identity and affect almost everything that men do. In the recent past, several initiatives using a masculinity framework have brought groups of men and boys together to form collectives or networks as a strategy to promote gender equality and prevent gender-based violence. It might be interesting to extend these efforts to include FP and demonstrate that gender equality and enhanced male responsibility can have a positive effect on FP issues.

Men belong to multiple associations and usually readily accept advice offered by these groups. Efforts may be made to identify organizations that could be approached as partners in this initiative to increase men's participation in SRH programmes. Such associations might include village governing bodies, religious associations, and other organizations of common interest to which men belong.

Lower rates of vasectomy are frequently due to lack of information or incorrect information. Many men believe that vasectomies reduce male sexual pleasure. The idea of pleasure during sexual intercourse is strongly related to perceptions of men's rights, and need to address men's concerns about vasectomies. Health institutions need to change their image as providers of "female-specific" health care to one of providers that also address the needs of couples, boys and married and unmarried men. Because of the complexity associated with male psychic environment, Family Planning professionals are also of opinion that the following areas need to be further researched and adequately explored to ensure full involvement of Male in taking family planning responsibility:

- 'We need a clearer understanding of how men conceptualize FP. Currently, FP is defined largely in terms of "spacing and limiting" the number of children; we should contextualize FP in a way that helps men learn how contraception addresses their own FP and sexual and reproductive health needs.
- What do men know and feel about various FP methods, and why don't men feel a responsibility to use contraception? To answer these questions, we must determine what the linkages are between FP and the sexual and reproductive health needs of men themselves. We certainly need context-specific information to address some of these attitudinal and knowledge-related barriers.
- Given the strong influence that men have in our family system, we also need to know more about several issues. For one, why and in how many ways do men continue to perpetuate the practices of early marriage? Also, with a strong preference for sons, are men explicitly or implicitly endorsing sex-selection and absolving themselves of the responsibilities of fatherhood? The answers to these questions are important in their role as determinants and also as consequences of

FP outcomes. The extent and nature of intra-familial communication would also be an important aspect to explore.
- There is little evidence from the supply perspective on what it means to work with men in FP. What barriers do front-line workers face accessing and working with men? In what ways can programmes better engage men? These questions need to be answered within an operations research framework to build evidence'.

3.4.2 Women in Family Planning and Other Related Women Issues

Gender is a multi-dimensional concept. It covers right to education and employment and have equal space in social and public life. From population control angle, the gender issue centres around reproductive rights. The International Conference on Women way back in 1995 (4–15 September), declared under the Beijing Platform for Action the following:

Good health is essential to leading a productive and fulfilling life, and the right of all women to control of all aspects of their health, in particular their own fertility, is basic to empowerment. Neglect of women's reproductive rights severely limits their opportunities in public and private life, including opportunities for education, economic and political empowerment. The ability of women to control their own fertility forms an important basis for the enjoyment of other rights. Government should therefore, pursue social, human development, education and employment policies to eliminate poverty among women in order to reduce their susceptibility to ill health and improve their own health.

The Broad components of Reproductive health care are as follows:

(a) Accessibility to good quality family planning services, counselling to suit the reproductive needs of individuals and couples, and prevention of unlawful pregnancy
(b) Provision of services safe motherhood services and infant care during and after pregnancy
(c) Provision of services related to infertility
(d) Provision and management of the consequences of unsafe abortion
(e) Prevention and management of reproductive disorders, including sexually transmitted disease, and prevention of HIV/AIDs
(f) Empowering adolescents by giving them reproductive and sexual health information and education in a comprehensive and sensitive way
(g) Ensure regular and uninterrupted availability of contraceptives, and quality family planning services, including counselling to individuals and couples.

The contraceptives availability for women is more than men and therefore women have far greater options to exercise a choice. However, before weighing option, it would be useful to look into pros and cons of each of the birth control methods and consult a doctor to guard against adverse impact on reproductive health. The following methods are usually available for contraception of Women:

Temporary Method:

Barrier Methods:
Barrier methods prevent sperm from entering the uterus and may be an option for women. It is reversible in nature.. Types of barrier methods include:

- **Female condoms**. These are thin, flexible plastic pouches. A portion of the condom is inserted into a woman's vagina before intercourse to prevent sperm from entering the uterus. The female condom also reduces the risk of STDs. Female condoms are disposed of after a single use.
- **Diaphragms**. Each diaphragm is a shallow, flexible cup made of latex or soft rubber that is inserted into the vagina before intercourse, blocking sperm from entering the uterus. Spermicidal cream or jelly should be used with a diaphragm. The diaphragm should remain in place for 6–8 h after intercourse to prevent pregnancy, but it should be removed within 24 h. Traditional latex diaphragms must be the correct size to work properly, and a health care provider can determine the proper fit. A diaphragm should be replaced after 1 or 2 years. The Woman needs to be measured again for a diaphragm after giving birth, having pelvic surgery, or gaining or losing more than 15 lb.
- **Cervical caps**. These are similar to diaphragms, somewhat smaller but relatively more rigid but less noticeable. The cervical cap is a thin silicone cup that is inserted into the vagina before intercourse to block sperm from entering the uterus. Like diaphragm, the cervical cap needs be used with spermicidal cream or jelly. The cap must remain in place for 6–8 h after intercourse to prevent pregnancy and needs to be removed within 48 h. Cervical caps happen to be of different sizes, and a health care provider determines the proper fit. With proper care, a cervical cap can be used for 2 years before replacement.
- **Contraceptive sponges**: These are soft, disposable, spermicide-filled foam sponges. It is required to be inserted into the vagina before intercourse. The sponge blocks sperm from entering the uterus, and the spermicide also kills the sperm cells. The sponge needs to be left in place for at least 6 h after intercourse and then removed within 30 h after intercourse.
- **Spermicides**: A spermicide destroys sperm. A spermicide can be used alone or in combination with a diaphragm or cervical cap. The most common spermicidal agent is a chemical called nonoxynol-9 (N-9). It is available in several concentrations and forms, including foam, jelly, cream, suppository, and film. A spermicide has to be inserted into the vagina close to the uterus no more than 30 min prior to intercourse and left in place 6–8 h after intercourse to prevent pregnancy. Spermicides do not prevent the transmission of STDs and may cause allergic reactions.

Hormonal Methods:
Hormonal methods of birth control use hormones to regulate or stop ovulation and prevent pregnancy. Ovulation is the biological process in which the ovary releases an egg, making it available for fertilization. Hormones can be introduced into the body through various methods, including pills, injections, skin patches, transdermal gels, vaginal rings, intrauterine systems, and implantable rods. Depending on the types of hormones that are used, these pills can prevent ovulation; thicken cervical mucus, which helps block sperm from reaching the egg; or thin the lining of the uterus. Health care providers prescribe, monitor, and administer hormonal contraceptives.

- **Combined oral contraceptives ("the pill")**. Combined oral contraceptive pills (COCs) contain different combinations of the synthetic estrogens and progestins and are given to interfere with ovulation. A woman takes one pill daily, preferably at the same time each day. Many types of oral contraceptives are available, and a health care provider helps to determine which type best meets a woman's needs. Use of COC pills is not recommended for women who smoke tobacco and are more than 35 years old or for any woman who has high blood pressure, a history of blood clots, or a history of breast, liver, or endometrial cancer.
- **Progestin-only pills (POPs)**. A woman takes one pill daily, preferably at the same time each day. Progestin-only pills may interfere with ovulation or with sperm function. POPs thicken cervical mucus, making it difficult for sperm to swim into the uterus or to enter the fallopian tube. POPs alter the normal cyclical changes in the uterine lining and may result in unscheduled or breakthrough bleeding. These hormones do not appear to be associated with an increased risk of blood clots.
- **Contraceptive patch**. This is a thin, plastic patch that sticks to the skin and releases hormones through the skin into the bloodstream. The patch is placed on the lower abdomen, buttocks, outer arm, or upper body. A new patch is applied once a week for 3 weeks, and no patch is used on the fourth week to enable menstruation
- **Injectable birth control**. This method involves injection of Depo-Provera, a progestin, in the arm or buttocks once every 3 months. This method of birth control can cause a temporary loss of bone density, particularly in adolescents. However, this bone loss is generally regained after discontinuing use of DMPA. Most patients using injectable birth control should eat a diet rich in calcium and vitamin D or take vitamin supplements while using this medication.
- **Vaginal rings**. The ring is thin, flexible, and approximately 2 inches in diameter. It delivers a combination of a synthetic estrogen and a progestin. The ring is inserted into the vagina, where it continually releases hormones for 3 weeks. The woman removes it for the fourth week and reinserts a new ring 7 days later. Risks for this method of contraception are similar to those for the combined oral contraceptive pills, and a vaginal ring is not recommended for any woman with a history of blot clots, stroke, or heart attack, or with certain types of cancer.

- **Implantable rods**. Each rod is matchstick-sized, flexible, and plastic. A physician surgically inserts the rod under the skin of the woman's upper arm. The rods release a progestin and can remain implanted for up to 5 years.
- **Emergency Contraceptive Pills** (**ECPs**). ECPs are hormonal pills, taken either as a single dose or two doses 12 h apart, that are intended for use in the event of unprotected intercourse. If taken prior to ovulation, the pills can delay or inhibit ovulation for at least 5 days to allow the sperm to become inactive. They also cause thickening of cervical mucus and may interfere with sperm function. ECPs should be taken as soon as possible after semen exposure and should not be used as a regular contraceptive method. Pregnancy can occur if the pills are taken after ovulation or if there is subsequent semen exposure in the same cycle.

Permanent Method:

Sterilization:

Sterilization is a permanent form of birth control that either prevents a woman from getting pregnant or prevents a man from releasing sperm. A health care provider must perform the sterilization procedure, which usually involves surgery. These procedures usually are not reversible.

- A **sterilization implant** is a nonsurgical method for permanently blocking the fallopian tubes. A health care provider threads a thin tube through the vagina and into the uterus to place a soft, flexible insert into each fallopian tube. No incisions are necessary. During the next 3 months, scar tissue forms around the inserts and blocks the fallopian tubes so that sperm cannot reach an egg. After 3 months, a health care provider conducts tests to ensure that scar tissue has fully blocked the fallopian tubes. A backup method of contraception is used until the tests show that the tubes are fully blocked.
- **Tubal ligation** is a surgical procedure in which a doctor cuts, ties, or seals the fallopian tubes. This procedure blocks the path between the ovaries and the uterus. The sperm cannot reach the egg to fertilize it, and the egg cannot reach the uterus.

3.5 Socio-Religious Issues

3.5.1 Minoritism

This is a fast changing social phenomena with strong spin-off effect. The Constitution of India has not defined minority nor has it enshrined any provision other than protection of Cultural and Educational Rights in Article 29, Right of minorities to establish and administer educational institutions in Article 30 and linguistic minority in Article 350B. Minority is definitely not a mathematical concept but presumed to have its constitutional meaning and definition for its correct usage or of its focus. However, Minoritism has emerged as a strong social phenomenon in Indian body-politic being championed by political class for vote-bank politics. Though all social sector data have now started being structured on religious lines including education, employment, poverty status and other components of human development deficit, efforts on showcasing family planning or population control of religious groups has not surfaced as yet. It is an area in which the dominant minority religious group is allergic and does not show any visible interest. The Sikhs, the Christens, the Jains and the Buddhists are found to be better responsive either on family Planning or on population control. The legal age of marriage, onset of pregnancy before 18 years of age, number of pregnancy, right of the women to have a say on pregnancy, per capita children of a married women, threat of dissolution of marriage for not agreeing to have more children etc. are some issues which have mixed up with religion. Since the status of family planning by the minority social groups is not captured in our country, the only option to understand the on-going family planning practices is by going through the census data on decadal growth of the social group. The approach of Indian Muslim towards family planning and population control is a big issue for sustainable population of our country as its decadal growth rate and decadal incremental addition are disproportionately higher. Family Planning for Muslim social group is almost a prohibited zone in our country least it injures their sentiment and adversely impacts the vote banks. It is a great irony of this country that any campaign on family planning is taken as attack on religion and the field workers feel insecure and threatened when a serious campaign is made. The role of the intelligentsia and civil society of the related social group is also less functional in this area. This is a challenging issue for the family planning in the states of India. Like dedicated programmes/schemes for minorities under Prime Minister's programme for minorities, there is a necessity to have a dedicated Eligible Couple List and Children Register for monitoring the family planning practices of the minority group as a part of national family planning and population control in India.

It would be fair and equitable to bring in the discourse religious issues of all descriptions to understand the linkage, either directly or indirectly, with the family planning and population control.

3.5.2 Religious Issues

Family planning or birth control is, in the final analysis, a collective decision of an eligible couple based on a number of factors including health of the mother, the financial and child caring support system available in the family, the societal norm or expectation of number of children in a family and a host of other factors as relevant for quality of life of family members in a civil society. Unlike animals, human beings do not indulge in breeding game of numbers. Normally, therefore, Religion has no definite role on family planning or birth control; it is very much a part of human psychic and human system. However, human beings from its ancient beginning have passed through phases of strife, stress and stains and overcame them with some divine code of conduct, belief and faith which led to the growth of different forms of religion and shaped its fundamental traits. Since Religions go by their own age old traditions, the apparently invisible religious mind-set plays a crucial role in influencing family size, its fertility behaviour and its birth control decisions including the use of its preventable methods. Way back in 1959, West off observed that 'the religious affiliation of the couple connotes a system of values which can affect family via several routes: (a) directly, by imposing sanctions on the practice of birth control or legitimising the practice of less effective methods only or (b) by indoctrinating its members with moral and social philosophy of marriage and family which emphasizes the virtues of reproduction'. Generally speaking, in all kinds of society, the knowledge of birth control methods, in some form or other has been in place since early times, especially with respect to temporary sterility. Though religious views on birth control vary widely, family planning is looked upon across the religious spectrum, by and large, as a moral good, a responsible choice, and a basic human right. Family planning is also looked upon as a means to protect the health of women and children and prevent unwanted pregnancies. Still then, religion does have its influence in the mental environment of an individual, in a given society or in the taking of collective decision of a particular faith. It would be worth trying to present, in brief, the linkage between fertility behaviour and the major religions in India:

Hinduism:
Unlike other religions of the world, Hinduism does not have any sole preceptor and a particular religious doctrine; in practice it is akin to way of life based on humans' existence in a transcendental world. Very recently, the Supreme Court of India observed that Hinduism was the collective wisdom and inspiration of the centuries. The Supreme Court further observed:

'It is a religion that has no single founder, no single scripture and no single set of teachings. It has been described as Sanatan Dharma, namely eternal faith, as it is the collective wisdom and inspiration of centuries that Hinduism seeks to preach and propagate.

The Times of India in its edition on December 17. 2015 also observed as follows:

Essentially, the Hinduism is the longest-running socially liberal program in the world. Unlike many other religions, Hinduism has never had a 'head', which is one of the reasons why it has survived countless attacks-those who have sought to kill it have failed to do so for the simple reason that they could never find a 'head' to chop off. It has thrived and evolved over thousands of years because of its secular, global, assimilative character-one where 'anything goes', be it food, dress, language or even belief. Indeed, Hinduism has exhibited an infinite capacity to contain contradictions and implicitly granted its 'subscribers' a democratic rights to create mythological hagiographies of their own.'

As usual in the true spirit of Hinduism, it does not have any religious doctrine of birth control or family planning. Most Hindus accept that there is a duty to have a family during the householder stage of life, and do believe in having children to continue the lineage. But there does not have any religious injunction against contraception. Therefore, for birth control or family planning, there is no obstruction from the Hinduism nor does exist any confusion whatsoever.

Wikipedia recorded the following note under the above subject:

"The Hinduism encourages procreation within marriage, yet there is no opposition against contraception. Most Hindus accept that there is a duty to have a family during that stage of one's life. So they are unlikely to use birth control to avoid having children altogether.

Traditional Hindu texts praise large families (which was normal in ancient times). Yet, Hindu scriptures that applaud small families also exist which emphasize the development of a positive social conscience. So family planning is seen as an ethical good. The Upanishads (texts delineating key Hindu concepts) describe birth control methods, and some Hindu scriptures contain advice on what a couple should do to promote contraception (thus providing a type of contraceptive advice). Contraception views vary widely among Hindu scholars. Although Gandhi advocated abstinence as a form of birth control, Radhakrishnan (a key Indian philosopher) and Tagore (the most prolific writer in modern Indian literature) encouraged the use of artificial contraceptive methods. Arguments in favour of birth control are drawn from the moral teachings of Hinduism. The Dharma (doctrine of the religious and moral codes of Hindus) emphasizes the need to act for the sake of the good of the world. Some Hindus, therefore, believe that producing more children than one or the environment can support goes against this Hindu code. Although fertility is important, conceiving more children than can be supported is treated as violating the *Ahimsa* (nonviolent rule of conduct).

In 1971, abortion was legalized in India, and there has very rarely been any objection to it. India has a high population, so discussion about contraception focuses more on overpopulation rather than moral or personal ethics. India was the first nation to establish a governmental population strategy based on birth control measures".

Buddhism:

In Buddhism, there is no established doctrine on birth control or contraception. 'The Traditional Buddhist teaching favours fertility over birth control. One view is that only after being a human can a soul reach Nirvana; so limiting the numbers of humans necessarily limits the numbers achieving Nirvana. The Buddhists, in general, do not favour in tampering with the natural development of life. A Buddhist may accept any form of contraception with a lot of hesitation. They usually do not like birth control by means of abortion or 'killing a human to be.'
'Buddhism is distinguished by wholesomeness which is the basis of all moral judgment. Thus, in a family set-up, the Buddhists believes they have duty to their parents. Similarly, Buddhism gives due importance to children and the duty to take care of them to grow up as decent human being with a good quality of life. Buddhist teachings, in this bigger context, support appropriate family planning when people feel that it would be too much of a burden on themselves or their environment to

have more children. Birth control allows couples to plan to have a certain number of children and prevent an excessive number of pregnancies. Buddhists believe that family planning should be allowed and that a good government should provide those services'.

Buddhists, in general, accept the Birth control options of pills and condoms but prefer condoms. According to a Buddhist activist, "the Buddhist scriptures say that many births cause suffering, so Buddhism is not against family planning. And we even ended up with monks sprinkling holy water on pills and condoms for the sanctity of the family."

Sikhism:
'The Sikh scripture, it is widely believed, does not condemn the use of birth control. Neither does the Sikh tradition condemn the same. Appropriate family planning is promoted by the community. The couple decides how many children they want and can support. They also decide whether or not to use contraception, and the type of birth control they are likely to use. Contraception decisions are centred on the needs of the family. Although Sikhs have no objection to birth control, they are not allowed to use it as a way to avoid a pregnancy resulting from adulterous behaviour. There is no religious mandate against abortion. Some don't support it because they believe that the fetus has a soul. But this decision is considered a personal choice'.

Jainism:
'The modern population problems have not been discussed in Jain scriptures. Jainism therefore, does not provide for any religious guidance on birth control or family planning. The Jains are aware of the alarming population problem of modern times and believe that they have to be resolved on the basis of the principles of non-violence, non-attachment and Jain-ethics. They argue that Jainism scriptures advocate sexual restraint and celibacy which fall in the category of family planning. They, however, admit that it cannot be a suitable method under modern complex situation. They hold the view, however, that contraception should be accepted but when accepted, it should be taken with regret by the acceptor as lack of self-control. Regarding abortion, since it involves both physical and psychic pain, the Jains's views are different. The Sthaananga Sutra has mentioned many forms of abortion in India before Mahabir's time. Jainism does not favour abortion as it involves murder of human being. Jainism believes that life begins in the day of conception'.

Christianity:
'The Bible, it is held, has said little about contraception and birth control. The Christian views on birth control in fact emanates from the church teachings and interpretations on marriage, sex, and family. The use of contraception by the Christians was a forbidden area and it had no religious sanction until the start of the 20th century, as it was perceived to be a barrier to God's procreative purpose of marriage. Many Christians consider sex as a gift from God and a positive force that strengthens the institution of marriage provided they do not feel threatened by having a number of children they cannot support. The majority of Protestant

denominations, theologians, and churches allow contraception to promote family planning as an important moral good. They stress that morality should come from the conscience of each person rather than from outside teachings'. It would be useful to capture separately, (from accepted websites and from its original version), the attitude of different Christian groups hereunder.

Roman Catholic:
'The Roman Catholic Church forbids sex outside marriage. Its teachings about contraception centres around the relationship between the husband and the wife. Catholicism strictly forbids the use of contraception. The Church teaches that sex must be both unitive and procreative and therefore, it is all against any kind of chemical and barrier methods of birth control and considers them morally unacceptable on the ground that artificial birth control methods impede the procreative aspect of sex, making contraception sinful. Natural family planning such as periodic abstinence is the only contraceptive method sanctioned by the Church. The catechism of the Catholic Church claims sex has a twofold purpose: "the good of the spouses themselves and the transmission of life (2363)."

The most Catholics, however, appear to disagree with the prohibition of birth control; in fact, surveys find that approximately 90 % of sexually active Catholic women of childbearing age use a birth control method forbidden by the church.

Incidentally, the Roman Catholic's aggressive ban on contraceptive is undergoing transformational change with the publication in 'Laudatosi' where Pope Francis wrote eloquently about our Mother Earth, "our common home who sustains and governs us." Allowing all women to plan for smaller families would give Mother Earth a much welcome respite.

Protestants:
'The Protestants in general uphold the use of some methods of family planning by married couples. They believe in creating a biblical model as a framework through which Christians can evaluate the moral and religious liberty issues confronting families in modern culture. The church believes that the use of birth control, as a means to regulating the number of children a couple has, and as a means to space out the ages of the children, is a moral decision that is left up to each couple. A section of them is, however, of the view that a couple may use any form of contraception that prevents conception'.

Evangelical Protestants:
The attitude of Evangelical groups, who rely more heavily on Catholic teachings, is not very uniform. Some of them oppose all forms of contraception short of abstinence while others allow natural family planning but oppose other methods. Some sects even support any form of birth control that prevents conception but are against any method that keeps a fertilized egg from implanting in the uterus. In 1954, The Evangelical Lutheran Church stated that "to enable them to more thankfully receive God's blessing and reward, a married couple should plan and govern their sexual

relations so that any child born to their union will be desired both for itself and in relation to the time of its birth."

The United Methodist Church:
'Methodists preach that every couple has the right and the duty prayerfully as well as the responsibility to control conception according to their circumstances. The United Methodist's Resolution on Responsible Parenthood dictates that as a means to uphold the sacred dimensions of personhood, all possible efforts should be made by the community and parents to ensure that every child enters the world with a healthy body and is born into an environment prepared to help the child to reach his/her fullest potential. That's why Methodists support public funding and participation in family planning services'.

The Presbyterian Church:
'Presbyterianism fully promotes equal access to birth control options. In fact, the Presbyterian Church advocates for legislation that would require insurance companies to cover the costs of birth control, asserting that contraceptive services are part of basic health care and warned that unintended pregnancies can lead to higher rates of infant mortality and maternal morbidity, and threaten the economic viability of families'.

Islam:
Family planning is a very sensitive subject under Islam and opinion sharply divides even among Muslim scholars. In such circumstances, it would be desirable to write down the correct position from an website of a Muslim academy or from writings of any Muslim scholar. Since an academy represents a collective view of Muslim scholars, it would be desirable to note down their views. Accordingly, a website of muslim-academy.com/islam-family-planning-birth-control/ which addresses the subject, has recorded the following (as on 19.12.2015):

"Islam tells its followers not to worry concerning the fact that if they would be able to raise and provide the children or not. Children are blessings from Allah, and He, Himself, provides for them.

Abortion in Islam is not promoted or encouraged, unless the life and health of the mother is in danger. Even in such circumstances, she was allowed to abort the fetus till a certain period of her pregnancy. If that certain period has passed, then abortion is no longer an option, even if the life of the mother is in danger.

To avoid aborting altogether, people refer to family planning and birth control strategies. Islam, under certain rules, allows Muslims to use them, as well.

Islamic Rulings on Birth Control:
Islam allows birth control methods to be used for the following conditions:

The methods or procedures do not provide any damage to the sexual organ and neither should they lead to permanent sterilization.

If the woman feels that she needs to opt for birth control methods/procedures, then she must first talk to her husband about it. With his approval, she is not to resort to anything, regarding this matter. The husband must be on board with his wife's decision.

There are different birth control procedures; some involve the actual placing of devices inside the person. Whatever procedure one opts for, should not involve the performance of any 'Haram' or forbidden act.

So, according to many Islamic scholars, birth control is permissible if the above conditions were kept in mind, no forbidden acts were performed to achieve it.

For those people who refer to abortion as a means of birth control, that was forbidden in Islam. But in severe cases if it must absolutely be done, then it should be before the soul was infused in the fetus.

Generally speaking, Islam does not prohibit the use of birth control methods. However, the reasons for why you are using birth control are to be kept in mind. With the population rate increasing by leaps and bounds, all over the world, and in such circumstances it only makes sense to do some family planning before you start a family. In countries and areas where managing a great population is a hassle, then information related to birth control and the various methods involved in it, should be disseminated.

The wife and husband should reach both a mutual decision. Islam however gives more right to the husband's approval or rejection regarding birth control strategies. Having his approval on this is mandatory for the wife."

3.6 Political Issues

Unlike other countries in the world, the population control and family planning is a major political issue in India. Paradoxically enough, the political parties do not have any population policy of their own nor do they have any serious interest on demographic challenges of the nation. Since the political parties do not have long term interest on sustainable population, they never bother them by indulging in periodical discussion or in stocktaking the population control programme of the government either in the parliament or in the concerned state legislature. They are mostly concerned with such population as are included in the voter list and have relevance in vote bank politics. The nation has not witnessed any serious contribution of the Parliament's committee on population on any discussion either in the parliament or outside. The role of political leaders in promoting the norm of small family and for taking required interest and care in preventing runaway population growth in the country, or in some of its parts, is not very praiseworthy. The underlying principle governing the political interest is only to see whether population stabilization programme would impact its vote bank politics and endanger political prospect. The courage of conviction to speak out the undesirable growth of population in any state or of any community or the possible limits of carrying such huge burden of population in a sustainable way is sadly missing from the contribution of national political parties of all descriptions. Even the role of regional parties to speak on the dangerous trend of population growth for the related state is also not visible and forthcoming. The ticking sound of additional population in every second in the population watch of our country does not disturb or bother our political leaders nor do they raise any serious question either in the parliament or outside as what to do in the face of ever increasing population size of the country.

Linked with number of population is the entry of illegal immigrants from across the borders where the voices of the political parties are seldom heard. The issue of illegal immigration in our border states also calls for a nationalist approach by the political spectrum. It is really sad and puzzling that the political voice of country, so shrill and loud, is silent on population issue and does not feel alarmed at the prospect of India overtaking China in the total count of population by 2025. The political parties of the country appears to be smug and comfortable with any number of population irrespective of the country's carrying capacity to sustain quality of human development of them. For the same reason, the political fraternity does not come forward with any blueprint of strong population policy to stop this runaway population menaces. Even they do not feel disturbed that the country is yet to adopt any population policy even after the expiry of the NPP, 2010. Given the track record, it may be a big wish that the clarion call for sustainable population in the SDGs summit in September, 2015 would really impact our political class and reactivate them.

In conclusion it may be said, however, that there is a strong case of role reversal of the political parties. All the national political parties need to revisit their current contribution and do justice to the goal of sustainable population in our country, in the SDGs framework, by way of enlisting their support through their published document, public utterances and otherwise.

3.7 Contraceptive Issues

Contraception means deliberate prevention of conception or impregnation. It prevents pregnancy by interfering with the normal process of ovulation, fertilization, and implantation through the use of various devices, sexual practices, chemicals, drugs, or surgical procedures. This means that something becomes a contraceptive if its purpose is to prevent a woman from becoming pregnant. For birth control in a family and/or population control in a country, contraception is the only medium to prevent unwanted pregnancy and thereby limiting the size of population.

The birth of "unwanted" additional number in a family is solely due to not using contraceptive by either of the couple. It is the single most important instrument for population control. Sadly enough, it does not get commensurate importance and space in the Family Welfare management in our country as the health issues overshadow the system. Be that as it may, the Eligible couple List (ECL) is the primary requirement and the most important requirement as well, to enable outreach of contraceptives to the acceptors. The preparation of this basic document is vested with the family welfare functionary in a given area. The ECL is required to keep records of all married persons in the age group of 18–45 with details of date of couple's respective age, date of marriage, the reproductive history including number of children born and its related date, the acceptance of any particular type of contraceptive practices by the couple and the like. The Ministry of Health & Family Welfare designed a comprehensive format in 2012–13 under the revised name 'Reproductive and Child Health (RCH) Register'. Unless this newly revised ECL is correctly prepared, updated, studied and thereafter taken up as a single reference document for service intervention and monitoring by all kinds of service providers, the gap in contraceptive use will ever remain unabridged.

The eligible couples need to take informed choice on types of contraceptives, whether temporary or permanent, which would suit them at any particular point of time. Reaching out to eligible couples with information on plus and minus of each of the contraceptives, as included in the National Family Welfare Programme, is a very important function. Way back in 2000, the National population Commission published the 'Report of the Working Group on Strategies to address Unmet Needs' which mentioned that 'India has around 29 million women having unmet needs for family planning'. It based its findings on NFHS I (1992–93) and NFHS II (1998–99) data which reported that the unmet needs for contraception were 19.5 % and 15.8 % respectively. The situation has not improved since then. The Annual Report of the Ministry of Health and Family Welfare, 2014–15 mentioned that as per DLHS III (2007–08), the unmet needs of contraception stood at 21.3 %. As a matter of fact, most of the family planning service providers do not pay as much attention as it deserves to meet the unmet needs of contraception which, incidentally, is also connected with the human right to have access to desired contraceptive. Since unmeet need of contraception happen to be a confused concept to a good number of service providers, it calls for conceptual clarification. Unmet need for contraception is the percentage of fertile married women of reproductive age who do not want to be pregnant and are not currently using any contraceptive method. In fact unmet need points to the gap between women's reproductive intentions and their contraceptive behaviour. There may be two types of such categories: (a) those with an unmet need for limiting, and (b) those with an unmet need for spacing. Women with an unmet need for limiting are those who desire no additional children and who do not currently use a contraceptive method. Women with an unmet need for spacing are those who desire to postpone their next birth by a specified length of time and who do not currently use a contraceptive method. These needs are to be met by service providers. Recently, Population Council researchers recommended three strategies for reducing unmet need for modern contraception, which are as follows:

- The first strategy is to expand the range of contraceptive methods available so that women have more options to choose from and thus less likely to discontinue contraceptive use;
- The second strategy is to improve quality of care and service provider–client interactions;
- The third strategy focuses on improving the characteristics of contraceptive technologies.

Access to available contraceptives is another very important area under contraceptive discourse. Access to family planning information and contraceptives can change lives of couples, especially of women. The far-away rural locations, tribal areas, urban slums and minority-concentrated areas often remain un-served and under-served to a great extent in the network of contraceptive services, and usually suffer from unwanted pregnancies. Such areas deserve appropriate extension of contraceptive services-network. Enabling women to make informed decisions about whether and when to have children reduces unintended pregnancies as well as maternal and new born deaths.

Another very important area that necessitates appropriate focus is the monitoring of contraceptive services. Monitoring the reach-out of contraceptive services would not only bridge the gap of service delivery, it will immensely prevent the additional births through unwanted pregnancies. Linked with monitoring, there is also an urgent necessity to undertake performance audit of methods of contraception and contraceptive prevalence rate (CPR) to verify the real outcome of services.

Further, there are hosts of other issues which need appropriate address, such as gender-based options for contraception and the issue of shared responsibility; the marketability of contraception and the role of private service providers; counselling set-up for family planning services; quality service care by service providers, affordable cost of contraception; research on new generation of contraception and its inclusion under national family planning services; infertility, HIV and AIDS, and the like.

While the issues of production and supply network of contraception fall under the primary responsibility of the department of family planning in the government, the meeting the unmet needs of contraception is the ultimate objective of contraceptive service providers. All the issues deserve independent attention and consideration for effective and functioning family planning set-up.

To sum up, the central focus of contraceptive issues is to address the availability of different kinds of contraceptives in the family welfare domain, the kind of contraceptives formally taken up under the National Family Planning Programme, the relative merits and role of temporary or permanent method and its suitability under client specific requirements. Further, family planning service providers need to keep account the choice of individual couple, the method best suited to them at the given situation, and to reach out to them. It is thus crucial that service providers must have complete knowledge of all the available methods together with those included under the National Family Planning Programme and are touched upon here under.

The different methods of family planning may be divided under six categories;

- Natural methods
- Barrier methods
- Hormonal methods
- Emergency contraception

- Intrauterine methods
- Sterilization.

(i) **Natural methods**:

Natural methods of contraception are considered "natural" because they are not mechanical and not a result of hormone manipulation. Instead, these natural methods prevent pregnancy during the time when an egg is available to be fertilized by a sperm. There are three types of natural methods:

(a) Natural Method:
(b) Withdrawal Method:
(c) Abstinence.

(a) **Natural Method**:

Under Natural birth control method, women determine the fertile phase (typically 7 to 10 days long) of their menstrual cycle and refrain from penetrative intercourse on fertile days to avoid pregnancy. There are many variations of natural birth control. The most effective method is for a woman to chart the signs of fertility that ebb and flow with the natural hormonal changes of each menstrual cycle. However, if a woman's cycle does not follow a typical pattern, using natural birth control will be more difficult.

(b) **Withdrawal Method**:

In this contraceptive method, the man withdraws his penis from a woman's vagina before ejaculation. It is premised on the basis of agreement of the partners. However, failure occurs which range in 1 out of 5 users. It also requires a lot of self-control and practice. It very often happens that withdrawal takes place too late.

(c) **Abstinence**:

Abstinence means choosing to **abstain from different levels** of sexual activity. However, for prevention of birth to be specific, sexual abstinence really means avoiding vaginal intercourse (penis to vagina sex), i.e. penetrative sex. Avoidance of vaginal intercourse is very effective for preventing unwanted pregnancy and still allows a couple to be involved in other forms of sexual expression. There are no risks involved with this method.

(ii) **Barrier Methods**

Barrier methods prevent sperm from entering the uterus and may be an option for women. It is reversible in nature. Types of barrier methods include:

- **Female condoms**. These are thin, flexible plastic pouches. A portion of the condom is inserted into a woman's vagina before intercourse to prevent sperm

from entering the uterus. The female condom also reduces the risk of STDs. Female condoms are disposed of after a single use.
- **Diaphragms**. Each diaphragm is a shallow, flexible cup made of latex or soft rubber that is inserted into the vagina before intercourse, blocking sperm from entering the uterus. Spermicidal cream or jelly should be used with a diaphragm. The diaphragm should remain in place for 6–8 h after intercourse to prevent pregnancy, but it should be removed within 24 h. Traditional latex diaphragms must be the correct size to work properly, and a health care provider can determine the proper fit. A diaphragm should be replaced after 1 or 2 years. The Woman needs to be measured again for a diaphragm after giving birth, having pelvic surgery, or gaining or losing more than 15 lb.
- **Cervical caps**. These are similar to diaphragms, somewhat smaller but relatively more rigid but less noticeable. The cervical cap is a thin silicone cup that is inserted into the vagina before intercourse to block sperm from entering the uterus. Like diaphragm, the cervical cap needs be used with spermicidal cream or jelly. The cap must remain in place for 6–8 h after intercourse to prevent pregnancy and needs to be removed within 48 h. Cervical caps happen to be of different sizes, and a health care provider determines the proper fit. With proper care, a cervical cap can be used for 2 years before replacement.
- **Contraceptive sponges**: These are soft, disposable, spermicide-filled foam sponges. It is required to be inserted into the vagina before intercourse. The sponge blocks sperm from entering the uterus, and the spermicide also kills the sperm cells. The sponge needs to be left in place for at least 6 h after intercourse and then removed within 30 h after intercourse.
- **Spermicides**: A spermicide destroys sperm. A spermicide can be used alone or in combination with a diaphragm or cervical cap. The most common spermicidal agent is a chemical called nonoxynol-9 (N-9). It is available in several concentrations and forms, including foam, jelly, cream, suppository, and film. A spermicide has to be inserted into the vagina close to the uterus no more than 30 min prior to intercourse and left in place 6–8 h after intercourse to prevent pregnancy. Spermicides do not prevent the transmission of STDs and may cause allergic reactions.

(iii) **Hormonal Methods**:

Hormonal methods of birth control use hormones to regulate or stop ovulation and prevent pregnancy. Ovulation is the biological process in which the ovary releases an egg, making it available for fertilization. Hormones can be introduced into the body through various methods, including pills, injections, skin patches, transdermal gels, vaginal rings, intrauterine systems, and implantable rods. Depending on the types of hormones that are used, these pills can prevent ovulation; thicken cervical mucus, which helps block sperm from reaching the egg; or thin the lining of the uterus. Health care providers prescribe, monitor, and administer hormonal contraceptives.

- **Combined oral contraceptives ("the pill")**. Combined oral contraceptive pills (COCs) contain different combinations of the synthetic estrogens and progestins and are given to interfere with ovulation. A woman takes one pill daily, preferably at the same time each day. Many types of oral contraceptives are available, and a health care provider helps to determine which type best meets a woman's needs. Use of COC pills is not recommended for women who smoke tobacco and are more than 35 years old or for any woman who has high blood pressure, a history of blood clots, or a history of breast, liver, or endometrial cancer.
- **Progestin-only pills** (**POPs**). A woman takes one pill daily, preferably at the same time each day. Progestin-only pills may interfere with ovulation or with sperm function. POPs thicken cervical mucus, making it difficult for sperm to swim into the uterus or to enter the fallopian tube. POPs alter the normal cyclical changes in the uterine lining and may result in unscheduled or breakthrough bleeding. These hormones do not appear to be associated with an increased risk of blood clots.
- **Contraceptive patch**. This is a thin, plastic patch that sticks to the skin and releases hormones through the skin into the bloodstream. The patch is placed on the lower abdomen, buttocks, outer arm, or upper body. A new patch is applied once a week for 3 weeks, and no patch is used on the fourth week to enable menstruation.
- **Injectable birth control**. This method involves injection of Depo-Provera, a progestin, in the arm or buttocks once every 3 months. This method of birth control can cause a temporary loss of bone density, particularly in adolescents. However, this bone loss is generally regained after discontinuing use of DMPA. Most patients using injectable birth control should eat a diet rich in calcium and vitamin D or take vitamin supplements while using this medication.
- **Vaginal rings**. The ring is thin, flexible, and approximately 2 inches in diameter. It delivers a combination of a synthetic estrogen and a progestin. The ring is inserted into the vagina, where it continually releases hormones for 3 weeks. The woman removes it for the fourth week and reinserts a new ring 7 days later. Risks for this method of contraception are similar to those for the combined oral contraceptive pills, and a vaginal ring is not recommended for any woman with a history of blot clots, stroke, or heart attack, or with certain types of cancer.
- **Implantable rods**. Each rod is matchstick-sized, flexible, and plastic. A physician surgically inserts the rod under the skin of the woman's upper arm. The rods release a progestin and can remain implanted for up to 5 years.

(iv) **Emergency contraception**:

- **Emergency Contraceptive Pills (ECPs)**. ECPs are hormonal pills, taken either as a single dose or two doses 12 h apart, that are intended for use in the event of unprotected intercourse. If taken prior to ovulation, the pills can delay or inhibit ovulation for at least 5 days to allow the sperm to become inactive. They also cause thickening of cervical mucus and may interfere with sperm function. ECPs should be taken as soon as possible after semen exposure and should not be used as a regular contraceptive method. Pregnancy can occur if the pills are taken after ovulation or if there is subsequent semen exposure in the same cycle.

(v) **Intrauterine Methods**

An IUD is a small, T-shaped device that is inserted into the uterus to prevent pregnancy. A health care service provider inserts the device. An IUD can remain and function effectively for many years at a time. After the recommended length of time, or when the woman no longer needs or desires contraception, a health care provider removes or replaces the device.

- A **copper IUD** releases a small amount of copper into the uterus, causing an inflammatory reaction that generally prevents sperm from reaching and fertilizing the egg. If fertilization of the egg does occur, the physical presence of the device prevents the fertilized egg from implanting into the lining of the uterus. Copper IUDs may remain in the body for 12 years. A copper IUD is not recommended for women who may be pregnant, have pelvic infections, or had uterine perforations during previous IUD insertions. It also is not recommended for women who have cervical cancer or cancer of the uterus, unexplained vaginal bleeding, or pelvic tuberculosis.
- A **hormonal IUD** releases a progestin hormone into the uterus. The released hormone causes thickening of the cervical mucus, inhibits sperm from reaching or fertilizing the egg, thins the uterine lining, and also may prevent the ovaries from releasing eggs. Hormonal IUDs can be used for up to 5 years.

Permanent Method:

(vi) **Sterilization**:

Sterilization is a permanent form of birth control that either prevents a woman from getting pregnant or prevents a man from releasing sperm. A health care provider must perform the sterilization procedure, which usually involves surgery. These procedures usually are not reversible.

- A **sterilization implant** is a nonsurgical method for permanently blocking the fallopian tubes. A health care provider threads a thin tube through the vagina and into the uterus to place a soft, flexible insert into each fallopian tube. No incisions are necessary. During the next 3 months, scar tissue forms around the inserts and blocks the fallopian tubes so that sperm cannot reach an egg. After 3 months, a health care provider conducts tests to ensure that scar tissue has

fully blocked the fallopian tubes. A backup method of contraception is used until the tests show that the tubes are fully blocked.
- **Tubal ligation** is a surgical procedure in which a doctor cuts, ties, or seals the fallopian tubes. This procedure blocks the path between the ovaries and the uterus. The sperm cannot reach the egg to fertilize it, and the egg cannot reach the uterus.

Against this background, the contraceptives services that are formally provided under the National Family Planning Programme in India, are as follows:

Temporary Method or Spacing Method:

(a) For Male

It is a temporary method of contraception to be used by a Male partner of the couple who do not wish to have any child in the immediate future

- Condoms: These are the barrier methods of contraception which offer the dual protection of preventing unwanted pregnancies as well as transmission of RTI/STI including HIV. The brand "Nirodh" is available free of cost at government health facilities and supplied at doorstep by ASHAs at minimal Cost.

Permanent Method: Male Sterilisation:

(a) For Male

Male Sterilisation:

Through a puncture or small incision in the scrotum, the provider locates each of the 2 tubes that carries sperms to the penis (vas deferens) and cuts or blocks it by cutting and tying it closed or by applying heat or electricity (cautery). The procedure is performed by MBBS doctors trained in these. However, the couple needs to use an alternative method of contraception for first three months after sterilization till no sperms are detected in semen.

Two techniques are being used in India:

(i) Conventional
(ii) Non-scalpel vasectomy– no incision, only puncture and hence no stiches.

For Female

Temporary Methods or Spacing Methods—These are the reversible methods of contraception to be used by couples who wish to have children in future. These include:

(i) Oral contraceptive pills: These are hormonal pills which have to be taken by a woman, preferably at a fixed time, daily. The strip also contains additional placebo/iron pills to be consumed during the hormonal pill free days. The method may be used by majority of women after screening by a trained provider. At present, there is a scheme for delivery of OCPs at the doorstep of beneficiaries by ASHA with a minimal charge. The brand "MALA-N" is available free of cost at all public healthcare facilities.

(ii) Intrauterine contraceptive devices (IUCD): Copper containing IUCDs are a highly effective method for long term birth spacing. It Should not be used by women with uterine anomalies or women with active PID or those who are at increased risk of STI/RTI (women with multiple partners). The acceptor needs to return for follow up visit after 1, 3 and 6 months of IUCD insertion as the expulsion rate is highest in this duration.

Two types: Cu IUCD 380A (10 yrs) and Cu IUCD 375 (5 yrs) are in place in India.

A new approach of postpartum IUCD insertion by specially trained providers to tap the opportunities offered by institutional deliveries has been in place.

(iii) Emergency Contraceptive Pill (ECP): ECP has to be consumed in cases of emergency arising out of unplanned/unprotected intercourse. The pill should be consumed within 72 h of the sexual act and should never be considered a replacement for a regular contraceptive.

Permanent Method: Female Sterilisation

These methods may be adopted by a married woman under certain physical conditions and are generally considered irreversible.

(i) Minilap:
Minilaparotomy involves making a small incision in the abdomen. The fallopian tubes are brought to the incision to be cut or blocked. It has to be performed by a trained MBBS doctor.

(ii) Laparoscopy:
Laparoscopy involves inserting a long thin tube with a lens in it into the abdomen through a small incision. This laparoscope enables the doctor to see and block or cut the fallopian tubes in the abdomen. It has to be done only by trained and certified gynaecologist/surgeon.

3.8 Health Issues

A good number of Health Issues are relevant and critical for promotion of family planning services. These are listed as below:

3.8.1 Pregnancy Care/Prenatal Care

Pregnancy is a beautiful phase in women's life. Life of a women is not complete until and unless she gets pregnant and gives births a child. Pregnancy is referred to a period when a woman is raising a fetus inside her which usually span a period of 40 weeks or a little more than 9 months. Some common pregnancy symptoms are missed-menstrual periods, morning sickness, sleepiness, food cravings, backache, nausea and vomiting and the like. Women deserve special care during pregnancy. As against the most common discomforts during pregnancy a list of basic pregnancy care tips for healthy pregnancy, worked out by medical professional, is captured here under:

"Morning sickness. Nausea or vomiting may strike anytime during the day (or night). The suggested tip is eating frequent, small meals and avoids greasy foods. Doctor has to be consulted if morning sickness lasts past the first 3 months of pregnancy or causes weight loss.

Tiredness. Sometimes tiredness in pregnancy is caused by anaemia, the suggested tip is to get enough rest and take a daytime nap if possible.

Leg cramps. The suggested tip is to gently stretch the calf of the leg by curling toes upward, toward knee.

Constipation. The suggested tip is to drink plenty of fluids and eat foods with lots of fibre, such as fruits, vegetables and bran cereal and not to take laxatives without medical advice.

Hemorrhoids. The suggested tip is not to strain during bowel movements and avoid becoming constipated.

Urinating more often. The need to urinate more often as baby grows because he or she will put pressure on bladder. This can't be helped.

Varicose veins. Avoid clothing that fits tightly around legs or waist. Rest and put feet up as much as you can. Move around if you must stand for long periods.

Moodiness. Hormones are on a roller coaster ride during pregnancy. Plus, life is undergoing a big change. Don't be too hard on yourself. If you feel very sad, talk to your doctor.

Heartburn. Eat frequent, small meals. Avoid spicy or greasy foods. Don't lie down right after eating. Ask doctor about taking antacids.

Yeast infections. The amount of discharge from the vagina increases during pregnancy. Yeast infections, which can also cause discharge, are more common during pregnancy. Consult doctor in any case of unusual discharge.

Bleeding gums. Brush and floss regularly, and see dentist for cleanings and be sure to tell your dentist that you're pregnant.

Stuffy nose. This is related to changes in the levels of the female hormone estrogen. You may also have nosebleeds.

Edema (retaining fluid). Rest with your legs up. Lie on your left side while sleeping so blood flows from your legs back to your heart better. Don't use diuretics (water pills). If you're thinking about cutting down on salt to reduce swelling, talk with your doctor first. Your body needs enough salt to maintain the balance of fluid and cutting back on salt may not be the best way to manage your swelling.

Skin changes. Stretch marks appear as red marks on your skin. Lotion with shea butter can help keep your skin moist and may help reduce the itchiness of dry skin. Stretch marks often can't be prevented, but they often fade after pregnancy."

Prenatal care is very important. As soon as pregnancy is confirmed, it is essential to make an appointment with doctor. The first prenatal visit needs to take place at the time of 6–8 weeks of pregnancy. The doctor is likely to take medical history, also weight and blood pressure which are likely tobe taken during each visit to the doctor. Additionally, on the first visit, pelvic examination will be made to check the size and shape of uterus and abnormalities, if any, of the cervix Urine and blood tests samples will be taken on the first visit and again at later visits. A pregnant needs to take at least three (Anti natal care) ANC check-ups and one PNC check-up after a child is born. A pregnant woman needs to take care of the following aspects in the interest of her own health and for the unborn baby:

- First Trimester Changes
- Second Trimester Changes
- Third Trimester Changes
- Exercise During Pregnancy
- Eating During Pregnancy
- Sleeping During Pregnancy
- Sex During Pregnancy.

Incidentally, the Maternal Health Programme which is a component of the Reproductive and Child Health (RCH) Programme aims at reducing maternal mortality. The major interventions under the programme includes essential obstetric care and the RCH Programme aims at providing at least 3 antenatal check-ups during which weight and blood pressure check, abdominal examination, immunization against tetanus, iron and folic acid prophylaxis as well as anaemia management are provided to the pregnant women. Under the RCH Programme emergency obstetric care, 24 h delivery services at primary health centres and safe abortion services are also provided. In this care services, ASHA or other link health worker associated with JSY facilitate registration for ANC, provide and/or help the women in receiving at least three ANC check-ups including TT injection, IFA tablets. ASHA or other link health worker associated with JSY also assist in receiving full benefits under JSY.

3.8.2 Post Partum Care/Post Natal Care

A postpartum period or postnatal period is the period beginning immediately after the birth of a child and extending for about six weeks. During this time, the mother is monitored for bleeding, bowel and bladder function, and baby care. The mother is examined for tears, and is sutured, if necessary. Her constipation and bladder status is also assessed at this stage.

The essence of postpartum care is to ensure that the mother is healthy and capable of taking care of her new-born. She is given at this time all information regarding breastfeeding, reproductive health and contraception along with the tips of life adjustment in new situation.

The World Health Organization (WHO) describes the postnatal period as the most critical and yet the most neglected phase in the lives of mothers and babies; most deaths occur during the postnatal period. WHO website states as follows:

"Of the 2.9 million new-born deaths that occurred in 2012, close to half of them occurred within the first 24 h after birth. Many of these deaths occurred in babies born too early and too small, babies with infections, or babies asphyxiated around the time of delivery. Labour, birth and the immediate postnatal period are the most critical for new-born and maternal survival. Unfortunately, the majority of mothers and new-borns in low- and middle-income countries do not receive optimal care during these periods.

Basic care for all new-borns should include promoting and supporting early and exclusive breastfeeding, keeping the baby warm, increasing hand washing and providing hygienic umbilical cord and skin care, identifying conditions requiring additional care and counselling on when to take a new-born to a health facility. New-borns and their mothers should be examined for danger signs at home visits. At the same time, families should be counselled on identification of these danger signs and the need for prompt care seeking if one or more of them are present. New-borns with who have preterm birth or low birth weight, who are sick or are born to HIV-infected mothers need special care.

New-borns born in health facilities should not be sent home in the crucial first 24 h of life, and postnatal visits should be scheduled. For all home births a visit to a health facility for postnatal care as soon as possible after birth is recommended. In high mortality settings and where access to facility based care is limited, WHO and UNICEF recommend at least two home visits for all home births: the first visit should occur within 24 h from birth and the second visit on day 3. If possible, a third visit should be made before the end of the first week of life (WHO-UNICEF Joint Statement on Home Visits for New-born Care)."

3.8.3 Birth Spacing

Birth spacing refers to the amount of time between births and/or pregnancies. In other words, a birth-to-pregnancy interval is the amount of time from the birth of a baby until the mother gets pregnant again. Usually, birth spacing is considered essential for the mother to need sufficient time to recover and rejuvenate after the ordeal of pregnancy and also time for caring of her infant child. It is presumed that normally after two years, the mother would be physically and mentally ready to undergo another spell of pregnancy without affecting the health care of her and also her infant child. This interval of at least two years is called healthy birth spacing.

WHO, in its Report on Technical Consultation and Scientific Review of Birth Spacing on 2005 recorded that after a live birth the recommended interval before attempting the next pregnancy is at least 24 months in order to reduce the risk of adverse maternal, peri-natal outcomes. The rationale for the recommendations was based on various maternal, infant and child outcomes. For each outcome, different birth to pregnancy intervals was associated with highest to lowest risks. To summarise, birth to pregnancy interval of 6 months or shorter are associated with elevated risk of maternal morbidity. Birth to pregnancy interval of about 18 months or shorter are associated with elevated risk of infant, neonatal or perinatal mortality, low birth weight, small size for gestational age and pre-term delivery. Some residual elevated risk might be associated with the interval 18–27 months but it was limited.

From family planning angle, birth spacing enables growth of healthy child and prevents unwanted pregnancies. By spacing birth it also takes care too many pregnancies in the reproductive period of a woman. Birth spacing in a way is the most effective contraceptive for promotion of family planning.

3.8.4 Breast Feeding

Breast feeding means feeding of a child with breast milk from mother's breast immediately after the birth and thereafter. The baby sucks and swallow mother's milk instinctively. It is the most natural human response for providing ideal food for the healthy growth and development of the infant. As per WHO norms, the baby must receive the benefits of early initiation of breastfeeding within an hour and exclusive breast feeding for first six months of life. Thereafter, the child should receive appropriate complementary foods with continued breastfeeding up to two years or beyond. It also provides nutritious food in a form more easily digested than any infant formula.

Breast milk provides the ideal nutrition for infants. It has a nearly perfect mix of vitamins, protein, and fat—everything a baby needs to grow that help fight off viruses and bacteria. It lowers baby's risk of having asthma or allergies and have fewer ear infections, respiratory illnesses, and bouts of diarrhoea. It minimizes the chance of visiting doctors or hospitalizations. It has very often been linked to higher IQ scores in later childhood. Breast-feeding sets physical closeness, skin-to-skin touching, and eye contact that help baby bonding with mother and the baby feels secure. It enables the baby to gain the right amount of weight as they grow rather than become overweight children.

Breastfeeding benefits the Mother as well. It is also an integral part of the reproductive process with important implications for the health of the mother. It burns extra calories and thus helps lose pregnancy weight faster. It is also held that breastfeeding releases the hormone oxytocin that helps uterus return to its pre-pregnancy size and may reduce uterine bleeding after birth. Additionally, it also lowers risk of breast and ovarian cancer and osteoporosis.

In an article In MCH Community Newsletter 'Breastfeeding, Special Issue, August, 2008, Faridi MMA in her wonderful article 'Health Care System In The Protection, Promotion and Support of Breastfeeding observed as follows:

'Human life begins in the womb, like in all mammals, where individual as fetus gets all that is required for growth and development from the mother. After birth all mammalian off-springs need mother for survival, so the human body. It includes prevention from diseases, optimal nutrition for growth and stimulation and bonding for quality development. Breastfeeding fulfils all the three attributes of survival to the maximum. Breast milk, full of species specific anti-infective factors, is the first vaccination to prevent diseases. It contains easily digestible nutrients in optimal amounts and in appropriate ratios to achieve maximum growth. Breast feeding is the best form of stimulation and bonding for the baby to guaranty quality development. Hence breast feeding is a part of reproductive cycle.'

3.8.5 *Reproductive Tract Infection (RTI) & Sexually Transmitted Infection (STI)*

The ICPD Program of Action gave a prominent space to Reproductive Tract Infections (RTI) and Sexually Transmitted Infections (STI). It highlighted the need to prevent and reduce the spread of reproductive tract infections (RTI) and sexually transmitted infections (STI) including HIV/AIDs, and to provide treatment for STIs and their complications, such as infertility, with special attention to increasing the ability of girls and women to protect themselves.

RTIs cover a variety of bacterial, viral, and protozoal infections of the lower and upper reproductive tract of both sexes. Many RTIs are sexually transmitted. Though most STIs falls in the category of RTIs, some STIs such as syphilis, Hepatitis B and AIDS, are also systemic diseases. A good number of STIs also affect the mouth, rectum, and urinary tract, the latter being part of the reproductive tract in males but not females.

Female RTIs originate in the lower reproductive tract (external genitals, vagina and cervix) and, in the absence of early treatment; they can spread to the upper tract (uterus, fallopian tubes and ovaries). Infections can ascend from the lower to the upper tract spontaneously to cause pelvic inflammatory disease (PID), but the risk of upper tract infection rise dramatically during procedures such as IUD insertion, abortion, and child birth when instruments are introduced through the cervix.

RTIs are comprised of: (a) iatrogenic infections, which are related to inadequate medical procedures, such as unsafe abortions; (b) endogenous infections, which may be associated with inadequate personal, sexual and menstrual hygiene practices, and (c) sexually transmitted infections (STIs).

RTIs and STIs represent a serious threat to the health and well-being of women and men in our country. They are exposed to the pain and discomfort of acute illness, and often experience long-term impairment of their reproductive function as a consequence of these infections. Some complications, such as infertility, are a source of psychological distress and family disruption. Others, such as ectopic pregnancy and cervical cancer, represent a significant source of mortality. Some infections may also cause fatal death or affect child survival by causing pre-term deliveries of low birth weight infants or by infecting newborns during delivery. RTIs and STIs are, therefore, responsible for a significant number of female. male and infant morbidity and mortality and form an enormous public health burden in India.

3.8.6 HIV and AIDS

HIV (Human Immunodeficiency Virus) attacks human's immune system that serves as our body's natural defence against illness. Such HIV is found in the body fluids of an infected person such as semen and vaginal fluids, blood and breast milk. The virus is passed from one person to another through blood-to-blood and sexual contact. In addition, infected pregnant women can pass HIV to their babies during pregnancy, at the delivering point during childbirth and through breast feeding. HIV is, however, not transmitted through sweat, saliva or urine, infection and disease and is very difficult to fight off; but with early diagnosis and effective antiretroviral treatment, people with HIV can live a normal, healthy life.

AIDS (Acquired immune deficiency syndrome or acquired immunodeficiency syndrome) is a syndrome caused by a virus called HIV (Human Immunodeficiency Virus). The illness alters the immune system, making people much more vulnerable to infections and diseases. This susceptibility worsens as the syndrome progresses. Someone with AIDS may develop a wide range of other infections including: pneumonia, thrush, fungal infections, TB, toxoplasmosis and cytomegalovirus and may ultimately lead to death.

The upshot of discussion of HIV and AIDS in the context of family planning is that unprotected sex may transmit HIV and AIDS, and that condom serves as a protective shield against such infection.

3.8.7 Adolescent Care

Adolescence denotes the period between the onset of puberty and the cessation of physical growth. It roughly spans from 11 to 19 years of age. This period is also popularly called the second decade of life, a period of rapid change and maturation when the child grows into the adult. It is located in between 'being children and being adults'. During this period, children undergo physiologic changes of their bodies, make efforts to keep pace with adjustments and search out their own sexual identification. They strive to find out personal identity and wanting freedom and independence of thought and action though they continue to remain under protective cover and dependence on their parents. Psychologically speaking, the very urge at this period to show off as adult, and no longer child, enable them to find out new peers-their new company for all possible help and guidance on challenges they face within their body's unknown behaviour. By and by, they tend to identify with their peers and usually yield to peer influence and conform to peer group values, behaviour, and tastes on matters such as clothing, food, and entertainment. Such association with peer group very often leaves longstanding impact on the personality and health of adolescents and the critical role of the peer groups need to be carefully taken into account.

From the premises of health care needs, there are specific needs for the Adolescents. It relates to new knowledge of growth of human body, health hygiene, nutritional requirement, sexually transmitted diseases and its prevention, matters concerning pregnancy and contraceptives options to prevent adolescent pregnancy. Another aspect in the health care is adolescence health education so that adolescents become knowledgeable about the relationship between their lifestyle and their physical and mental health. They also need help in achieving the maturity essential to choosing a healthy lifestyle and accepting responsibility for their personal health.

Providing Adolescents health care is indeed a very challenging task and requires skill to be able to communicate with them in a manner they can understand The service providers have to be warm and compassionate, needs to have a strong sense of presence of mind and humour, and be able to show emotional responsiveness at the required level. Such care providers need to be objective and non-judgmental as it deals with sensitive adolescent health problems. Besides, they should have required flexibility, tolerance, and comfortability in working with adolescents in addition to being knowledgeable about the special needs of adolescents.

National Population Policy 2000 in our country earlier identified adolescents as an under served group for which health needs and appropriate reproductive and sexual health interventions are to be put in place. The National Youth Policy 2003 also recognized 13–19 years as a distinct age group which had to be covered by special programmes in all sectors including health. The National Curriculum Framework 2005 for school education highlights the need for integrating adolescent reproductive and sexual health messages into school curriculum. Based on this, the National Adolescence Education Programme of NACO along with the Ministry of Human Resources Development has been developed.

Under National Rural Health Mission as a part of RCH, a Programme called the Adolescent Reproductive and Sexual Health Programme (ASRH) has been launched. This focus on ARSH and special interventions for adolescents include prevention of under-aged marriage, reduction of incidence of teenage pregnancies, meet unmet contraceptive needs and reduction of number of maternal deaths and incidence of sexually transmitted diseases and reduce the proportion of HIV positive cases in the 10–19 years age group. Other items covered under this initiative include improvement of calorie intake, vaginal discharge, headache, painful menstruation, irregular and excessive bleeding, urinary tract infection, poor menstrual hygiene and the like.

Another programme linked with adolescent care is Kishori Balika scheme under ICDS of the Department of Women and Child Development where 100 mg iron Folic Acid supplementation per week is given to all adolescent girls through schools and anganwadi centres.

3.9 Child Care Issues Including Child Nutrition

The word 'child' has different connotations; it has its constitutional meaning, it has different meanings under different legislations but it has quite separate meaning under child care.

Before addressing child care as such, it would be relevant to look at it from child rights angle.

The United Nations and United Nations Convention on the Rights of the Child (UNCRC) declared that Child Rights are minimum entitlements and freedoms that should be afforded to all persons below the age of 18 regardless of race, colour, gender, language, religion, opinions, origins, wealth, birth status or ability and therefore apply to all people everywhere. There are four broad classifications of these rights. These four categories cover all civil, political, social, economic and cultural rights of every child.

- Right to Survival: A child's right to survival begins before a child is born. According to Government of India, a child life begins after twenty weeks of conception. Hence the right to survival is inclusive of the child rights to be born, right to minimum standards of food, shelter and clothing, and the right to live with dignity.
- Right to Protection: A child has the right to be protected from neglect, exploitation and abuse at home, and elsewhere.
- Right to Participation: A child has a right to participate in any decision making that involves him/her directly or indirectly. There are varying degrees of participation as per the age and maturity of the child.
- Right to Development: Children have the right to all forms of development: Emotional, Mental and Physical. Emotional development is fulfilled by proper

care and love of a support system, mental development through education and learning and physical development through recreation, play and nutrition.

With this background, the child health care in India is briefly addressed from the broad linkage of family welfare under two heads:

(a) **Child health**

Child health care is an essential requisite for developing childhood development and building social health of the nation. All health sector intervention programmes including Reproductive and Child Health programme (RCH) II under the National Rural Health Mission (NRHM) have been addressing interventions that improve child health and addresses factors contributing to Infant and under-five mortality. Currently, the Twelfth Five Year plan (2012–2017) and National Health Mission (NHM) laid down the Goal to Reduce Infant Mortality Rate (IMR) to 25 per 1000 live births by 2017.

A set of issues relevant to child health care is addressed hereunder:

(i) **Essential new-born care**

The Essential New-born Care aims to ensure mother/family members/health workers to get the required knowledge and skills for appropriate care at the most vulnerable period in a baby's life. For the new-born care a range of needs like Air, Warmth, Food, Hygiene and love, as summarised in the Unicef website, are required to put in place which are listed below:

- Resuscitate and maintain an airway
- Keep the new-born warm and avoid unnecessary hypothermia or cold stress
- Encourage early breast feeding, and feed high-risk new-borns more frequently
- Maintain hygiene during delivery and cord cutting; treat infections promptly
- Ensure the new-born infant stays close to its mother, and mothers have open access to their new-born infant if he or she requires special care

(ii) **Exclusive breast feeding and weaning**

Breast feeding means feeding of a child with breast milk from mother's breast immediately after the birth and thereafter. The baby sucks and swallow mother's milk instinctively. It is the most natural human response for providing ideal food for the healthy growth and development of the infant. As per WHO norms, the baby must receive the benefits of early initiation of breastfeeding within an hour and exclusive breast feeding for first six months of life. Thereafter, the child should receive appropriate complementary foods with continued breastfeeding up to two years or beyond. It also provides nutritious food in a form more easily digested than any infant formula.

Breast milk provides the ideal nutrition for infants. It has a nearly perfect mix of vitamins, protein, and fat—everything a baby needs to grow that help fight off viruses and bacteria. It lowers baby's risk of having asthma or allergies and have fewer ear infections, respiratory illnesses, and bouts of diarrhoea. It minimizes the chance of visiting doctors or hospitalizations. It has very often been linked to higher IQ scores in later childhood. Breast-feeding sets physical closeness, skin-to-skin touching, and eye contact that help baby bonding with mother and the baby feels secure. It enables the baby to gain the right amount of weight as they grow rather than become overweight children.

Breastfeeding benefits the Mother as well. It is also an integral part of the reproductive process with important implications for the health of the mother. It burns extra calories and thus helps lose pregnancy weight faster. It is also held that breastfeeding releases the hormone oxytocin that helps uterus return to its pre-pregnancy size and may reduce uterine bleeding after birth. Additionally, it also lowers risk of breast and ovarian cancer and osteoporosis.

In an article In MCH Community Newsletter 'Breastfeeding, Special Issue, August, 2008, Faridi MMA in her wonderful article 'Health Care System In The Protection, Promotion and Support of Breastfeeding observed as follows:

'Human life begins in the womb, like in all mammals, where individual as fetus gets all that is required for growth and development from the mother. After birth all mammalian off-springs need mother for survival, so the human body. It includes prevention from diseases, optimal nutrition for growth and stimulation and bonding for quality development. Breastfeeding fulfils all the three attributes of survival to the maximum. Breast milk, full of species specific anti-infective factors, is the first vaccination to prevent diseases. It contains easily digestible nutrients in optimal amounts and in appropriate ratios to achieve maximum growth. Breast feeding is the best form of stimulation and bonding for the baby to guaranty quality development. Hence breast feeding is a part of reproductive cycle.'

(iii) Appropriate management of diarrhoea

Diarrhoea is one of the major causes of mortality among under five children. It contributes to 11 % deaths in age group beyond neonatal period. Diarrhoea is defined as passage to three or more loose tools per day with or without blood. On an average a child usually suffers from 1.7 episodes of diarrhoea per year.

The diarrhoeal disease control programme has been started in our country since 1978 under different programme heads. The main objective of the programme is to prevent death due to dehydration. In order to control Diarrheal diseases Government of India has adopted the WHO guidelines on Diarrhoea management.

- India introduced the low osmolality Oral Rehydration Solution (ORS), as recommended by WHO for the management of diarrhoea.
- Zinc has been approved as an adjunct to ORS for the management of diarrhoea. Addition of Zinc would result in reduction of the number and severity of episodes and the duration of diarrhoea.

The protocol and advices as suggested for management of diarrhoea are the following:

3.9 Child Care Issues Including Child Nutrition

1. The child has to be given plenty of home based fluids during episode of diarrhoea. ORS is the preferred fluids. In its absence, Rice water, Lassi, Soup, Daal, water, Nimbupani, Tea etc. may be given.
2. ORS is supplied in the kits to all sub-centres in the country for free distribution against any occurrence of diarrhoea. In case such ORS are not readily available, it need to be purchased from a medicinal shop and prepare the same as below:
   - Wash your hands with soap.
   - Take a litre of clean drinking water in a clean container.
   - Take a packet of ORS and add all its contents in the water.
   - Stir thoroughly so that the powder is completely mixed.
   - Cover the vessel.
3. ORS is to be given to the infants every one or two minutes. The amount of ORS to be given to the infant is based on the infant's age, as given here:
   - Up to 2 months—5 spoons of ORS.
   - From 2 months up to 2 years—One-fourth to half a cup.
   - More than 2 years—half a cup to one full cup.
   - Give enough ORS for the patient pass pale, yellow urine, four or five times a day.

(iv) **Appropriate management of Acute Respiratory Infection (ARI)**

ARI is one of the major causes of mortality among under five children. It accounts for 19 % deaths in age group beyond neonatal period. Every child very often suffers 4 to 6 episodes of acute respiratory infections in one year. This includes common cold cough and pneumonia. On an average 44Lac pneumonia cases take place in India. Early diagnosis and appropriate use of antibiotics remains one of the most effective interventions to prevent deaths due to pneumonia. The medical advice and protocols for the management of ARI is the following:

- If child is having cold cough and fever but respiratory rate is normal and there is no evidence of chest recession, then the child can be managed at home with paracetamol for fever. The child has to be given plenty of fluids, keep warm, and give normal diet. Home remedies like honey, ginger and tulsikada can be given.
- If child is having cold cough, fever and respiratory rate is increased but there is no evidence of chest recession, then the child is suffering from pneumonia. The child has to be taken to the nearest health centre and give him/her medicine for five days as advised by the health functionary. The child has to be given plenty of fluids, keep warm, and give normal diet. The child has to be Kept watch on his/her respiratory rate and chest recession.
- If child is having cold cough, fever, respiratory rate is increased as well as there is evidence of chest recession along with colour change on tongue and lips to

blue then child is suffering from severe pneumonia. The child has to be taken to the nearest hospital because child needs admission and injections for antibiotics.

(v) **Universal Immunization Program**

Immunization Programme is one of the key interventions for protection of children from life threatening conditions, which are preventable. Immunisation programme in our country was initiated in 1978 with limited outreach under Expanded Programme of Immunisation (EPI). In 1985 with the launch of Universal Immunization Programme (UIP), it was extended to the entire country and became a part of Child Survival and Safe Motherhood Programme in 1992. The UIP now falls under National Rural Health Mission (NRHM) since 2005. Under the Universal Immunization Programme, Government of India is providing vaccination to prevent eight vaccine preventable diseases nationally, i.e. Diphtheria, Pertussis, Tetanus, Polio, Measles, severe form of Childhood Tuberculosis and Hepatitis B and meningitis & pneumonia caused by Haemophilus influenza type B, and against Japanese Encephalitis in selected districts. An Immunisation Schedule drawn by the Ministry of Health and Family Welfare is shown below:

Immunisation Schedule under the Ministry of Health and Family Welfare

Sl. No.	Vaccination	Protection	Number of doses	Vaccine schedule
1	BCG (Bacillus Calmette Guerin	Tuberculosis	1	At birth (up to 1 year if not given earlier)
2	OPV (Oral Polio Vaccine)	Polio	5	Birth dose for institutional deliveries within 15 days; primary dose at 6, 10 and 14 week and booster dose at 16–24 months of age. Given orally
3	Hepatitis B	Hepatitis	4	Birth dose for institutional deliveries within 24 h and primary dose at 6, 10 and 14 week
4	DPT (Diphtheria Pertussis Tetanus Toxoid)	Diphtheria Pertussis Tetanus Toxoid	5	Three doses at 6, 10 and 14 week and two booster dose at 16–24 month and 5 years of age
5	Measles	Measles	2	9–12 months of age and 2nd dose at 16–24 months

(continued)

(continued)

Sl. No.	Vaccination	Protection	Number of doses	Vaccine schedule
6	TT (Tetanus Toxoid)	Tetanus	2	10 Years and 16 years of age
7	JE Vaccination (in selected 177 JE blocks			1st dose in 9–12 months in age and 2nd dose at 16–24
	177 endemic districts in 19 States)	Japanese encephalitis	2	16–24 months of age at JE endemic districts immediately after completion of campaign
8	Hib containing Pentavalent vaccine (Hib+DPT+Hep B) presently in eight States (Tamil Nadu, Kerala, Gujrat, Haryana, Karnataka, Goa, J&K and Puducherry	Diphtheria, Pertussis, Tetanus, Hepatitis B and Haemophilus Influenza type B associated Pneumonia, Meningitis	3	6, 10 and 14 week of age

The Ministry of Health and Family Welfare launched the Rotavirus vaccine as part of universal immunisation programme very recently on 26th March, 2016 among children less than five years of age to prevent diarrhoeal deaths, which claims the lives of nearly 1 lakh kids every year. With this addition, nine vaccinations are now available under universal immunisation programme.

(vi) **Vitamin A deficiency (VAD)**

Vitamin A deficiency (VAD) is a major contributor to child mortality. There is increasing evidence that it also raises significantly the risk of maternal death. The VAD is the leading cause of preventable severe visual impairment and blindness in children and increases the risk of disease and death from severe infections from common childhood infections as diarrhoeal disease and measles. For pregnant women, vitamin A deficiency occurs especially during the last trimester when demand by both the unborn child and the mother is highest. The mother's deficiency is demonstrated by the high prevalence of night blindness during this period.

The website of WHO has mentioned that 'the basis for lifelong health begins in childhood. Vitamin A is a crucial component. Since breast milk is a natural source of vitamin A, promoting breastfeeding is the best way to protect babies from VAD. For deficient children, the periodic supply of high-dose vitamin A in swift, simple, low-cost, high-benefit interventions has also produced remarkable results, reducing mortality by 23 % overall and by up to 50 % for acute measles sufferers. Planting these "seeds" between 6 months and 6 years of age can reduce overall child mortality by a quarter in areas with significant VAD. The impact of this single supplementation on childhood mortality is as great or greater than that of any one vaccine—and it costs a very insignificant sum for a dose However, because breastfeeding is time-limited and the effect of vitamin A supplementation capsules lasts only 4–6 months, they are only initial steps towards ensuring better overall nutrition and not long-term solutions.

Cultivating the garden is the next phase necessary to achieve long-term results. Food fortification takes over where supplementation leaves off. Food fortification, maintains vitamin A status, especially for high-risk groups and needy families. For vulnerable rural families, growing fruits and vegetables in home gardens complements dietary diversification and fortification and contributes to better lifelong health.'

(vii) **Treatment of Anaemia**

Anaemia refers to any deficiency in the blood, either in quality or quantity. Though there are different forms of anaemia, but the term usually refers to a decrease in the number of red blood cells or a reduction in haemoglobin. It is not a disease, but is a symptom of a disease. It can be caused by bleeding, poor diet, reaction to certain drugs or diseases of bone marrow. Treatment depends on the cause of the anaemia.

Iron deficiency anaemia is the most common type of anaemia and frequently occurs in girls and women during the female reproductive years. While excessive blood loss most commonly causes iron deficiency anaemia, increasing dietary intake of iron-rich foods can both help prevent and treat low iron levels in blood. Further, intake of folic acid and vitamin C in foods helps body absorb iron better.

(viii) **Child nutrition**

Nutrition is recognised as a basic pillar for physical and mental development. Adequate nutrition is essential in early childhood to ensure growth, proper organ formation and function, a strong immune system, and neurological and cognitive development. Good nutrition is the bedrock of child survival, health and development. Proper nutrition in childhood can reinforce lifelong eating habits that contribute to children's overall wellbeing and help them to grow up to their full potential and a healthy life. A child needs to be well-nourished to be able to grow as a natural human being and be in a position learn properly and acquire skill efficiently so that she/he can participate in the bigger outer world and contribute meaningfully. Nutrition also provides resilience in the face of disease, disasters, and other crises. Thus, from a country's perspective of economic growth and human development, well-nourished population can contribute meaningfully for the development of their country.

The state of child nutrition in India is far from ideal. Unicef website summarises the state of child nutrition in India under caption 'Fast Facts' which are as follows:

"In India 20 % of children under five years of age suffer from wasting due to acute under-nutrition. More than one third of the world's children who are wasted live in India.

Forty three per cent of Indian children under five years are underweight and 48 % (i.e. 61 million children) are stunted due to chronic under-nutrition, India accounts for more than 3 out of every 10 stunted children in the world.

Under-nutrition is substantially higher in rural than in urban areas. Short birth intervals are associated with higher levels of under-nutrition.

The percentage of children who are severely underweight is almost five times higher among children whose mothers have no education than among children whose mothers have 12 or more years of schooling.

Under-nutrition is more common for children of mothers who are undernourished themselves (i.e. body mass index below 18.5) than for children whose mothers are not undernourished.

Children from scheduled tribes have the poorest nutritional status on almost every measure and the high prevalence of wasting in this group (28 %) is of particular concern.

- India has the highest number of low birth weight babies per year at an estimated 7.4 million.
- Only 25 % of new-borns were put to the breast within one hour of birth.
- Less than half of children (46 %) under six months of age are exclusively breastfed.
- Only 20 % children age 6–23 months are fed appropriately according to all three recommended practices for infant and young child feeding.
- 70 % children age 6–59 months are anaemic. Children of mothers who are severely anaemic are seven times as likely to be severely anaemic as children of mothers who are not anaemic.
- Only half (51 %) of households use adequately iodized salt.
- Only one third (33 %) Indian children receive any service from an anganwadi centre; less than 25 % receive supplementary foods through ICDS; and only 18 % have their weights measured in an AWC".

In this given background, it is desirable to have fair idea about ICDS programme and also to know what should be the age-appropriate diet for children. An age-appropriate diet is one that provides adequate nutrition and is appropriate for a child's state of development. In view of great diversity in respect of economic status, customs, cultural practices and plural dietary habits in the country side, no national standard for age-appropriate diet for children of the country has been put out by the Ministry of Health and Family Welfare though under ICDS programme supplementary nutrition is provided. WHO, Unicef and medical professionals are all in agreement that Breast milk contains all the nutrients that a child is needed and hence the child has to be exclusively on breastfeed for the first six months. Post six months, breast milk alone doesn't meet other nutrients requirements, like iron, and hence the child has to be introduced slowly with other foods. In other words, age-appropriate diet for children for post six months is very crucial for building up the nutrition profile of a child. As a general principle, nutrition must be given through a judicious combination of a variety of foodstuffs from different food groups. Dietary intakes lower or higher than the body requirements can lead to under-nutrition. A just quantum of diet, providing all nutrients, is needed during this period.

There are no uniform prescriptions for age-appropriate diet for children However, after a child has grown six months of age he/she needs to be introduced semi-solid food along with breast milk. Before such cereal is introduced, the mother needs to check up whether the baby is ready to eat semi-solid foods by looking at the following milestones:

- The birth weight has doubled
- The baby has good control of head and neck
- The baby can sit up with some support
- The baby can show fullness by turning the head away or by not opening the mouth
- The baby begins showing interest in food when others are eating.

Normally, the pedestrians suggest that diet for 6–8 months is semi-solid feedings with iron-fortified baby rice cereal mixed with breast milk or formula. Initially, cereal may be offered 2 times per day in servings of 1 or 2 tablespoons which may be increased gradually to 3 or 4 tablespoons of cereal. However, breast milk or formula has to be offered three to four times per day at this age as well. After a baby is accustomed to baby cereals, fruits and vegetables may be introduced one at a time, waiting 2–3 days in between to check for any allergic reaction. Thereafter, gradually small quantum of locally available seasonal plain vegetables such as green peas, potatoes, carrots, sweet potatoes, squash, beans, beets and locally available seasonal plain fruits such as bananas, guava, papia, melon etc. may be given. At 8–12 months of age, a baby may be ready to try iron-rich foods such as finely chopped meats but not more than once in a week and too with very small amount. Other option like Eggs may be given 3–4 times per week, but only the yolk until the baby is 1 year old, as some babies are sensitive to egg whites. By the age of 1, most children are likely to be off the bottle. If the child still uses a bottle, it should contain water only. After a baby is 1-year old, whole milk may replace breast milk or formula. Children under the age of 2 should not be given low-fat milk (2 %, 1 %, or skim) as they need the additional calories from fat to ensure proper growth and development. As a general rule, Toddlers and small children will usually eat only small amounts at one time, but will eat frequently (4–6 times) throughout the day. After the age of 2, the diet has to be moderately low in fat, as diets high in fat may contribute to heart disease, obesity, and other health problems later in life. Providing a variety of foods from each of the food groups (breads and grains, meats, fruits and vegetables, and dairy) will help to ensure enough vitamins and minerals and prevent nutrition deficiencies. Further, children should rather get all their nutrients from foods rather than vitamin supplements.

3.9 Child Care Issues Including Child Nutrition

Integrated Child Development Services Scheme (ICDS)

Against the above background, the Government of India launched the Integrated Child Development Services (ICDS) Scheme in 1975 with the following objectives:

- to improve the nutritional and health status of children in the age group 0–6 years;
- to lay the foundation for proper psychological, physical and social development of the child;
- to reduce the incidence of mortality, morbidity, malnutrition and school dropout;
- to achieve effective coordination of policy and implementation amongst the various departments to promote child development; and
- to enhance the capability of the mother to look after the normal health and nutritional needs of the child through proper nutrition and health education.

The above objectives are sought to be achieved through a package of six services comprising:

- supplementary nutrition,
- immunization,
- health check-up,
- referral services,
- pre-school non-formal education and
- Nutrition & health education.

ANM identifies the target group with the assistance of AWW from among under-privileged families. The details of six services are as follows:

(1) **Nutrition including Supplementary Nutrition**

This includes supplementary feeding and growth monitoring; and prophylaxis against vitamin A deficiency and control of nutritional anaemia. All families in the community are surveyed, to identify children below the age of six and pregnant & nursing mothers. They avail of supplementary feeding support for 300 days in a year. By providing supplementary feeding, the Anganwadi attempts to bridge the caloric gap between the national recommended and average intake of children and women in low income and disadvantaged communities. Growth Monitoring and nutrition surveillance are two important activities that are undertaken. Children below the age of three years of age are weighed once a month and children 3–6 years of age are weighed quarterly. Weight-for-age growth cards are maintained for all children below six years. This helps to detect growth faltering and helps in assessing nutritional status. Besides, severely malnourished children are given special supplementary feeding and referred to medical services.

Type of Supplementary Nutrition:

(i) Children in the age group 0–6 months:
- For Children in this age group, States/UTs may ensure continuation of current guidelines of early initiation (within one hour of birth) and exclusive breast-feeding for children for the first 6 months of life.

(ii) Children in the age group 6 months to 3 years:
- For children in this age group, the existing pattern of Take Home Ration (THR) under the ICDS Scheme continues. However, in addition to the current mixed practice of giving either dry or raw ration (wheat and rice) which is often consumed by the entire family and not the child alone, THR is given in the form that is palatable to the child instead of the entire family.

(iii) Children in the age group 3–6 years:
- For the children in this age group, State/UTs are to make arrangements to serve Hot Cooked Meal in AWCs and mini-AWCs under the ICDS Scheme. Since the child of this age group is not capable of consuming a meal of 500 calories in one sitting, the States/UTs are to consider serving more than one meal to the children who come to AWCs. Since the process of cooking and serving hot cooked meal takes time, and in most of the cases, the food is served around noon, States/UTs may provide 500 calories over more than one meal. States/UTs may arrange to provide a morning snack in the form of milk/banana/egg/seasonal fruits/micronutrient fortified food etc.

(2) **Immunization**

Immunization of pregnant women and infants protects children from six vaccine preventable diseases, namely-poliomyelitis, diphtheria, pertussis, tetanus, tuberculosis and measles are organised. These are major preventable causes of child mortality, disability, morbidity and related malnutrition. Immunization of pregnant women against tetanus also reduces maternal and neonatal mortality.

(3) **Health Check-ups**

This includes health care of children less than six years of age, antenatal care of expectant mothers and postnatal care of nursing mothers. The various health services provided for children by Anganwadi workers and Primary Health Centre (PHC) staff includes regular health check-ups, recording of weight, immunization, management of malnutrition, treatment of diarrhoea, deworming and distribution of simple medicines etc.

(4) **Referral Services**

During health check-ups and growth monitoring, sick or malnourished children, in need of prompt medical attention, are referred to the Primary Health Centre or its sub-centre. The Anganwadi worker has also been oriented to detect disabilities in young children. She enlists all such cases in a special register and refers them to the medical officer of the Primary Health Centre/Sub-centre.

(5) **Non-formal Pre-School Education (PSE)**

The Non-formal Pre-school Education (PSE) component of the ICDS may well be considered the backbone of the ICDS programme, since all its services essentially converge at the Anganwadi—a village courtyard. Anganwadi Centre (AWC) is the main platform for delivering of these services. These AWCs have been set up in every village in the country. In pursuance of its commitment to the cause of India's Children, the government has decided to set up an AWC in every human habitation/settlement. This is also the most joyful play-way daily activity, visibly sustained for three hours a day. It brings and keeps young children at the Anganwadi centre-an activity that motivates parents and communities. PSE, as envisaged in the ICDS, focuses on total development of the child, in the age up to six years, mainly from the under-privileged groups. Its programme for the three-to six years old children in the Anganwadi is directed towards providing and ensuring a natural, joyful and stimulating environment, with emphasis on necessary inputs for optimal growth and development. The early learning component of the ICDS is a significant input for providing a sound foundation for cumulative lifelong learning and development. It also contributes to the universalization of primary education, by providing to the child the necessary preparation for primary schooling and offering substitute care to younger siblings, thus freeing the older ones—especially girls—to attend school.

(6) **Nutrition and Health Education**

Nutrition, Health and Education (NHED) are a key element of the work of the Anganwadi worker. This forms part of BCC (Behaviour Change Communication) strategy. This has the long term goal of capacity-building of women, especially in the age group of 15–45 years so that they can look after their own health, nutrition and development needs as well as that of their children and families.

(b) **Child Growth Standards**

The WHO and the Unicef jointly worked out growth standard for identification of 6–60 month old infants and children for the management of severe acute malnutrition (SAM). Two most important standards are weight-for-height and mid-upper arm circumference (MUAC). The WHO and the Unicef recommend the use of a cut-off for weight-for-height of below-3 standard deviations (SD) of the WHO standards to identify infants and children as having SAM. Children with a MUAC less than 115 mm have a highly elevated risk of death compared to those who are above.

3.10 Educational Issues

Education, the spread of education and its quality are hardly given its due importance on population control and family planning, as it deserved, as a positive contributor in containing the run-away population scenario in a country. As a matter of fact, the standard of general education of a country is very crucial to facilitate acceptance of any new educational input, ideas or practices which benefits any individual or the society. It has all pervading influences in understanding sustainable population and its linked responsibility to have a limited size of a family. Education broadens the mental horizon, sharpens the intellectual skill and enables any individual to appreciate the perspective to take reasoned decision. The inherent strength of education does not allow the walls of prejudice or any age-old beliefs etc. to put up any obstruction whatsoever. Further, education provides a platform of seeing, feeling and acting which he/she could not have done it otherwise and that too, spontaneously. The main function of the educative process is also to pass down knowledge from generation to generation—a process that is essential to the development of culture for the good of the society. Formal education helps to inculcate crucial skills and values central to the survival of the society. Inherent in education, in all period of man's history is a stimulus to creative thinking and action which accommodates needs for cultural change from a generation to another. Education also helps citizens to develop common values important for integrative role in the social life. The role of education as an agent or instrument of social change is recognized when it brings about a change in outlook and attitude of man/woman.

Similarly, the role of mass education is also important in shaping and up-scaling the average education status of its non-school citizens and enables them to take appropriate decision in a given social context. It also promotes equality in the average standard of general education and perception in social life by way of generating a kind of common consciousness of their duties, obligations, responsibilities both in relation to work and society. In the context of population control programme, the role of mass education is thus very important as it educates and prepares the social mind-set to respond to issues connected with family planning.

Finally, in a broader sense, Education has the function of cultural transmission in all societies. Culture is a growing whole. There can be no break in the continuity of culture. The cultural elements are passed on through the agents like family, school and other associations. All societies maintain themselves through their culture which includes a set of beliefs, skills, art, literature, philosophy, religion, music etc. These social heritages are also usually transmitted through social education.

3.11 Population Education

Population education is a new programme under the discipline of education, an innovative programme designed to address population balance in a country. About population education, there is no universally accepted definition as such but it varies the way one looks at it. While some define it as to what population education should aim at, some others look at it by giving a listing of its contents. Way back in 1970, in a regional workshop in UNESCO at Bangkok, Population education was conceived as 'an educational programmes which provides for a study of population situation in the family, community, nation and world with the purpose of developing in the students rational and responsible attitudes and behaviour toward that situation'. Another definition that emerged by way of refinement of the earlier definition in 1972 in Philippines was that 'population education is the process of developing awareness and understanding of population situations as well as a rational attitude and behaviour toward those situations for the attainment of quality of life for the individual, the family, the community, the nation and the world'.

An UNESCO document defines 'Population education as designed to help people to understand the nature and particularly the causes and consequence of population events. It is directed at a people as individuals or as members of groups, as decision-makers or potential decision-makers within their families, as citizens within a community, as leaders within a society and as policy makers within a nation. All people are population actors making population-related decisions throughout their lives....

...The sum of their decisions shapes the nature of the population forces (fertility, mortality, migration) which operate within a society and which then affect other social, political and economic forces. In turn, the social, political and economic decisions made by the larger mass of people (the society or nation) influence the behaviour of the individual, the family and the small communities to which they belong. Population education, in essence, is an educational response to contemporary economic, social and political issues....

...Within the population context, population education may help people recognize and the nature of problems which have population components. It can help them realize better how problems arise and what consequences their decisions and actions will have. Within a development context, population education may be designed to help people comprehend that social and economic development is, to some variable extent, influenced by population processes and that their decisions may depend upon the social and economic status of a society or nation. Within an educational context, population education processes can contribute in many ways as to the general development of education and to innovation and renovation. Since most population issues have economic and social and political components, population education, in seeking to deal with these interactions, can contribute to the reorganization of curricula along interdisciplinary lines. Population education can aid in selecting contents which is relevant to the lives of learners and can contribute to the development and learner-centred method of instruction".

The above definition covers comprehensively the goal and objectives of population education. In short, the goal of population education is to involve people in a learning process that will broaden their understanding of population-related issues and develop in them appropriate skills to define and analyse these issues so that ultimately they will be able to make rational and responsible decisions regarding these issues in a way that is personally and socially relevant.

Just like education, population education need to reach out to students through structured curricula in a formal system, and to out-of-school students through informal system. It is usually held that population concepts spread too thinly in too many school subjects are not very effective. However, when population content is treated as an integral part of the school curricula and textbook, the teachers would teach topics with all seriousness; even the sensitive issues can be properly reached out. The core message and textual material are very crucial in a formal system of population education. Of many items, the core message relating to family size, marriage at appropriate permissible age, responsible parenthood, population related beliefs and faith, population load impact on quality of life of family members and the society, population change and effect on resources and environment etc. are important. The subjects through which the aforesaid message need to be conveyed belongs to Science, Geography, Civics, Economics, Social studies, Health and home economics etc. As a matter of fact, the UNESCO Regional office for Education in Asia and the Pacific published such learning materials on population education long ago in 1984.

For the out-of-school Education programme of population education, the education components are most critical. The items need to be the minimum but focussed with core message of population education imbedded in them. The presentation could be in entertainment format but with message on fast changing demographic scenario, the linkage of social, cultural, economic and psychological determinants with the population process, the consequences of population changes upon the environment and also on quality of life of the individual, the family and the nation. The National policy and programme in relation to population control and family planning need also to be included in it.

The National Population Policy, 2000 of India has not mentioned conspicuously about the role of population education in furthering the cause of population stabilisation efforts in the country. However, the long term effects of population education in arresting the growth of population is well recognized. In the context of India's commitment to achieve SDGs by 2030, the requisite of sustainable population has to be put in place. For achieving sustainable population, all available options need to be fully utilised. Population education can play its crucial role in restoring population balance in our country.

3.12 Media Issues

The term media is derived from Medium, which means carrier. Media denotes a links specifically designed to reach large viewers. Media is decisively the most powerful instruments of communication. It helps to promote the right things in right perspective and on right time. It helps to add on education and knowledge to enable people at large to take informed decision as to what is right and what is wrong. In our complex life, it is hardly possible to find out enough time or resources to have access to materials which give an objective analysis on any important social issue or any issue that has potential for any far reaching impact on our social life, cultural life or on our environmental existence. The media plays that role in a very constructive way in increasing public awareness, sharpening perceptions, formatting right attitudes and aptitude, and coalesces them in forming or shaping collective views toward certain issue having common social goal for today or for the tomorrow.

Media itself is like a universe. It has several segments of formats, each serving its jurisdiction of information and knowledge dissemination. Thus there are mass media, broadcast media, print media and the web media. The television and the radio belong to broadcast media while newspapers, magazines and journals fall under print media and the internet news and services are labelled as the web media. Each of the media links has its own clientele for its news segments, entertainment segments, and interaction forum for exchange of ideas, suggestions and views. The Media, serving its national, regional or local areas and jurisdiction, as the case may be, connect the people with a new message to undergo change in a silent mode to its extant mental environment and then enable it to graduate to accept new ideas for a better tomorrow.

Ours is a knowledge society in our current times. The society is in search of new ideas and new knowledge. This quest for new knowledge is almost universal irrespective of its rural and urban settings. However, the format need to be client-friendly; for some particular location or for some particular social group, repackaging of knowledge inputs might be necessary to meet the needs for such local areas on ground of extant culture and/or on consideration of local, regional or social mind-sets. In some cases, intensive inputs might be necessary for focussed media intervention at an interval on ground of issue attention cycle.

Population control and family planning entails a lot of societal, cultural, educational, health, gender, environmental issues and the like for which public support and participation are essential. Such public support and participation depends to a great extent on their awareness and collective views. Media has the potentiality to effect strong social and cultural impact upon the society. Because of its inherent ability to reach large number of public, it is the best forum to convey message to build public opinion and awareness. It is also less expensive to reach out large number of people at minimal time.

In our country's context, the vast potential of the role of media and its strong overarching power remains largely unutilised. Except on annual coverage of World

Population Day, Media hardly covers all related issues in a focused way as a part of its social commitment and also as a social partner for safeguarding against run-away population growth in our country. Admittedly, the media has the potentiality and inherent strength to assess from outside the grey areas of programme implementation and the capacity to outreach any national level advocacy group, region-based social and community leaders, local level social groups, decentralised local bodies and volunteers to revisit their responsibility. Media may serve as a social monitoring format as well by way of issue-specific coverage, special supplements, annual performance report on the population control front and the like. In the context of India's journey towards sustainable development by 2030, the role of the media in ensuring sustainable population is immense and the media has to be brought in a big way to realise it.

3.13 IEC Issues

Knowledge is a significant driving force for promoting growth and development. Knowledge and transfer of functional knowledge is very important for social development as it prepares the social mind to appreciate new ideas to overcome extant challenges. Information, education and communication (IEC) serves as vehicle for knowledge empowerment by broadening social perception and skill on the related issues. In the context of family planning and related public health, IEC often serves to change or reinforce health-related behaviour in a target audience. It strives to address a specific problem within a defined period of time, through communication methods and principles. This boils down the need for IEC initiatives to have a clear objective to ensure focussed outcome.

The most important component of IEC relates to information. The aspect of information becomes meaningful when it is communicated. On the other hand, communication can take place only when it contains some component of message of education. In family planning area, the role of communication is very important. The social environment obtaining in the family planning area is characterised by the prevalence of wall of ignorance and resistance. A good part of social groups and individuals perceive family planning from different perspectives with different levels of social focus. Such perception emanates from economic status, educational backwardness, ignorance and certain hard and inflexible religious, cultural and subcultural beliefs. IEC aims at addressing such mind-sets and effect behavioural change.

For designing an effective IEC, understanding the basic objective and the related concern is very relevant. Mere elucidating the importance of IEC and introducing some ideas to a motley group of persons do not go far in achieving intended objectives. It is an irony that avowed goal of IEC very often goes astray for not having clear and focused ideas of what have to be achieved. In other words, the clarity of what the IEC is supposed to intervene and achieve is very important, and have to be settled at the outset. For effective Population Control and family planning, a good number of issues have to be covered and addressed through the instrument of IEC for behavioural change. Such items range from medical issues to non-medical issues falling in the category of educational, socio-cultural, legal and infrastructural issues. Among medical and non-medical issues, there are a variety of sub-issues which have to be addressed individually. A short-list of them is given below by way of illustration:

Medical issues	Non-medical issues
(•) Child health: – Essentials of new born care – Immunization – Breast feeding – ARI – Diarrhoea – Preventive health care for the child	(•) Educational: – Female education – Family life education – Literacy and adult education – Population education
(a) Maternal health • Prenatal, ante natal care • Neonatal and peri-natal care • Post partum care • Birth spacing • Maternal morbity • STI • RTI • MTP • Methods of family planning • Infertility	(a) Socio-cultural • Age of marriage • Child marriage • Son preference • Girl child • Attitude of elderly members in the family • Male attitude and participation including non-scalpel vasectomy
(b) AIDS	(b) Legal Issues • Marriage laws • Adoption laws • PNDT Act • Marriage registration • Birth and death registration
(c) Adolescents care including anaemia	(c) Socio-infrastructural issues • Eligible couple list • Contractive prevalence survey/KAP studies • Community needs assessment • Training of family planning workers/Opinion leaders • Decentralized family planning set-up • Employment based FP set-up • Counselling set-up • Role of Civil Society including Women organisations
(d) School health care	(d) IEC intervention programme
(e) Nutrition care for all	(e) Monitoring of performance criteria

Linked with the focused objectives of the IEC, the most crucial function is to identify clients and reach them. The targeted clients happen to be located in different geographical situations having diverse role and plural background. There would, therefore, be varying IEC needs for each of them. A common IEC format for such diverse IEC needs is not likely to have any meaningful impact and trigger behavioural change. There has to have linkage between the IEC needs and IEC intervention. For that it is essential to undertake an assessment of IEC needs for each of the target groups to begin with. In order that such need assessment is correctly done, it would be necessary to have a fair idea of the category of clients and shortlist them for programme coverage. An illustration is given for possible clients and their probable IEC needs:

List of possible clients and their probable IEC needs

Possible clients	Probable IEC needs
(a) Council of ministers State planning board State women commission State minority commission State backward commission State government secretaries Divisional commissioners	• Demographic status of the State and its all-India relative position • Demographic status of the districts and its state-specific relative position • Religion-wise decadal growth in the districts of the State • Vital statics of the State and the districts, where ever available • Family planning related performance data • National population policy/State population policy • National health policy • Population control and FP related programmes in the state and its efforts index
(b) Home department	• Demographic status of the State and its all-India relative position • Demographic status of the districts and its state-specific relative position • Religion-wise decadal growth in the districts of the State • National population policy/State population policy • Nature of possible infiltration and migration
(c) Health and family planning department	• Demographic status of the State and its all-India relative position • Demographic status of the districts and its state-specific relative position • Religion-wise decadal growth in the districts of the State • Vital statics of the State and the districts, where ever available • Family planning related performance data • National population policy/State population policy • National health policy • District specific population control and FP related programmes and its efforts index • Supportive and complementary programmes of other line departments, local bodies (Panchayats and municipalities) and civil societies etc.

(continued)

3.13 IEC Issues

(continued)

Possible clients	Probable IEC needs
(d) Other associated departments (Rural & Panchayat, Urban development/, Municipal affairs, social welfare, minority affairs, mass education, labour, Information & culture affairs, planning etc.	• Demographic status of the State and its all-India relative position • Demographic status of the districts and its state-specific relative position • National population policy/State population policy • National health policy • District specific population control and FP related programmes and its supportive and complementary programmes of the department
(e) School education department	• Integration of population education in school curriculum
(f) GoI's units at the State HQ including AIR, Doordarshan and other media units	• Highlights of State Government's performing data on population control and family planning
(g) Civil Society including opinion leaders/advocacy group/social workers/religious leaders/women organizations etc.	• Un-served, under-served, resistant groups and areas of concern
(h) Minority organizations/social groups	• List of resistant families for family planning
(i) Municipal corporations/municipalities	• Demographic status of the area and its relative position • Religion-wise decadal growth in its area • National population policy/State population policy • National health policy • Programmes of the government
(j) MPs/MLAs/Secretariat of all political parties	• Demographic status of the district and its relative position • Religion-wise decadal growth in the districts of the State • Vital statics of the State and the districts, where ever available • Family planning related performance data • National population policy/State Population policy • National health policy • Programmes of the government • Areas of concern
(k) State unit of the Indian medical association	• Programmes of the government • Areas of concern
(l) Central government's medical units/ESI/railway hospitals/private medical clinics/nursing homes/private medical practitioners/health units of private sector organizations/residential colony	• Programmes of the family planning of the State government

(continued)

(continued)

Possible clients	Probable IEC needs
(m) Newly Married Couples	• Knowledge of plus and minus points of all contraceptive methods and its availability • Value of quality of life in a small family • Birth spacing • Family life education
(n) Eligible couples	• Knowledge of plus and minus points of all contraceptive methods and its availability • Birth spacing • Value of quality of life in a small family
(o) Problematic eligible couples having more than two children	• Knowledge of terminal method of contraceptive services for the male and also for the female • Impact of unwanted pregnancies on mother, on family and the environment
(p) Eligible couples located at unstructured situation (e.g. pavement dwellers etc.)	• Knowledge and availability of contraceptive methods including permanent methods • Immunization of children
(q) Adolescent girls	• Knowledge of health and hygiene care • Counselling care and its availability
(r) College students	• Family life education • Knowledge of contraceptive methods
(s) Others (Pharmaceutical shop)	• Knowledge of display of contraceptives in a customer friendly manner
Zilla Panchayat/District administration including DPC, District health & family welfare Samity	• Demographic status of the districts and its state-specific relative position • Religion-wise decadal growth in the districts of the State • Vital statics of the State and the districts, where ever available • Family planning related performance data • National population policy/State population policy • National health policy • District action plan on population control and FP related programmes in the district for all programme partners • Monitoring indicators and periodical performance report
District school education department/board	• Monitoring of population education in primary and secondary schools
Sub-divisional/Taluk level administration including sub-divisional/Taluk level health & family welfare administration	• Demographic status of the Subdivision and its state-specific relative position • Religion-wise decadal growth of its area • Vital statics of the district, where ever available • Family planning related performance data

(continued)

3.13 IEC Issues

(continued)

Possible clients	Probable IEC needs
	• District action plan on population control and FP related programmes falling in the district for all programme partners
Block development office/block Panchayat office/block primary health centre/municipality	• Demographic status of the block and its state-specific relative position • Religion-wise decadal growth of its area • Vital statics of the district, where ever available • Family planning related performance data • Block action plan/municipality action plan on population control and FP related programmes in the district for all programme partners • Updated EC&CR, local body wise • Monitoring indicators and periodical performance report based on EC&CR
Gram Panchayat/sub-centre/municipal ward committee/ICDS centre/	• Demographic status of the local area • Religion-wise decadal growth of its area • Block action plan on population control and FP related programmes falling in the local area • List of all programme partners • Updated EC&CR, local body wise • Monitoring indicators and periodical performance report based on EC&CR

IEC Medium

The critical role of need assessment for IEC in shaping behavioural change does not call for further elaboration. However, the nature of the medium through which the needs are to be met is important.. As a matter of fact, the power of medium as a change agent is very critical. It sets on motion new wave for effecting and internalising behavioural change. The selection of medium does play crucial role in meeting identified needs of varying clients located at different stations of programme schedule. Internal message transmission through this process, however, moves rather slowly, indeed very slowly, step by step, and, therefore, be never attempted to rush forward. The initial Awareness promotes KAP status which ultimately matures to trigger behavioural change. The sensibility and delicate nature of community mind-set need to be kept in view during these stages. Further, situation analysis is an essential requisite before embarking on IEC intervention on diverse and differential clients. It has to be culture specific and be flexible. The medium strategy itself may call for a change with the first order intervention as the dynamic human mind-set may need an altogether new IEC intervention from a particular stage on to another. Additionally, a combination of one or more medium may be needed to reach out to difficult social groups or individuals.

The available of possible IEC medium for the purpose of intervention to different sets of clients may be shown as below:

- Electronic Media—TV/Doordarshan
    - Musical logo on Family Planning
    - Spots
    - Films
    - Panel discussion
    - Quiz on Family Planning
    - Song on Family Planning
    - Question and Answer session

- Radio
    - Musical logo on Family Planning
    - Spots
    - Films
    - Panel discussion
    - Quiz on Family Planning
    - Song on Family Planning
    - Drama
    - Question and Answer session

- Other Electronic media

    Help Line
    Cinema slides
    News/advertisement through cable network
    Video
    Video cassette

- Print Media

    Advertisement
    Family Planning related news
    Family Planning related editorial
    Special issues

    Other Media

- Poster/Bill board/Display board
- Leaflet
- Booklet
- Fact sheet
- Flash card
- Flip chart
- Hoarding
- Banner/Festoons

3.13 IEC Issues

- Arms band
- Wall chart
- Wall painting/Transport cabin painting
- Slide
- Mobile IEC van
- Audio cassette
- Miking
- Beat of drums
- Folk song
- Door to door visit
- Group discussion
- Seminar
- Workshop
- Counselling centre
- Greetings card for newly married couple from Family Planning unit
- Family Planning football cup
- Inter collegiate quiz on adolescent health
- Inter school quiz on adolescent health for girls
- Inter school quiz on population education
- Family Planning fair/festivals
- Exhibition
- Rally
- Cartoon
- Puppet show
- Mime
- Drama
- Street drama
- Role play
- News Bulletin
- Population tower/clock
- Baby show
- Interactions with media personality/Social personality/Film Personality/Sports personality etc.

Given the plethora of alternative choices, it would be worthwhile to work out a simple matrix which could be used as a possible guide map for intervention of IEC medium. It is reiterated that three stages are crucial before the onset of any behavioural change, namely, awareness generation, knowledge development and behavioural change. In keeping with these change format, IEC medium need to be appropriately linked and put in place. Normally, mass media is very useful for creation of awareness. Select leaflets, booklets and group communication make the perspective clear, and facilitate in acquiring the related knowledge and deepen it while person to person communication and counselling promotes behavioural change. Further, IEC messages should have positive focus and non-argumentative in presentation to kick up emotional response in an entertainment format. Focus attention can be drawn by creating humorous sex related situation leaving suggestion of the intended core message.

List of possible clients and their probable IEC medium

Possible clients	Probable IEC medium
(a) Council of ministers State planning board State women commission State minority commission State backward commission State government secretaries Divisional commissioners	Fact sheet
(b) Home department	Fact sheet
(c) Health and family planning department	Fact sheet and Booklet
(d) Other associated departments (Rural & Panchayat, Urban development/, Municipal affairs, social welfare, minority affairs, mass education, labour, information & culture affairs, planning etc.	Fact sheet and Booklet
School education department	UNESCO publications on population education
GoI's units at the State HQ including AIR, Doordarshan and other media units	Fact sheet
Civil society including opinion leaders/advocacy group/social workers/religious leaders/women organizations etc.	Fact sheet, Booklet, group discussion, seminar, workshop, rally etc.
Minority organizations/social groups	Fact sheet, counselling, group discussion, door to door visit
NGOs working on population and family planning	Leaflet, Booklet, wall chart, posters, cassettes, door to door visit, group discussion, seminar, workshop, rally, role play, drama, Street dram, Mime etc.
Municipal corporations/municipalities	Fact sheet and Booklet
MPs/MLAs/secretariat of all political parties	Fact sheet, leaflet, news bulletin etc.
State unit of the Indian medical association	Fact sheet and news bulletin
Central government's medical units/ESI/railway hospitals/private medical clinics/nursing homes/private medical practitioners/health units of private sector organizations/residential colony	Fact sheet and news bulletin
Newly married couples	Greetings folder along with flip chart
Eligible couples	Leaflet, door to door visit, group discussion, counselling, family planning greetings card, rally, role play, drama, street drama, Mime, puppet show, cartoon, exhibition etc.

(continued)

(continued)

Possible clients	Probable IEC medium
Problematic eligible couples having more than two children	Booklet on non-scalpel vasectomy, door to door visit, group discussion, counselling, role play, puppet show, cartoon etc.
Eligible couples located at unstructured situation (e.g. pavement dwellers etc.)	Door to door visit, group discussion, counselling, street drama, Mime, puppet show, cartoon, contraceptive available points etc.
Adolescent girls	Booklet, quiz, seminar, question and answer session, essay competition, cassette etc.
College students	Booklet, quiz, seminar, question and answer session, essay competition, cassette etc.
Others (Pharmaceutical shop)	Guidelines for display contraceptives in a customer = friendly manner and its quiet sale
District administration including DPC, District health & family welfare Samity etc.	Fact sheet, leaflet, Booklet, wall chart, stickers, seminar, news bulletin, departmental periodicals etc.
District school education department/board	Booklet and seminar
Sub-divisional administration including sub-divisional health & family welfare administration	Fact sheet, booklet, news bulletin, seminar, workshop etc.
Block development office/block Panchayat office/block primary health centre	Fact sheet, Booklet, leaflet, seminar, workshop, wall chart, news bulletin etc.
Gram Panchayat/sub-centre/municipal ward committee/ICDS centre/	Fact sheet, Booklet, leaflet, seminar, workshop, wall chart, cartoon, news Bulletin, group discussion, drama, street drama, door to door visit

3.14 Legal Issues

The Population control programme is more of a social issue management programme than a legal one. The area of social mind-set, social awareness and social ownership ultimately decides the societal approved population size and number. However, the administration and management of population control entails a lot of issues which have legal implications and have, therefore, to be addressed legally. It would, therefore, be desirable to enlist them formally before discussing each of them. They are listed as follows:

3.14.1 Constitutional Provisions

The Constitution of India, adopted in 1950, did not have any provision, direct or otherwise, on items relating to the population problem of the country. In 1976, however, a new entry, 20A to List III-Concurrent List, had been added which reads as "Population Control and Family Planning". The entry of this item to the Concurrent list empowered the Parliament to legislate on this very important social issue at its own. Further, in the Constitution, a proviso in Article 81 was subsequently introduced by the 42nd Constitutional Amendment providing that until the Census 2001, the increasing number of population in any state would not be reckoned for the purpose of further number of members to the Parliament from any state or in the state legislatures in order to guard against rising number of population in some states, poor performance in family planning and clamour for higher representation in Parliament.

Similar provisions had also been added to Article 55 (Manner of Election of President), Article 82 (Readjustment after each census), Article 170 (Composition of Legislative assemblies of States), and Article 330 (Reservation of Seats for scheduled Castes and Scheduled Tribes in the House of the People). In 2000, the current freeze on the number of seats in Lok Sabha and State assemblies has been extended till 2026. Thus, the 42nd Amendment, 1976 and 91st Amendment, 2000 of the Constitution of India has brought in certain important changes with its bearing on family planning.

3.14.2 Legal age of Marriage

Earlier it has been discussed in details about child marriage in the context of population control. The related Act is now shown below:

- **The Prohibition Of Child Marriage Act, 2006(6 of 2007)**:

Child Marriage is one of the prime reasons for number of child births, poor health of the Children and the mother and is also important causal factor for high IMR and MMR in our country. Earlier, in the pre-independence era, an Act namely, The Child Marriage Restraint Act, 1929 was in vogue for restraining solemnisation of child marriages. The Act was subsequently amended in 1949 and 1978 in order to raise the age limit of the male and female persons for the purpose of marriage. The Act had no provision to declare Child marriage invalid. Following the recommendations of the National Human Right Commission and the National Commission on Women, The Prohibition Of Child Marriage Act, 2006 was enacted by the Union Government. It extends to the whole of India and it applies also to all citizens of India without and beyond India.

3.14 Legal Issues

The salient features of this Act are as follows:

- A provision to declare child marriage as voidable at the option of the contracting party to the marriage, who was a child (sec-3)
- A provision requiring the husband or, if he is a minor at the material time, his guardian to pay maintenance to the minor girl until her remarriage (sec-4)
- A provision for the custody and maintenance of children born of child marriages (sec-5)
- A provision that notwithstanding a child marriage has been annulled by a decree under section 3, every child born of such marriage, whether before or after the commencement of the Act, shall be legitimate for all purposes (sec-6)
- Empowering the district court to add to, modify or revoke any order relating to maintenance of the female petitioner and her residence and custody or maintenance of children, etc. (sec-7)
- A provision for declaring the child marriage as void in certain circumstances (sec-12)
- Empowering the courts to issue injunctions prohibiting solemnisation of marriages in contravention of the provisions of this Act (sec-13)
- A provision making the offences under the Act to be cognizable for the purposes of investigation and for other purposes (sec-15)
- A provision for appointment of Child Marriage Prevention Officers by the State Governments (sec-16)
- Empowering the State Governments to make rules for effectively administration of the Act (sec-19).

The crucial provision in this Act is the role of the State Government in that it has to frame Rules under provisions of the Act for its proper administration. This is a grey area as many State governments have neither framed Rules nor appointed Child Marriage Prevention Officers for effective administration of this important legislation.

3.14.3 Registration of Marriage

Registration of marriage is a format needed to enable the husband or the wife to get a legal document from the appropriate authority as to the marital status of the couple in the eye of law. It is a useful reference document for all incidental issues connected with the marriage including right and obligations of the husband or the wife, as the case may be. Earlier registration of marriage was optional but not compulsory. It is now required that all marriages should be registered under the related Act of the couple.

A marriage which has already been solemnised can be registered either under the Hindu Marriage Act, 1955 or under the Special Marriage Act, 1954. The Hindu Marriage Act is applicable in cases where both husband and wife are Hindus,

Buddhists, Jains or Sikhs or where they have converted into any of these religions. Where either of the husband or wife or both are not Hindus, Buddhists, Jains or Sikhs, the marriage is registered under The Special Marriage Act, 1954.

Further, marriage can be solemnised between any two persons under The Special Marriage Act, 1954.

Delhi (Compulsory Registration of Marriage) Order, 2014:

The Hon'ble Supreme Court of India in a case titled Smt. Seema vs Aswini Kumar in transfer petition (c) No, 291 of 2005 directed compulsory registration of marriages solemnized in the respective State/Territories. As a consequence thereof, registration of marriage is now compulsory and such States/Union Territories are required to issue their own marriage registration order. The Government of Delhi has since issued the Delhi (Compulsory Registration of Marriage) Order, 2014 vide Notification No. F.1(12)DC/MC/2014/4392/dt 24.04.2014 which is briefly captured hereunder to enable other state governments/Union Territories to have similar regulation in their areas.

1. Short title extent and commencement

    (a) This order is called the Delhi (Compulsory Registration of Marriage) Order, 2014
    (b) This order shall extend to all marriages solemnized in Delhi irrespective of caste, creed and religion professed by the parties to the marriage.
    (c) This order shall come into effect on the date of its notification in the official gazette.

2. Compulsory Registration of marriage:

    (a) On the commencement of this order, any marriage solemnized in Delhi between a male having completed 21 years of age and a female having completed 18 years of age on the date of solemnization of marriage and of which one of the parties is an Indian citizen, such marriage shall be compulsorily registrable in accordance with this order, irrespective of caste, creed and religion professed by any party or parties to such marriage.

    Provided that if the marriage has already been solemnized under any existing law, the same shall not be required to be registered under this order.

    (b) Notwithstanding anything in any custom or practice, which may be derogatory with the aforesaid provision regarding compulsory registration of marriage, the provision of sub-clause (a) of this order shall have an overriding effect.

3. Marriage Officer:

    For the purpose of compulsory registration of marriage in accordance with this order, the marriage officer already appointed for registration of marriage under the Hindu Marriage Act, Special Marriage Act or any other law for the time being in force in Delhi, shall have the jurisdiction to register marriage under this order also.

Provided that the Government of NCT of Delhi may appoint additional marriage officer for the purpose of this order and may also delegate the powers of marriage officer to any retired gazetted officer appointed for the purpose for such period not exceeding 5 years after his retirement and having attained the age of 65 years.

4. Procedure:

   (a) Within a period of 60 days, excluding the day on which marriage is solemnized, the parties to the marriage shall apply jointly in the Prescribed Form-A for registration of their marriage addressed to the marriage officer having jurisdiction to register the same.
   (b) Such prescribed application shall be accompanied by documentary proof of age of both the parties, place of residence of the parties to the marriage, Solemnization of marriage, identification of parties to the marriage, citizenship of the parties, if any along with requisite fee of Rupees two hundred.
   (c) On receipt of such application along with requisite documents as prescribed above and satisfaction of the marriage officer as regard authenticity of such proof, the same shall be entered in the register of marriage prescribed for the purpose as per Form-B.
   (d) After having received such application complete in all aspect and having entered the same in the prescribed register, the marriage officer shall fix a date for the parties to appear in person along with the witnesses who shall certify to the solemnization of such marriage and having proof of permanent resident of Delhi.
   (e) The marriage officer thereafter on personal appearance of the party with witnesses on such appointed date or any other extended date and on satisfaction of solemnization of such marriage in Delhi shall issue the requisite certificate of registration of such marriage as Form-C.

5. Jurisdiction of the marriage officer:

   The marriage officer within whose district the marriage has been solemnized shall have the jurisdiction to register the marriage.

   Provided that the Sub divisional Magistrate (Head Quarter), Additional District Magistrate and District Magistrate of Delhi shall have the concurrent jurisdiction over entire Delhi to register any marriage solemnized in Delhi within any Revenue District of Union Territory of Delhi.

6. Condonation of Delay:

   In case of default to get the marriage registered within the prescribed period of 60 days, the marriage officer shall have the power to condone the delay not exceeding further 60 days subject to additional fee of Rupees two hundred and thereafter register the marriage.

7. Consequence of non-registration marriage within the period prescribed:
   Any party to the marriage having not registered their marriage within the prescribed period/and or extended period shall suffer a penalty of Rupees one thousand, imposed by the marriage officer.
   Provided that such penalty or any part of it may be remitted by the Additional District Magistrate of such district or the District Magistrate, Delhi on application made in respect thereof by any party explaining reasonable cause for not getting the marriage registered. The Additional District Magistrate or the District Magistrate shall thereafter direct the marriage officer having jurisdiction, to register such marriage, on payment of requisite fee and the penalty so imposed.
8. Validity:
   Registration of marriage under this order will not tantamount to validity of marriage as the same would be the subject matter of the respective law, custom and practices professed by the parties as applicable to such marriage.
9. E. registration:
   The Govt. of NCT of Delhi shall endeavour o create a dedicated portal for the purpose of online submission of application and prior appointment to facilitate compulsory registration of marriage. The application form alternatively shall be available on such portal which may be downloaded by the parties and be submitted along with requisite documents manually at the respective counters of the marriage registration offices. On such submission manually or online, a computerized priority number along with appointed date for registration of marriage shall be made available to the applicants to be produced at the time of personal appearance before the marriage officer to register the marriage.
10. Tatkal registration of marriage:
    To facilitate registration of marriage in case of urgency on priority basis, corresponding optional facility shall also be available subject to payment of additional fee of Rupees Ten thousand.

3.14.4 Dowry Prohibition Act, 1961

The object of the Dowry Prohibition Act, 1961 is to prevent the evil practice of giving and taking dowry by the family of the bridegroom from the family of the bride. The dowry system is a black spot in the beginning of a new family life linked with marriage. It reduces the very institution of marriage to an auction exchange. The linkages between dowry and population control is that dowry gives premium concession on age of the bride which leaves the under aged bride with prolonged reproductive period and in the process gives birth to a number of children. The Dowry Prohibition Act, 1961 is therefore, a valuable instrument to control number of births in an indirect manner. The extract of the Dowry Prohibition Act, 1961 is now given below:

Extracts of the Dowry Prohibition Act, 1961

1. Short title, extent and commencement.—(1) This Act may be called the Dowry Prohibition Act, 1961.

    It extends to the whole of India except the State of Jammu and Kashmir.

    It shall come into force on such date as the Central Government may, by notification in the official Gazette, appoint.

2. Definition of 'dowry'.—In this act, 'dowry' means any property or valuable security given or agreed to be given either directly or indirectly—

    (a) by one party to a marriage to the other party to the marriage; or

    (b) by the parents of either party to a marriage or by any other person, to either party to the marriage or to any other person; at or before or any time after the marriage in connection with the marriage of said parties but does not include dower or mahr in the case of persons to whom the Muslim Personal Law (Shariat) applies.

    ***

3. Penalty for giving or taking dowry.—(1) If any person, after the commencement of this Act, gives or takes or abets the giving or taking of dowry, he shall be punishable with imprisonment for a term which shall not be less than five years, and with the fine which shall not be less than fifteen thousand rupees or the amount of the value of such dowry, whichever is more:

    Provided that the Court may, for adequate and special reasons to be recorded in the judgment, impose a sentence of imprisonment for a term of less than five years.

    (2) Nothing in sub-section (1) shall apply to or, in relation to,—presents which are given at the time of a marriage to the bride (without nay demand having been made in that behalf):

    Provided that such presents are entered in list maintained in accordance with rule made under this Act; presents which are given at the time of marriage to the bridegroom (without any demand having been made in that behalf):

    Provided that such presents are entered in a list maintained in accordance with rules made under this Act;

    Provided further that where such presents are made by or on behalf of the bride or any person related to the bride, such presents are of a customary nature and the value thereof is not excessive having regard to the financial status of the person by whom, or on whose behalf, such presents are given.

4. Penalty for demanding dowry.—If any person demands directly or indirectly, from the parents or other relatives or guardian of a bride or bridegroom as the case may be, any dowry, he shall be punishable with imprisonment for a term which shall not be less than six months but which may extend to two years and with fine which may extend to ten thousand rupees:

    Provided that the Court may, for adequate and special reasons to be mentioned in the judgment, impose a sentence of imprisonment for a term of less than six months.

5. **Agreement for giving or taking dowry to be void.**—Any agreement for the giving or taking of dowry shall be void.
6. **Dowry to be for the benefit of the wife or heirs.**—(1) Where any dowry is received by any person other than the woman in connection with whose marriage it is given, that person shall transfer it to the woman

   (a) if the dowry was received before marriage, within three months after the date of marriage; or

   (b) if the dowry was received at the time of or after the marriage within three months after the date of its receipt; or

   (c) if the dowry was received when the woman was a minor, within three months after she has attained the age of eighteen years, and pending such transfer, shall hold it in trust for the benefit of the woman.

   (2) If any person fails to transfer any property as required by sub-section (1) within the time limit specified therefor or as required by sub-section (3), he shall be punishable with imprisonment for a term which shall not be less than six months, but which may extend two years or with fine which shall not be less than five thousand rupees, but which may extend to ten thousand rupees or with both.

   (3) Where the woman entitled to any property under sub-section (1) dies before receiving it, the heirs of the woman shall be entitled to claim it from the person holding it for the time being:

   Provided that where such woman dies within seven years of her marriage, otherwise than due to natural causes, such property shall-

   if she has no children, be transferred to her parents, or

   if she has children, be transferred to such children and pending such transfer, be held in trust for such children.

   (3-A) Where a person convicted under sub-section (2) for failure to transfer any property as required by sub-section (1) or sub-section (3) has not, before his conviction under that sub-section, transferred such property to the women entitled thereto or, as the case may be, her heirs, parents or children, the Court shall, in addition to awarding punishment under that sub-section, direct, by order in writing, that such person shall transfer the property to such woman, or as the case may be, her heirs, parents or children within such period as may be specified in the order, and if such person fails to comply with the direction within the period so specified, an amount equal to the value of the property may be recovered from him as if it were a fine imposed by such Court and paid to such woman, as the case may be, her heirs, parents or children.

   (4) Nothing contained in this section shall affect provisions of Sec. 3 or Sec. 4.

7. Cognisance of offences.—(1) Notwithstanding anything contained in the Code of Criminal Procedure, 1973 (2 of 1974),—no Court inferior to that of a Metropolitan magistrate or a Judicial Magistrate of the first class shall try any offence under this Act;

8-A. Burden of proof in certain cases.—Where any person is prosecuted for taking or abetting the taking of any dowry under Sec. 3, or the demanding of dowry under Sec. 4, the burden of proving that he had not committed an offence under those sections shall be on him.

8-B. Dowry Prohibition Officers.—(1) The State Government may appoint as many Dowry Prohibition Officers as it thinks fit and specify the areas in respect of which they shall exercise their jurisdiction and powers under this Act.

(2) Every Dowry Prohibition Officer shall exercise and perform the following powers and functions, namely,—

(a) to see that the provisions of this Act are complied with;
(b) to prevent, as far as possible, the taking or abetting the taking of, of the demanding of, dowry;
(c) to collect such evidence as may be necessary for the prosecution of persons committing offences under the Act; and
(d) to perform such additional functions as may be assigned to him by the State Government, or as may be specified in the rules made under this Act.

(3) The State Government may, by notification in the official Gazette, confer such powers of a police officer as may be specified in the notification, the Dowry Prohibition Officer who shall exercise such powers subject to such limitations and conditions as may be specified by rules made under this Act.

(4) The State Government may, for the purpose of advising and assisting the Dowry Prohibition Officers in the efficient performance of their functions under this Act, appoint an advisory board consisting of not more than five social welfare workers (out of whom at least two shall be women) from the area in respect of which such Dowry Prohibition Officer exercises jurisdiction under sub-section (1).

9. Power to make rules.—(1) The Central Government may, by notification in the official Gazette, make rules for carrying out the purposes of this Act.
**

10. Power of the State Government to make rules.—The State Government may, by notification in the official Gazette, make rules for carrying out the purposes of this Act.

THE DOWRY PROHIBITION (MAINTENANCE OF LISTS OF PRESENTS TO THE BRIDE AND BRIDEGROOM) RULES, 1985

G.S.R. 664 (E), dated 19th August, 1985—In exercise of the powers conferred by Sec. 9 of the Dowry Prohibition Act, 1961 (28 of 1961), the Central Government hereby makes the following rules, namely:

1. Short title and commencement.—(1) These rules may be called the Dowry Prohibition (Maintenance of Lists of Presents to the Bride and Bridegroom) Rules, 1985.

   (2) They shall come into force on the 2nd day of October, 1985, being the date appointed for the coming into force of the Dowry Prohibition (Amendment) Act, 1984 (63 of 1984).

2. Rules in accordance with which lists of presents are to be maintained.—(1) The list of presents which are given at the time of the marriage to the bride shall be maintained by the bride.

   (2) The list of present which are given at the time of the marriage to the bridegroom shall be maintained by the bridegroom.

   Every list of presents referred to in sub-rule (1) or sub-rule (2),—

(a) shall be prepared at the time of the marriage or as soon as possible after the marriage:
(b) shall be in writing;
(c) shall contain,—
   (i) a brief description of each present;
   (ii) the approximate value of the present;
   (iii) the name of the person who has given the present; and
   (iv) where the person giving the present is related to the bride or bridegroom, a description of such relationship;
(d) shall be signed by both the bride and the bridegroom.
   3. The bride or the bridegroom may, if she or he so desires, obtain on either or both of the lists referred to in sub-rule (1) or sub-rule (2) the signature or signatures of any relations of the bride or the bridegroom or of any other person or persons present at the time of the marriage.

3.14.5 The Preconception and Pre-natal Diagnostic Test Act

The demographic balance of any healthy society depends on the existence/prevalence of a natural sex ratio. Sex ratio means the ratio of females to males in a given population. Such sex ratio may be looked upon from three angles: primary, secondary and tertiary. The primary sex ratio is the ratio at the time of conception; secondary sex ratio is the ratio at time of birth while tertiary sex ratio is the ratio of mature organisms. The Preconception and Prenatal Diagnostic Test Act aims at safeguarding the primary and the secondary sex ratio. Incidentally, the secondary sex ratio is commonly assumed, in human system, to be 105 boys to 100 girls and is often referred as "a ratio of 105'.

In the census presentation in India, the sex is captured under two formats: (a) Sex ratio-the number of females per 1000 of male population; (b) Child Sex ratio-the number of female children per 1000 of male child population in the age group of 0–6 years of age. The Child Sex Ratio indicates the gap between the natural Child Sex Ratio vis-à-vis the actual sex ratio at the early years of age group. The trend of Sex Ratio and Child Sex Ratio in India since 1951 to 2011, which appears disturbing, has been shown at Table 3.6.

The census findings over the decades point to the adverse female sex ratios in the country foretelling the onset of an alarming demographic imbalance, with the help of unscrupulous medical practitioners and modern technology to avert the birth of female children. The pre-birth elimination was not limited to any group; it happened to be a nationwide phenomenon. This practice initially started with the availability of amniocentesis in 1974 and then with the coming in of the ultrasound scanning, it rapidly gained momentum in the 80 s and 90 s. Popular upsurge against this evil practice led the Government of Maharashtra to pass legislation at the first instance to regulate the misuse of prenatal diagnostic techniques. The Government of India passed the Prenatal Diagnostic Techniques (Regulation and Prevention of Misuse) Act—(PNDT Act) in 1994.

The purposes of the PNDT Act were to provide for the regulation of the use of pre-natal diagnostic techniques for the purpose of detecting genetic or metabolic disorders or chromosomal abnormalities or certain congenital disorders and for the prevention of the misuse of such techniques for the purpose of pre-natal sex determination leading to female foeticide, and for matters connected there with or incidental thereto.

The salient features of the PNDT Act are described hereunder:

- Sec 3(1) provided for regulation of Genetic Counselling Centre, Genetic Laboratory or Genetic Clinic and none shall conduct, associate or help in activities relating to pre-natal diagnostic techniques.
- Sec 3(2) stipulated that no such Genetic Counselling Centre, Genetic Laboratory or Genetic Clinic shall employ any person who does not possess the prescribed qualifications
- Sec 3 (3) mentioned that no medical geneticist, gynaecologist paediatrician, registered medical practitioner or any other person shall conduct or aid in

Table 3.6 Sex ratio and child sex ratio of India since 1951 to 2011

Census year	1951	1961	1971	1981	1991	2001	2011
Sex ratio	946	941	930	934	929	933	940
Child sex ratio	983	976	964	962	945	927	914

conducting by himself or through any other person any pre-natal diagnostic techniques at a place other than a place registered under this Act.

Section-4 of the Act provides details on regulation of pre-natal diagnostic techniques

- Sec 4 (1) stipulates that no place including a registered Genetic Counselling Centre, Genetic Laboratory or Genetic Clinic shall be used by any person for conducting pre-natal diagnostic techniques except for the purposes specified in clause (2) and after satisfying any of the conditions specified in clause (3)
- Sec 4 (2) mentions that no pre-natal diagnostic techniques shall be conducted except for the purposes of detection of any of the following abnormalities, namely:

    (i) Chromosomal abnormalities;
    (ii) Genetic metabolic diseases
    (iii) Haemoglobinopathies
    (iv) Sex-linked genetic diseases
    (v) Congenital anomalies
    (vi) any other abnormalities or diseases as may be specified by the Central Supervisory Board;

Sec 4 (3) provides that no pre-natal diagnostic techniques shall be used or conducted unless the person qualified to do so is satisfied that any of the following conditions are fulfilled, namely:

(i) Age of the pregnant woman is above thirty-five years;
(ii) The pregnant woman has undergone of two or more spontaneous abortions or foetal loss;
(iii) The pregnant woman had been exposed to potentially teratogenic agents such as drugs, radiation, infection or chemicals;
(iv) The pregnant woman has a family history of mental retardation or physical deformities such as spasticity or any other genetic disease;
(v) Any other condition as may be specified by the Central Supervisory Board; Sec 4(4) stipulates that no person, being a relative or the husband of the pregnant woman shall seek or encourage the conduct of any pre-natal diagnostic techniques on her except for the purposes specified in clause (2).

Sec 5(1) provides that no person referred to in clause (2) of section 3 shall conduct the pre-natal diagnostic procedures unless—

He has explained all known side and after effects of such procedures to the pregnant Section 5 mentions that written consent of pregnant woman is necessary. Further it also prohibits communication of sex of foetus as mentioned below:

(a) Woman concerned;
(b) He has obtained in the prescribed form her written consent to undergo such procedures in the language which she understands; and

(c) A copy of the written consent obtained under clause (b) is given to the pregnant woman.

Sec (6) is about determination of sex prohibition which provides that on and from the commencement of this Act,—

(a) no Genetic Counselling Centre or Genetic Laboratory or Genetic Clinic shall conduct or cause to be conducted in its centre, Laboratory or Clinic, pre-natal diagnostic techniques including ultrasonography, for the purpose of determining the sex of a foetus;
(b) no person shall conduct or cause to be conducted in its centre, Laboratory or Clinic, pre-natal diagnostic techniques including ultrasonography, for the purpose of determining the sex of a foetus.

Sec-7 provides that the Central Government shall constitute a Central Advisory Board with the Minister in charge of the Ministry or Department of Family Welfare as ex officio Chairman with members as provided under sec 7(2). The Board shall have the following functions, namely:

(i) to advise the Government on policy matters relating to use of pre-natal diagnostic techniques;
(ii) to review implementation of the Act and rules made there under and recommend changes in the said Act and rules to the Central Government;
(iii) tocreat public awareness against the practice of pre-natal determination of sex and female foeticide;
(iv) to lay down code of conduct to be observed by persons working at Genetic Counselling Centre or Genetic Laboratory or Genetic Clinic;
(v) any other functions as may be specified under the Act.

Sec-17 empowers the Central Government to appoint one or more Appropriate Authority for each of the Union Territories. It has also empowered the State Governments to appoint one or more similar Appropriate Authority for the whole or part of the state. Sec 17(3) provides that when such Appropriate Authorities are appointed for the whole of the State or the Union Territory, they shall be of or above the rank of the Joint Director of Health and Family Welfare, but when appointed for any part of the State or Union Territory it shall be of such rank as the State Government or the Central Government as may deem fit.

The Appropriate Authority shall have the following functions, namely:

(a) to grant, suspend or cancel registration of a Genetic Counselling Centre or Genetic Laboratory or Genetic Clinic;
(b) to enforce standards prescribed for the Genetic Counselling Centre or Genetic Laboratory or Genetic Clinic;
(c) to investigate complaints of breach of the provisions of this Act or the rules made there under and take immediate action; and

(d) to seek and consider the advice of the Advisory Committee, constituted under sub-section (5), on application for registration and on complaints for suspension or cancellation for registration.

Sec-17 (5) empowers respectively the Central Government and the State Government to appoint an Advisory Authority for each of Appropriate Authority to aid and advise the Appropriate Authority in the discharge of its functions, and shall appoint one of its members of the Advisory Committee to be its Chairman. The Advisory Committee shall consist of—

(a) three medical experts from amongst gynaecologists, obstetricians, pediatricians and medical geneticists;
(b) one legal expert;
(c) one officer to represent the department dealing with information and publicity of the State Government or the Union Territory, as the case may be;
(d) three eminent social workers of whom not less than one shall be from amongst representatives of women's organisations;

No person having been associated with the use or promotion of pre-natal diagnostic techniques for determination of sex shall be appointed as a member of the Advisory Committee.

The functions of such Committee shall be such as may be prescribed in Rules by the State Government/Union Territory.

Sec 18 stipulates that no person shall open any Genetic Counselling Centre or Genetic Laboratory or Genetic Clinic after the commencement of this Act unless such Centre, Laboratory or Clinic is duly registered separately or jointly under this Act.

Sec 19 provides that the Appropriate Authority, after holding an inquiry and after satisfying the compliance of all requirements, shall grant certificate of registration to any Genetic Counselling Centre or Genetic Laboratory or Genetic Clinic separately or jointly.

Sec 20 provides that the Appropriate Authority may suo moto, or on complaint, issue a notice to any Genetic Counselling Centre or Genetic Laboratory or Genetic Clinic to show cause why its registration should not be suspended or cancelled for the reasons mentioned in the notice. Such Authority may, after proper application of mind, suspend the registration of such unit with recorded reasons.

Sec 21 stipulates that the concerned Genetic Counselling Centre or Genetic Laboratory or Genetic Clinic may prefer an appeal against such order to the Central Government or the State Government, as the case may be, within 30 days from the date of receipt of such order.

Sec 22 prohibits advertisement in any form (e.g. notice, circular, label wrapper or any other visible representation) relating to pre-natal determination of sex. Such contravention shall be punishable with imprisonment for a term which may extend to 3 years and with fine which may extend to ten thousand rupees.

Sec 23 provides that any medical geneticist, gynaecologist, registered medical practitioner etc. or any owner of such Genetic units or its employees who

3.14 Legal Issues

contravenes any provisions of this Act or rules there under shall be punishable with imprisonment for a term which may extend to 3 years and with fine which may extend to ten thousand rupees. The name of the registered medical practitioner shall be reported by the Appropriate Authority to the respective State Medical Council for removal of his name from the register of the council for a period of two years for the first offence and permanently for subsequent offence.

Any person who seeks the aid of any form of Genetic Centre or any category of medical practitioners etc. for conducting pre-natal diagnostic technique on any pregnant woman (including such woman unless she was compelled to undergo such diagnostic techniques) for purposes other than those specified in clause (2) of sec 4 shall be punishable with imprisonment for a term which may extend to 3 years and with fine which may extend to ten thousand rupees. And on subsequent conviction with imprisonment which may extend to 5 years and with fine which may extend to fifty thousand rupees.

Sec 27 provides that every offence under this Act shall be cognizable, non-bailable and non-compoundable.

Following a public interest litigation on non-enforcement of the law and Supreme Court's directive on it, the Ministry of Health and Family Welfare, Government of India amended the original Act. The Act is now called The Pre-Conception and Pre-Natal Diagnostic Techniques (Prohibition of Sex Selection) Act. 2003.

The status of implementation of The Pre-Conception and Pre-Natal Diagnostic Techniques (Prohibition of Sex Selection) Act. 2003 may be seen from the Report Card as below:

Report card on PNDT act

	Cases	Convictions	Sealing	Licence cancellation/suspension
May 2011	869	55	409	–
Jan 2012	1048	85	869	16
June 2012	1212	111	866	33
Jan 2013	1327	111	989	33
July 2013	1521	116	1180	53
Sept 2013	1833	143	1242	65

Source Annual report, 2013–14, ministry of health and family welfare, Govt of India

The Report Card with fewer number of cases, low convictions and lower licence suspensions shows the trend of poor implementation of the Act in a big country like India.

Incidentally, the Beti Bachao, Beti Padhao, a pilot programme introduced in October, 2014 to address the issue of declining Child Sex Ratio (CSR) in 100 selected districts with low in CSR, was launched nationwide by the Prime minister on 22 January 2015 to address the dipping child sex ratio and empower girl child in the country.

3.14.6 The Medical Termination Of Pregnancy Act, 1971

Abortion is one of the options for any woman to exercise her reproductive right and decision. It is, however, not an unfettered right but is subject to several biological and legal restrictions. Uppermost consideration in this regard is the risk of life of the pregnant woman including the possible grave injury of physical and mental health. Further, the possible physical or mental abnormalities in the event any child is born has to be seriously taken into consideration. Additionally, such abortion is legal only up to twenty weeks of pregnancy under specific conditions. No adult woman requires any other person's consent except her own for such abortion.

The Medical Termination of Pregnancy Act was enacted in Parliament in 1971 to *provide for the termination of certain pregnancies by registered medical practitioners and for matters connected therewith or incidental thereto.* It extends to the whole of India except the State of Jammu and Kashmir.

The salient features of this Act are as follows:

Sec 3. When Pregnancies may be terminated by registered medical practitioners.

(1) a registered medical practitioner shall not be guilty of any offence under Indian Penal Code or under any other law for the time being in force, if any pregnancy is terminated by him in accordance with the provisions of this Act.
(2) Subject to the provisions of sub-section (4), a pregnancy may be terminated by a registered medical practitioner,—

   (a) where the length of the pregnancy does not exceed twelve weeks, or
   (b) where the length of the pregnancy exceeds twelve weeks but does not exceed twenty weeks, if not less than two registered medical practitioners are of opinion, formed in good faith, that,—

      (i) the continuance of the pregnancy would involve a risk to the life of the pregnant woman or of grave injury physical or mental health; or
      (ii) there is a substantial risk that if the child were born, it would suffer from such physical or mental abnormalities as to be seriously handicapped.

Explanation 1.—Where any, pregnancy is alleged by the pregnant woman to have been caused by rape, the anguish caused by such pregnancy shall be presumed to constitute a grave injury to the mental health of the pregnant woman.

Explanation 2.—Where any pregnancy occurs as a result of failure of any device or method used by any married woman or her husband for the purpose of limiting the number of children, the anguish caused by such unwanted pregnancy may be presumed to constitute a grave injury to the mental health of the pregnant woman.

(3) In determining whether the continuance of pregnancy would involve such risk of injury to the health as is mentioned in sub-section (2), account may be taken of the pregnant woman's actual or reasonable foreseeable environment.

3.14 Legal Issues

(4) (a) No pregnancy of a woman, who has not attained the age of eighteen years, or, who, having attained the age of eighteen years, is a lunatic, shall be terminated except with the consent in writing of her guardian.
   (b) Save as otherwise provided in Cl. (a), no pregnancy shall be terminated except with the consent of the pregnant woman.

Sec 4. Place where pregnancy may be terminated.—
No termination of pregnancy shall be made in accordance with this Act at any place other than,—

(a) a hospital established or maintained by Government, or (b) a place for the time being approved for the purpose of this Act by Government.

Sec 5. Sections 3 and 4 when not to apply.—

(1) The provisions of Sec.4 and so much of the provisions of sub-section (2) of Sec. 3 as relate to the length of the pregnancy and the opinion of not less than two registered medical practitioner, shall not apply to the termination of a pregnancy by the registered medical practitioner in case where he is of opinion, formed in good faith, that the termination of such pregnancy is immediately necessary to save the life of the pregnant woman.
(2) Notwithstanding anything contained in the Indian Penal Code (45 of 1860), the termination of a pregnancy by a person who is not a registered medical practitioner shall be an offence punishable under that Code, and that Code shall, to this extent, stand modified.

Sec 6. Power to make rules.

(1) The Central Government may, by notification in the Official Gazette, make rules to carry out the provisions of this Act.

Sec 7. Power to make regulations.—

(1) The State Government may, by regulations,—

   (a) require any such opinion as is referred to in sub-section (2) of Sec. 3 to be certified by a registered medical practitioner or practitioners concerned in such form and at such time as be specified in such regulations, and the preservation or disposal of such certificates;
   (b) require any registered medical practitioner, who terminates a pregnancy to give intimation of such termination and such other information relating to the termination as maybe specified in such regulations;
   (c) prohibit the disclosure, except to such persons and for such purposes as may be specified in such regulations, of intimations given or information furnished in pursuance of such regulations.

(2) The intimation to be given or the information to be furnished in pursuance of regulations made by State shall be to the Chief Medical Officer of the State.

(3) Any person who wilfully contravenes or wilfully fails to comply with the requirements of any regulation shall be liable to be punished with fine which may extend to one thousand rupees.

Sec 8. Protection of action taken in good faith.—No suit for other legal proceedings shall lie against any registered medical practitioner for any damage caused likely to be caused by anything which is in good faith done or intended to be done under this act.

3.14.7 The Maharashtra Family Act, 1976

The Maharashtra Family Act was passed in 1976 to limit size of any family as per governmentally determined size. The main objective behind the Act was that contraception delivery system has not succeeded in reducing population. The Maharashtra Family Act requires that if a couple have 3 living children one of the parents be sterilized, unless the children are of the same sex.

This Act has been criticized on ground that it fails to protect the individual's right prior to sterilization. There is no provision in the Act to ensure that the government will inform the people of the available methods of family planning prior to the imposition of sterilization. The existing structure of the Act, which has not been implemented as yet, need to be completed with a scheme for compulsory family planning education and also providing opportunities to opt for other birth control methods before resorting to compulsory sterilization as it has to be compatible with the human right of family planning.

3.14.8 Others

Disqualification Laws:
Some States like Rajasthan, Orissa, Haryana, Delhi and Andhra Pradesh have enacted a law in order to have an effective check on the tendency of growing population to give a boost to the national family planning program. A disqualification is introduced with certain exceptions for being elected as a member of the local self-government. The validity of the impugned provisions has been upheld by the Division Bench of Rajasthan High Court Orissa High Court, Andhra Pradesh High Court and by the Supreme Court of India as well.

3.15 Organisational Issues and Organisational Set-up

An organization is an entity comprising multiple people with a particular purpose, be it in the business or government department. An *organizational structure in the government* defines how activities such as task allocation, coordination and supervision are directed towards the achievement of *organizational goal among the layers of government*. All organizations in the appropriate government level have a management structure that determines relationships between the different activities and the members, and subdivides and assigns roles, responsibilities, and authority to carry out different tasks. The organizational issues in population control and family planning at the government level are seldom addressed independently since it has become altogether an appendix of the health care system in our country and is thus away of the arc light of focus and public scrutiny. For the same reason, the goal of population stabilisation never becomes the key objective of the composite organisation of Health and family welfare. This is one of the important factors why population control and family planning is not eminently visible as a programme in the country.

From our discussion above, it would appear that population control and family planning is not an one dimensional management issue relating to health aspect only and that a good number of social, political, religious, gender and other issues are involved and inter-linked. Ideally management architecture for population control and family planning need to be so structured as to address each of the issues directly, preferably under the umbrella of a single organizational unit.

Right now, the Ministry of Health & Family Welfare, under Constitutional provision, is entrusted with the following broad sections of work for the Department of Health and Family Welfare, namely:

I. Union Business
II. List of Business for Legislative and Executive purposes in respect of Union Territories
III. List of Business with which the Central Government deal in a legislative capacity only for the Union and in both legislative and executive capacity for all Union Territories.
IV. Miscellaneous Business
V. Family Welfare Matters.

All sections of work falling under I to IV relate to health issues of the Union Government or those of Union Territories. Only section V deals with family welfare details of which are mentioned hereunder:

- Policy and organisation of Family Welfare
- All matters relating to:

    (a) National Rural Health Mission
    (b) National Commission on Population
    (c) Reproductive and Child Health

- Intersectoral Coordination in accordance with National Population Policy
- Matters related to Janasankhya Sthiarta Kosh and Empowered Action Group
- Organisation and direction of education, training and research on all aspects of family welfare including higher training abroad
- Production and supply of Aids to Family Welfare
- Liaison with foreign countries and international bodies as regards relating to family welfare
- Family Welfare Schemes and Projects with external assistance
- International Institute of Population Science, Mumbai
- Development and Production of audio visual aids, extension education and information in relation to population and family welfare
- Promoting Public Private Partnership for the Family Welfare Programme.
- All matters relating to following institutions:

    (a) Hindustan Latex Limited, Thiruvananthapuram
    (b) National Institute of Health and Family Welfare

- Implementation of Pre-conception and Pre-natal Diagnostic Technique (Prohibition of sex selection) Act, 1994 (57 of 1994)-Medical Termination of Pregnancy Act, 1971 (34 of 1971)

3.15.1 Organisational Set-up

As per website of the Ministry of Health and family Welfare, as on 12.12.2015, it has the following broad organisational set-ups:

- Departments
- Directorate General Health Services
- Autonomous Bodies
- State Health Departments
- State Health Societies
- Hospitals
- Teaching Institutions
- Research Institutions.

In the said website of the Ministry, Departments have the following two parts, namely,

- Departments of Health and Family Welfare
- Department of Health Research

It is rather unusual that even within departments in the Ministry, there is at present no separate department of Family Welfare reflecting the kind of priority the family welfare enjoys in the Ministry and thereby in the country. The Department of Family Welfare, that had an independent existence earlier, was abolished with the launching of National Health Mission in 2005 during the Tenth Five Year Plan. There is no separate directorate either for Family Welfare or Population Control and Family Planning making it impossible to undertake issue-based management of population control and family planning in the country. The entire Family Welfare programme has been subsumed under the National Health Mission with health focus only. This is a big area for policy reform in the country.

Looking around the issues as captured in Chap. 2, those listed under this Chap. 3 and the sustainable development issues as in Chap. 13, there are reasons to presume that Population Control and Family Planning need a bigger organisational framework with a much bigger organisational role than what it has been discharging at present, The limited forum of family welfare within a limited space in the Ministry of Health and Family Welfare is too inadequate and too meagre to play any pro-active role and discharge its Constitution ordained function. In fact, a good number of issues remains un-addressed and even when some of the issues are so addressed, those are purely notional, and merely touched upon, without any visible and commensurate outcome. Organisationally speaking, bigger problems need robust and focussed handling which the existing organisational structure is unable to cope with. It calls for a quantum jump from its existing status of being an appendix to the Health Care Services to the big functioning forum of Population Control and Family Planning in the country with authority of implementing, coordinating, monitoring and evaluating all issues connected with Population Control and Family Planning in the country. Further, such a visualised organisation of independent ministry would require to inform the citizens of the country periodically about what interventions are being made to stabilize population and seek their support to bring it back to its sustainable level. The population control programme of India is now threatened with an alarming scenario of overtaking even the huge population size of China by 2025. Incidentally, India, as an important member of UN and as a signatory to UN Summit of Sustainable Goals, 2015, is supposed to reach the SDGs targets by 2030. Unless India is in a position to put in place sustainable population, it will be well-nigh impossible to reach SDGs by 2030. The existing organisational set-up needs to be revamped to meet such new challenges as well.

3.15.2 Counselling Set-up and Counselling Issues

Motivation and awareness building are essential requisites to condition the mind-set of individuals to accept new ideas. While mere information through a service provider or otherwise may be good enough to self-motivate any individual, it may not be strong enough for some group of individuals to act upon it. Such individuals may need additional support and assistance to enable them to take personal decision for any informed choice. A good number of such eligible couples may remain in a state of confusion till all their silent queries are adequately addressed by a different set of people at another level of clarification. It is here that the role of counselling comes into facilitate to take final decision on the subject.

Counselling in the context of family planning is a process where face-to-face discussion takes place with a family planning service provider or a team of service providers to take informed decision on importance of family planning in the quest for the quality of life style of the family including the consideration of health of wife and the children. Moreover, counselling services provide related information on the available options for family planning method for the male and for the female, the plus and minus points of each of the methods and finally what would be the best option for the eligible couple in their particular situation having regard to age, reproductive behaviour and other factors. Additionally, it provides information on access to obtain such contraceptives and how to make use of them. The core intention of counselling is to enable the couple to take decision voluntarily based on informed choice.

The importance of counselling centres as a referral point for motivation is not in place for promotion of family planning services in our country. The institutional gap in awareness building and motivation may be filled into a great extent by a network of counselling points. The critically alarming demographic region deserves such counselling centres for breaking the shell of resistance and for scaling up of outreach of family planning acceptance. Additionally, in the context of unmet need of information of adolescents for right kind of hygienic practices, the importance of contraceptive, its availability and other services for prevention of unwanted pregnancies, the counselling centres can play very important role. There is also a strong case for establishing a counselling point near a cluster of colleges and universities for meeting the educational need of contraception for the upcoming eligible couples, preferably by civil society group under shared responsibility.

At the ground level, the word counselling is used in popular sense and not in terms of its professional meaning and services. Mere a talk or an advice by a health professional at a family planning clinic is not enough to graduate it as counselling. For a family planning counselling to be meaningful, it must have the following components, usually referred as GATHER:

G   Greet clients in a friendly way
A   Ask clients about their family planning needs
T   Tell clients about the available family planning methods
H   Help clients decide which method they decide
E   Explain how to use the method chosen
R   Return visit should be planned.

In other words, counselling can take place when both the husband and the wife remains present and participate in open and frank discussion in an exclusive and enabling environment in a particular location. The plus and minus points of all the methods are to be objectively discussed there and options of method chosen are to be validated with reference to age and reproductive behaviour. In so far as service providers are concerned, they must have a lot of empathy to understand the matter from the client's perspective and that due respect should also be given to client's feeling and attitude. Furthermore, privacy of the discussion need to be respected and honest guidance, as appropriate to the relevant client, need to be accorded. The counselling service providers must have the required skill to generate automatic confidence of the clients.

3.15.3 Incentives Versus Disincentives

Incentive is a kind of inducement or supplemental reward that serves as a motivational device for a desired action or behaviour. Disincentives, on the other hand, work as a factor, especially a financial disadvantage, that discourages a particular action. Promotion of family planning through Incentives and disincentives has been a management option in countries having adverse population scenario. Depending on the nature of crisis, a country falls back upon either on Incentives or disincentives or a combination of both. Incentives are usually given by a country facing critical population load and eligible couples are encouraged to have a small family and to accept certain category of permanent method of birth control in lieu of cash award for pocket expenses, transport cost, special leave during the period of medical services etc. In the reverse scenario, as in some western European countries facing very poor fertility growth, incentives are given to encourage more child births. In some country, such as China, severe disincentives are in place for couples flouting state dictated one-chid norm in terms of career growth, housing accommodation, social security and the like. In some country, as in India, disincentive in the form of maternity leave is not accorded for third pregnancy and beyond.

The advocates of Incentives and disincentives are very vocal and each of them have strong arguments for and against incentive and disincentive-based family planning. The arguments centre around various aspects including the desirability and sustainability of incentive and disincentive-based family planning programme. Be that as it may, let us see the kind of Incentives and disincentives that are in place in India. Incidentally, the NDC committee on population captured the scenario up to the period of Eight Plan which may be taken up as reference and discussed hereunder:

The Incentive scheme came into being in India as a compensation for the loss of wages to the poor undergoing sterilisation operation but expanded gradually to cover monetary benefits for the motivators, doctors and the other below:

(a) Government Employees: Government employees who undergo sterilisation with three or less number of children are eligible for the following benefits:

   (i) A special increment in the form of personal pay not to be absorbed in future increase of pay
   (ii) Reduction of half per cent on the rate of interest on house building advance
   (iii) Special casual leave up to seven days in respect of male employee and fourteen days in respect of female employee

As regards disincentives for Government employees, maternity leave is not available for the birth of the third or subsequent child. Similarly, children's Education Allowance and reimbursement of tuition fees, allowed to government employee, is restricted to two children only.

These incentives are also available for employees of public sector undertakings

(b) General Public: As far as the Central Government is concerned, compensation for loss of wages for sterilisation/IUD insertion is given to the acceptors. States/UTs are paid Rs 200/per tubectomy, Rs 180/per vasectomy and Rs 12/for IUD insertion in each case. Out of these, Rs 100/for sterilisation and Rs 9/for IUD insertion is passed on to acceptors (the rates have since been revised).

The Report of the NDC committee also mentioned some other incentive schemes introduced by a number of states, namely,

(i) Green Card Scheme: Under the Green Card Scheme, acceptors of sterilisation having up to two children are issued Green Cards which entitle them for preferential treatment for sanction of loans, grant of subsidies, house allotment, medical benefits, education facilities for children etc.
(ii) Lottery Scheme: Under the Lottery Scheme, acceptors of sterilisation having up to two children are entitled to participate in lotteries with large monetary prize from time to time.
(iii) Long term maturity bond scheme: Under the Social Security Certificate scheme, acceptors of sterilisation after one or more girl children get a lump-sum amount after a maturity period of seven years.

Besides, enhanced monetary incentives for sterilisation operations are given by some state governments from their State's budget.

Additionally, the State Governments were given 'incentives' by way of Additional Central Assistance from the Planning Commission under a structured Gadgil-Mukherjee formula where population control is accorded 1 % weightage within the category of performance of 'National Objective' to the tune of 3.0 %.

3.16 Social Security Issues

The uncertainty in life, dependence on livelihood support and the security in old age are invisibly linked with fertility behaviour for number of children in a family in a traditional society like us. The great expectation of financial support from the earning offspring is the principal contributing factor for a social mind-set to have number of children. The need for universal social security, especially in old age, can hardly be overemphasised in this very context. With the fast changing family values and the break-up of the traditional joint family system, the need for old age security support, especially the institutional security, has gone up tremendously. The organized sector including the employees of state government, central government, public sector institutions and the like enjoy various forms of old age security. The Central government and the various state governments have initiated a good number of old age pension schemes on social security for a select group of beneficiaries. However, considering the vast multitude of un-served unorganised sector, the issue of old age security dependence on children remains still very potent. This is a serious social issue that calls for appropriate resolution through institutional mechanism.

The social security schemes are of different nature with client-specific benefits. Since it may not be very relevant to go in for details of individual schemes, only the broad name of such schemes are mentioned below:

- Three components of Retirement Benefits for Central Government Employees—
  (a) Civil Services Pension Schemes
  (b) Civil Services Provident Fund Scheme
  (c) Gratuity to Civil Servants
- New Pension System (NPS) for Central Government Employees since January, 2004(also to all parts of the governments except for three State Governments and the Armed Forces)
- Military Pension Schemes
- Workers in the Organized Sector—
  (i) Employee Provident Fund (EPF)
  (ii) Employee Pension Scheme (EPS)
  (iii) Gratuity to Employees
  (iv) Special Provident Funds such as Coal Miners Provident Fund Act, Assam Tea Plantation Provident Fund Act etc.
- Voluntary Schemes:
  (a) Public Provident Fund
  (b) Private Pension Plans
  (c) Personal Pension Plan from Annuity Providers
- National Social Assistance Programme (NASP) for Poor and the Elderly:
  4 Components—
  (i) National Old Age Pension Scheme (NOAPS)
  (ii) National Family Benefit Scheme (NFBS)
  (iii) National Maternity Benefit Scheme (NMBS)
  (iv) Elderly Persons (over 65 years) under NOAPS.

Very recently the Prime Minister of India has introduced the following programme which is connected with financial support and retirement benefits, almost akin to pension schemes.

- Jan Dhan (providing bank accounts to 15 crore unbanked people)
- Pradhan Mantri Jeevan Jyoti Bima Yojana (PMJJBY),
- Pradhan Mantri Suraksha Bima Yojana (PMSBY)
- Atal Pension Yojana (APY).

In addition to the schemes listed above, the various state governments have also been implementing their own scheme on Old age pension, widow pension, Farmers' pension, Artisan's pension, Fishermen's pension etc. for the residents of such states.

In view of huge number of elderly population in the country, the country has to address, in no distant future, to a uniform social security scheme for all the elders of the country on consideration of equity and social justice. The destination point of such old age security scheme has to be universal in character without having any linkage to past services and occupational standards so as to insulate it from dependence syndrome on children. In the ultimate analysis, this would pre-empt any desire for additional child on ground of old age support.

Chapter 4
Issues Arising Out of the Health Policy and National Population Policy for Control of Population

Abstract The National policy of Health and the National Population Policy of the country focuses on issues that are critical and falls under priority concern for programme intervention. The recognition of any issue in the national policy of the country bestows it the seal of universal acceptance and stamp of priority. It automatically becomes an actionable item in work programme. Additionally, issue based policy pre-empts cloud of confusion that very often surfaces during the programme implementation stages. In so far as the Health policy is concerned, it basically aims at achieving specific health care goals within a society. With this broad objective, the first National Health Policy (NHP) of the country was adopted in 1983 and was subsequently revised in 2002. The Government of India is now in the process of finalising the National Health Policy, 2015. The National Health Policy looks the population control issues from the prism of family planning and specially those linked to reproductive health care services while the National Population Policy addresses broader issues linked with control of population. The highlights of issues as addressed by the Health Policy and National Population Policy of the country have been captured here.

Keywords Age of marriage · Birth spacing · Non-coercive approach to family planning · Unmet needs · Universal immunization · Small family norm · Replacement levels of TFR

Identification of issues is an important requisite for any effective policy intervention. Earlier, in Chap. 2 and in Chap. 3, all possible issues connected with population control and management have been sought to be identified. Further, it is also important to know how the National policy of Health and the National Population Policy of the country look on issues that are critical and which have been included for programme intervention. The recognition of any issue in the national policy of the country bestows it the seal of universal acceptance and stamp of priority. It automatically becomes an actionable item in work programme. Additionally, issue based policy pre-empts cloud of confusion that very often surfaces during the programme implementation stages. In so far as the Health policy is concerned, it

© Springer International Publishing Switzerland 2017
B. SyamRoy, *India's Journey Towards Sustainable Population*,
DOI 10.1007/978-3-319-47494-6_4

basically aims at achieving specific health care goals within a society. With this broad objective, the first National Health Policy (NHP) of the country was adopted in 1983 and was subsequently revised in 2002. The Government of India is now in the process of finalising the National Health Policy, 2015. The National Health Policy looks the population control issues from the perspective of family planning and specially those linked to reproductive health care services. The highlights of issues as addressed by the Health Policy of the country revolve around the following:

(a) Issues arising out of Health Policy:

- Better and safer contraceptive choice
- Age of Marriage
- Birth spacing
- Family planning services on any fixed day at non-camp site
- Increased male sterilisation
- Non-coercive approach to family planning with improved access

(b) Issues arising out of National Population Policy:

The Population Policy of the country covers the entire approach of the government in relation population control and family planning. It addresses demographic issues and formulates guidelines how to go forward. It addresses vital statistics of the country and designs way out for its course correction. Finally, it probes and diagnoses the medical and non-medical issues connected with family planning and suggest ways to address each of them. To be precise, Population policy generally addresses issues connected with fertility, births, deaths, geographical distribution, and immigration, population composition of sex, age structure and activities directly and indirectly influencing population variables. Population policy also focuses on society's culture and values.

The National Population Policy, 2000 identified a host of issues which were to be addressed to bring the TFR to replacement levels by 2010 and stable population by 2045. The items identified and listed are as below:

1. Unmet needs for basic reproductive and child health services, supplies and infrastructure
2. School education up to age 14 to be made free and compulsory along with reduction of drop outs at primary and secondary school levels to below 20 % for boys and girls.
3. Reduction of infant mortality rate to below 1000 live births
4. Reduction of maternal mortality ratio to below 100 per 100,000 live births
5. Universal immunization of children against all vaccine preventable diseases
6. Delayed marriage for girls not earlier than age 18 and preferably after 20 years of age.
7. 80 % institutional deliveries and 100 % deliveries by trained persons
8. Universal access to information/counselling, and services for fertility regulation and contraception with a wide basket of choices

9. 100 % registration of births, deaths, marriage and pregnancy
10. Containment of the spread of Acquired Immunodeficiency Syndrome (AIDS), and integration between the management of reproductive tract infections(RTI) and sexually transmitted infections(STI) and the National AIDS Control Organisation.
11. Prevention and control communicable diseases
12. Integration of Indian Systems of Medicine (ISM) with the reproductive and child health services and in reaching out to households.
13. Promotion of small family norm to achieve replacement levels of TFR
14. Convergence in implementation of related social sector programs so that family welfare becomes a people centred program

Chapter 5
Issues Emerging Out of the International Conference on Population and Development (ICPD) and Millennium Development Goals (MDGs)

Abstract The most eventful conference of the 20th century under the United Nations was in the forum of the International Conference on Population and Development held in Cairo, Egypt from 5 to 13 September 1994. The future shaping 115-page document embodying general agreement on the Programme of Action, adopted on 13 September, ushered a new approach and a new phase to population and development among 179 countries of which India happened to be a key nation. Central to this new approach is empowering women and providing them with more choices through extended access to education and health services and promoting skill development and employment. The Programme of Action includes a set of important population and development objectives, including both qualitative and quantitative goals that are mutually supportive and are of critical importance to these objectives. Among these objectives and goals are: sustained economic growth in the context of sustainable development; education, especially for girls; gender equity and equality; infant mortality and maternal mortality reduction; and the provision of universal access to reproductive health services, including family planning and sexual health.

Keywords Sustainable development · Gender equality · Empowerment of women · Reproductive health · Family planning · Sexually transmitted diseases · International migration · HIV/AIDS · Global partnership

5.1 International Conference on Population and Development (ICPD)

The most eventful conference of the 20th century under the United Nations was in the forum of the International Conference on Population and Development held in Cairo, Egypt from 5 to 13 September 1994. The future shaping 115-page document embodying general agreement on the Programme of Action, adopted on 13 September, ushered a new approach and a new phase to population and development among 179 countries of which India happened to be a key nation. Central to this new approach is empowering women and providing them with more choices through extended access to education and health services and promoting skill development and employment.

The Programme of Action includes a set of important population and development objectives, including both qualitative and quantitative goals that are mutually supportive and are of critical importance to these objectives. Among these objectives and goals are: sustained economic growth in the context of sustainable development; education, especially for girls; gender equity and equality; infant mortality and maternal mortality reduction; and the provision of universal access to reproductive health services, including family planning and sexual health.

A set of principles were set for guidance for the international community to collectively address the critical challenges and interrelationship between population and development as a part of a new global partnership which were as follows:

Principle-1

All human beings are born free and equal in dignity and rights. Everyone is entitled to all the rights nd freedoms set forth in the Universal Declaration of Human Rights, without distinction of any kind, such as race, colour, sex, language, religion, political or other opinion, national or social origin, property, birth, or other status. Everyone has the right to life, liberty and security of person.

Principle-2

Human beings are at the centre of concerns for sustainable development. They are entitled to healthy and productive life in harmony with nature. People are the most important and valuable resources of any nation. Countries should ensure that all individuals are given the opportunity to make the most of their potential. They have the right to an adequate standard of living for themselves and their families including adequate food, clothing, housing, water and sanitation.

Principle-3

The right to development is a universal and inalienable right and an integral part of human rights and the human person is the central subject of development. While development facilitates the enjoyment of all human rights, the lack of development may not be invoked to justify the abridgement of internationally recognised human rights. The right to development must be fulfilled so as to equitably meet the population, development and environment needs of present and future generations.

Principle-4

Advancing gender equality and equity and the empowerment of women, and all kinds of violence against the women, and ensuring women's ability to control their own fertility, are cornerstones of population and development-related programmes. The human rights of women and the girl child are an inalienable, integral and indivisible part of universal human rights. The full and equal participation of women in civil, cultural, economic, political and social life, at the national, regional, and international levels, and the eradication of all forms of discrimination on grounds of sex, are priority objectives of the international community.

Principle-5
Population-related goals and policies are integral parts of cultural, economic and social development, the principal aim of which is to improve the quality of life of all people.

Principle-6
Sustainable development as a means to ensure human wellbeing equitably shared by all people today and in the future, requires that the interrelationship between population, resources, the environment and development should be fully recognised, properly managed and brought into harmonious, dynamic balance. To achieve sustainable development and a higher quality of life for all people, states should reduce and eliminate unsustainable patterns of production and consumption and promote appropriate policies, including population related policies, in order to meet the need of current generations without compromising the ability of future generations to meet their own needs.

Principle-7
All states and all people shall cooperate in the essential task of eradicating poverty as an indispensable requirement for sustainable development, in order to decrease the disparities in standard of living and better meet the needs of the majority people of the world. The special situation and needs of developing countries, particularly the least developed, shall be given special priority. Countries with economics in transition, as well as all other countries, need to be fully integrated into the world economy.

Principle-8
Everyone has the right to the enjoyment of the highest attainable standard of physical and mental health. States should take all appropriate measures to ensure, on a basis of equality of men and women, universal access to health-care services, including those related to reproductive health care, which includes family planning and sexual health. Reproductive health-care programmes should provide the widest range of services without any form of coercion. All couples and individuals have the basic right to decide freely and responsibly the number and spacing of their children and to have the information, education and means to do so.

Principle-9
The family is the basic unit of society and as such should be strengthened. It is entitled to receive comprehensive protection and support. In different cultural, political and social systems, various forms of the family exist. Marriage must be entered into with the free consent of the intending spouses, and husband and wife should be equal partners.

Principle-10
Everyone has the right to education, which shall be directed to the full development of human resources, and human dignity and potential, with particular attention to women and girl child. Education should be designed to strengthen respect for human rights and fundamental freedoms, including those relating to population and

development. The best interest of the child shall be the guiding principle of those responsible for his or her education and guidance; that responsibility lies in the first place with parents.

Principle-11
All states and families should give the highest priority to children. The Child has the right to standards of living adequate for its well-being, and rights to the highest attainable standards of health, and the right to education. The child has the right to be cared for, guided and supported by parents, families and society and to be protected by appropriate legislative, administrative, social and educational measures from all forms of physical or mental violence, injury or abuse, neglect or negligent treatment, maltreatment or exploitation, including sale, trafficking, sexual abuse, abuse, and trafficking in its organs.

Principle-12
Countries receiving documented migrants should provide proper treatment and adequate social welfare services for them and their families, and should ensure their physical safety and security, bearing in mind the special circumstances and needs of countries, in particular developing countries, attempting to meet these objectives or requirements with regard to undocumented migrants, in conformity with the provisions of relevant conventions and international instruments and documents. Countries should guarantee to all migrants all basic human rights as included in the Universal Declaration of Human Rights.

Principle-13
Everyone has the right to seek and to enjoy in other countries asylum from persecution. States have responsibilities with respect to refugees as set forth in the Geneva Convention on the Status of Refugees and its 1967 Protocol.

Principle-14
In considering the population and development needs of indigenous people, states should recognize and support their identity, culture and interests, and enable them to participate fully in the economic, political and social life of the country, particularly where their health, education and well-being are affected.

Principle-15
Sustained economic growth, in the context of sustainable development, and social progress require that growth be broadly based, offering equal opportunities to all people. All countries should recognize their common but differentiated responsibilities. The developed countries acknowledge the responsibility that they bear in the international pursuit of sustainable development, and should continue to improve their efforts to promote sustained economic growth and to narrow imbalances in a manner that can benefit all countries, particularly the developing countries.

The adoption of holistic guidance on population and development by the ICPD in such a broad plane is by itself a landmark event. However, the high valued canvass required to be further broken down into activities and transformed them in a cogent programme of actions to be followed up. Such programme of actions, to all intents and purposes, are intended to be adapted in the programme design on population and development of participating countries. It is with this mission that the ICPD identified the general issues connected with population and development and adopted the following components on items of Programme of Action, as indicated herein:

5.1.1 Programme of Actions

1. Interrelationships between Population, Sustained Economic Growth and Sustainable Development

    A. Integrating population and development strategies
    B. Population, sustained economic growth and poverty

2. Gender equality, equity and empowerment of women

    A. Empowerment and status of women
    B. The girl child

3. The Family, its Roles, Rights, Composition and Structure

    A. Diversity of family structure and composition
    B. Socio-economic support to the family

4. Population Growth and Structure

    A. Fertility, mortality and population growth rates
    B. Children and youth
    C. Elderly people
    D. Indigenous people
    E. Persons with Disabilities

5. Reproductive Rights and Reproductive Health

    A. Reproductive rights and reproductive Health
    B. Family Planning
    C. Sexually transmitted diseases and prevention of human immunodeficiency virus (HIV)
    D. Human sexuality and gender relations
    E. Adolescents

6. Health, Morbidity and Mortality

    A. Primary health care and the health–care sector
    B. Child survival and health
    C. Women's health and safe motherhood

D. Human immunodeficiency virus (HIV) infection and acquired immunodeficiency syndrome (AIDS)

7. Population Distribution, Urbanization and Internal Migration

    A. Population distribution and sustainable development
    B. Population growth in large urban agglomerations
    C. Internally displaced persons

8. International Migration

    A. International migration and development
    B. Documented Migrants
    C. Undocumented migrants
    D. Refugees, asylum-seekers and displaced persons

9. Population, Development and Education

    A. Education, population and sustainable development
    B. Population, information, education and communication

10. Technology, Research and Development

    A. Basic data collection, analysis and dissemination
    B. Reproductive health research
    C. Social and economic research

11. National Action

    A. National policies and plans of action
    B. Programme management and human resource development
    C. Resource mobilization and allocation

12. International Cooperation

    A. Responsibilities partners in development
    B. Towards a new commitment to funding population and development

13. Partnership with the Non-Government Sector

    A. Local, national, international non-governmental organizations
    B. The Private sector

14. Follow–up to the Conference

    A. Activities at the national level
    B. Sub regional and regional activities
    C. Activities at the International level

The detailed objectives of ICPD document required actions for each of these broad issues and needed to be acted upon by programme professionals down the lines.

5.2 Millennium Development Goals (MDGs) and the Population Control Issues

The MDGs is an agreed set of goals that international community set for itself at the UN Millennium Summit 2000. It encompassed universally accepted human values and rights such as freedom from hunger, the right to basic education, the right to health and responsibility to future generations. The MDGs, as set for implementation and monitoring, were as follows:

Goal 1 Eradicate extreme poverty and hunger
Goal 2 Achieve universal primary education
Goal 3 Promote gender equality and empower women
Goal 4 Reduce child mortality
Goal 5 Improve maternal health
Goal 6 Combat HIV/AIDS, malaria and other diseases
Goal 7 Ensure environmental sustainability
Goal 8 Develop a global partnership for development

Along with these eight goals, the MDGs had set targets for each of the sub-goals, and also its monitoring indicators. There were nothing in it directly on population control and family planning; however, indirectly it strived to address the same through a number of indicators, which may be summarized hereunder. Such indicators might as well be identified and called as issues for population control under MDGs.

It is to mention in this connection that Millennium Development Goals Indicators and its country performance have been released by the United Nations in the MDG Report, 2015 on 6th July, 2015. The related indicators and its performance status over time for India are indicated below.

- Series Name: Children under five moderately or severely under-weight percentage
  Goal 1. Eradicate extreme poverty and hunger
  Target 1.C Halve, between 1990 and 2015, the proportion of people who suffer from hunger
  Indicator 1.8 Prevalence of underweight children under five years of age
  Last updated 6, July 2015

Country	1992	1993	1997	1999	2006
India	52.8	51.2	38.4	46.3	43.5

Source unstat/Millennium Indicators

- Series Name: Children under five severely under-weight, percentage
  Goal 1. Eradicate extreme poverty and hunger
  Target 1.C Halve, between 1990 and 2015, the proportion of people who suffer from hunger

Indicator 1.8 Prevalence of underweight children under five years of age
Last updated 6, July 2015

Country	1992	1993	1997	1999	2006
India	–	24.3	–	20.6	17.4

Source unstat/Millennium Indicators

- Series Name: Population under nourishment, percentage
  Goal 1. Eradicate extreme poverty and hunger
  Target 1.C Halve, between 1990 and 2015, the proportion of people who suffer from hunger
  Indicator 1.9 Proportion of population below minimum level of dietary energy consumption
  Last updated 6, July 2015

Country	1991	1992	1993	1994	1995	1996	1997	1998	1999	2000	2001	2002
India	23.7	22.2	22.4	22.1	21.6	20.5	19.2	18.0	17.3	17.0	17.5	18.6
	2003	2004	2005	2006	2007	2008	2009	2010	2011	2012	2013	
	19.9	20.9	21.2	20.5	18.9	17.2	16.2	17.2	15.7	15.6	15.4	

Source unstat/Millennium Indicators

- Series Name: Children 1 year old Immunization against measles
  Goal 4. Reduce Child mortality
  Target 4.A: Reduce by two-thirds, between 1990 and 2015, the under-five mortality
  Indicator 4.3 Proportion of 1 year old children immunized against measles
  Last updated 6, July 2015

Country	1990	1991	1992	1993	1994	1995	1996	1997	1998	1999	2000	2001
India	56	43	51	59	67	72	66	55	57	58	99	57
	2002	2003	2004	2005	2006	2007	2008	2009	2010	2011	2012	2013
	56	62	68	59	71	69	74	74	74	74	74	74

Source unstat/Millennium Indicators

- Series Name: Children under five mortality rate per 1000 live births
  Goal 4. Reduce Child mortality
  Target 4.A: Reduce by two-thirds, between 1990 and 2015, the under- five mortality
  Indicator 4.1 Under five-mortality rate
  Last updated 3, October 2014

5.2 Millennium Development Goals (MDGs) ...

Country	1990	1991	1992	1993	1994	1995	1996	1997	1998	1999	2000	2001
India	125.9	122.4	119	115.7	112.3	109	105.5	102	98.5	94.9	91.4	87.9
	2002	2003	2004	2005	2006	2007	2008	2009	2010	2011	2012	2013
	84.5	81.2	77.9	74.7	71.6	68.7	65.8	62.9	60.2	57.5	55	52.7

Source unstat/Millennium Indicators

- Series Name: Infant mortality rate (0–1) per 1000 live births
  Goal 4. Reduce Child mortality
  Target 4. A: Reduce by two-thirds, between 1990 and 2015, the under- five mortality
  Indicator 4.2 Infant mortality rate
  Last updated 3, October 2014

Country	1990	1991	1992	1993	1994	1995	1996	1997	1998	1999	2000	2001
India	88.4	86.2	84	81.9	79.8	77.7	75.5	75.3	71	68.7	66.5	64.3
	2002	2003	2004	2005	2006	2007	2008	2009	2010	2011	2012	2013
	62.1	60	57.9	55.9	53.9	52	50.1	48.3	46.4	44.7	42.9	41.4

Source unstat/Millennium Indicators

- Series Name: Maternal maternity ratio per 100000 live births
  Goal 5. Improve Maternal health
  Target: 5.A: Reduce by three quarter, from 1990 to 2015, maternal mortality ratio
  Indicator 5.1 Maternal mortality ratio

Country	1990	1995	2000	2005	2010	2015
India	560	460	370	280	220	190

Source unstat/Millennium Indicators

Chapter 6
Current Issues in the Post Census, 2011

Abstract The Census 2011 lays bare the aggregate concern that the population control programme of the day needs to address with focused attention. This has assumed enormous importance now because of the absence of any population policy right at the moment after the expiry of NPP 2000 at 2010. Usually, the population policy of any country undergoes generational changes in keeping with the changing status of its demographic profile; India is, however, still to have any new population policy after 2011. Be that as it may, the country needs to address the demographic challenges that opened up subsequent to publication of the final report of the Census 2011.

Keywords Demographic challenges · Size of population of the country · Density of population · Sex ratio · Child sex ratio · Muslim population

Earlier several issues relevant to population control and family planning have been discussed from various perspectives and angles giving a fair idea of the length and breadth of the relevant issues that call for programme intervention. While such issues are indeed very critical from population control management point of view, the Census 2011 lays bare the aggregate concern that the population control programme of the day needs to address with focused attention as well. This has assumed enormous importance now because of the absence of any population policy right at the moment after the expiry of NPP 2000 at 2010. Usually, the population policy of any country undergoes generational changes in keeping with the changing status of its demographic profile; India is, however, still to have any new population policy after 2011. Be that as it may, the country needs to address the demographic challenges that opened up subsequent to publication of the final report of the Census 2011.

The most alarming concern of the findings of census, 2011 is the size of population of the country. The total population of the country has reached a phenomenal figure of 121, 05, 69, 573 as on 1st March, 2011 adding a huge number of 18, 19, 59, 458 over 2001. The decadal (2001–2011) growth rate of population of India, though declined over the previous decade, was as high as 17.7 % with

© Springer International Publishing Switzerland 2017
B. SyamRoy, *India's Journey Towards Sustainable Population*,
DOI 10.1007/978-3-319-47494-6_6

varying growths of the states and Union Territories. The states which had the decadal growth rate of more than national average are Himachal Pradesh (23.6 %), Uttarakhand (18.8 %), Haryana (19.9 %), Rajasthan (21.3 %), Uttar Pradesh (20.2 %), Bihar (25.4 %), Arunachal Pradesh (26.0 %), Manipur (18.6 %), Mizoram (23.6 %), Meghalaya (27.9 %), Jharkhand (22.4 %), Chattisgarh (22.6 %), Madhya Pradesh (20.3 %), Gujarat (19.3 %), Poducherry (28.1 %), NCT of Delhi (21.2 %), Daman& Diu (53.8 %) and D &Haveli (55.9 %). Decadal growth rates for such a good number of States and Union Territories are matters of serious concern and call for appropriate policy response and action.

Density of population is in a way the final indicator of population load of a country as it conveys the ultimate message of carrying capacity of a country. The density of population for India has shot up from 325 in 2001 to 382 in 2011. Bihar occupies the first position with a population density of 1106, surpassing West Bengal of 1028. The position of other states and Union Territories which surpassed the national average include Punjab (551), Chandigarh (92580, Haryana (573), NCT of Delhi (11320), Uttar Pradesh (829), Bihar (1106), Assam (398), West Bengal (1028), Jharkhand (414), Daman &Diu (2191), D&N Haveli (700), Goa (394), Lakshadweep (2149), Kerala (860), Tamil Nadu (555) and Poducherry (2547).

Another area of concern is the Sex Ratio of the country. It is a fact that the Sex Ratio of the country has gone up by 10 points from 933 in 2001 to 943 in 2011. However, there are States and Union Territories where the current Sex Ratio is less than the national average implying scope for appropriate response from programme management angle. Such States and Union Territories include Jammu& Kashmir (889) Punjab (895), Chandigarh (818), Haryana (879), N&CT of Delhi (868), Rajasthan (928), Uttar Pradesh (912), Bihar (918), Sikkim (890), Arunachal Pradesh (938), Nagaland (931), Madhya Pradesh (931), Gujarat (919), Daman &Diu (618), D&N Haveli (774), Maharashtra (929) and A &N Islands (876).

The gravity of the Sex Ratio is understood when the Child Sex Ratio (0–6 years of age) is taken into consideration. As per Census 2011, the child population in the age group of 0–6 years stands at 16, 44, 78, 150 representing 13.6 % of the total population of the country. The Census 2011 also reveals that there has taken place a considerable fall of Child Sex Ratio from 927 in 2001 to 919 in 2011—a clear 8 points decline in India. The fall has been to the tune of 11 points (934 to 923) in rural areas and in urban areas the decline has been to the extent of 1 point (906 to 905) over the last decade. While the child sex ratio is much below the natural sex ratio, there are States and Union Territories where the current Child Sex Ratio is even less than the national Child Sex Ratio of 919. This signals a very grave and alarming situation for family welfare point of view. Such States and Union Territories include Jammu& Kashmir (862), Himachal Pradesh (909), Punjab (846), Chandigarh (880), Uttarahkand (890), Haryana (834), N&CT of Delhi (871), Rajasthan (888), Uttar Pradesh (902), Madhya Pradesh (918), Gujarat (890), Daman &Diu (904), Maharashtra (894) and Lakshadweep (911).

Caste has become an important issue in today's political economy. In view of the usual practice to highlight the demographic data among Schedule Castes and

Schedule Tribes population of the country, it is appropriate to present the demographic features of SC and ST population of India. The decadal growth rate of the SC population of India in 2011 is 20.8 % as against the country's decadal growth of 17.7 %. In other words, the decadal growth rate of SC population is higher by 3.1 % over the national average. Among the states and Union Territories where such decadal growth is higher than the national average of 17.7 % are Jammu &Kashmir (20.1 %), Punjab (26.1 %), Chandigarh (26.3 %), Uttarakhand (24.7 %), Haryana (25.0 %), NCT of Delhi (20.0 %) Rajasthan (26.1 %), Bihar (27.0 %), Manipur (61.6 %), Mizoram (347.8 %), Tripura (17.8 %) Meghalaya (55.8 %), Assam (22.2 %), Jharkhand (25.0 %), Odisha (18.2 %), Chattisgarh (35.4 %), Madhya Pradesh (23.9 %), Poducherry (24.4 %), Daman &Diu (26.6 %) and D &Haveli (50.7 %), Maharashtra (34.3 %), Karnataka (22.3 %) and Tamil Nadu (21.8 %).

As to the decadal growth of ST population in India in 2011, it is at 23.7 %, higher by 6 % of the national average of 17.7 %. The decadal growth of ST population is also higher than the decadal growth rate of 20.8 % of the SC population. Among the states and Union Territories where the decadal growth rate of ST population is higher than the national average of 17.7 % are Jammu &Kashmir (35.0 %), Himachal Pradesh (60.3), Rajasthan (30.2 %), Uttar Pradesh (950.6 %), Bihar (76.20 %), Sikkim (85.2 %), Arunachal Pradesh (35.0 %), Manipur (21.8 %), Mizoram (23.4 %), Meghalaya (28.3 %),West Bengal (20.2 %), Jharkhand (22.0 %), Chattisgarh (18.2 %), Madhya Pradesh (25.2 %), Gujarat (19.2 %), D &Haveli (30.1 %), Andhra Pradesh (17.8 %), Maharashtra (22.5 %), Karnataka (22.7 %), Goa (26,273.7 %) Kerala (33.1 %) and Tamil Nadu (22.0 %).

The Sex Ratio of the SC population of India is 945 in 2011, which is higher by 2 points over the national average of 943. However, there are States and Union Territories where the Sex Ratio is less than the national average. Such States and Union Territories include Jammu& Kashmir (902), Punjab (910), Chandigarh (872), Haryana (887), N&CT of Delhi (889), Rajasthan (923), Uttar Pradesh (908), Bihar (925), Mizoram (509), Meghalaya (895), Madhya Pradesh (920), Gujarat (931), Daman &Diu (944), D&N Haveli (853), Maharashtra (929) and A &N Islands (876).

The Sex Ratio of ST population for the country is 990 in 2011, which is 47 point higher than the country average of 943. It is also 45 point higher over the country average of SC population. Except in case of Jammu &Kashmir (924) and A&N Islands (937), in all States and Union Territories, the Sex Ratio of ST population is much higher than the corresponding state average, and in many cases those are very near to natural Sex Ratio.

Another concern area of the country is the size and growth of the Muslim population. The size of the Muslim population has increased from 13.81 crore in 2001 to 17.22 crore in 2011, registering decadal growth of 24.6 %, as against the decadal growth of Hindus at 16.8 %. Thus, the decadal growth of Muslim population is higher by 6.9 % over the national average of 17.7 %.It is also higher by 7.8 % over the Hindu growth of population at 2011. The growth rate of population of other religious communities in the same period was Christian: 15.5 %; Sikh: 8.4 %; Buddhist: 6.1 % and Jain: 5.4 %. It is only desirable that Muslims in India do conform to growth rate like other religious faith in the country.

Part II
Requisites of the Population Control Programme

Chapter 7
Conceptual Clarifications on Data Used in Population Control Area

Abstract The Family planning programme managers in the field and/or Population control professionals in the intermediate level very often uses some catchy phrases without sometimes understanding its proper meaning and implications. Such phrases belong to the discipline of demography and happen to be part of the lexicography of Population control subject area. Such concepts include Crude Birth Rate, Total Fertility Rate, Crude Death Rate, Infant Mortality Rate, Child Mortality Rate, Maternal Mortality Ratio, Maternal Mortality Rate, Still Birth, Sex Ratio at Birth, Sex Ratio, Growth Rate, Dependency Ratio, Stable Population, Couple Protection Rate, Effective Couple Protection Rate, Contraceptive Prevalence Rate, Birth Interval etc. The use of such concepts in a popular sense does not help in promoting intended outcome. On the contrary, it breeds confusion in the programme implementation area. Conceptual clarifications on data used in population Control area have been made in this chapter.

Keywords Conceptual clarifications · Crude birth rate · Total fertility rate · Infant mortality rate · Maternal mortality ratio · Sex ratio · Couple protection rate etc

The Family planning programme managers in the field and/or Population control professionals in the intermediate level very often uses some catchy phrase without sometimes understanding its proper meaning and implications. Such phrases belong to the discipline of demography and happen to be part of the lexicography of Population control subject area. Some of such concepts have been listed in the IIPS hand-outs and discussed as below:

Crude Birth Rate Crude Birth Rate (CBR) is defined as a ratio of the total number of births during given year and a given geographical area to the average (or mid-year) population ever lived in that year and geographical area.

It is the simplest method used to measure fertility in a given area in a given period.

Total Fertility Rate Total Fertility Rate (TFR) is the number of children which a woman of a hypothetical cohort would bear during her life time if she were to bear children throughout her life at the Age specific Fertility Rates for given year and if none of them dies before crossing the age of reproduction.

The TFR of level 2.1 children per woman in a population is generally taken as replacement level fertility. It is usually considered to be a precondition for population stabilization. However, if a population attains TFR of level 2.1, it does not mean the size of population get stabilized immediately. This takes some time to take effect because acceleration in population growth persists due to larger number of couples already existing in the reproductive ages in that population.

Crude Death Rate Crude Death Rate (CDR) is defined as the total number of deaths in a given year and a given geographical area to the mid-year population per thousand in that year and geographical region.

Infant Mortality Rate Infant Mortality Rate (IMR) is number of infants dying under one year of age in a year in a given geographical region per thousand live births of the same year and geographical region.

IMR is considered to be the most sensitive indicator of general health and medical facilities in a given community.

Child Mortality Rate Child Mortality Rate (CMR) refers to total number of deaths between age 0–5 in given year and geographical region and to total number of live births that year and in same geographical regions.

Neonatal Mortality Rate Neonatal Mortality Rate means that number of infants dying within the first month (4 weeks) i.e. within 28 days of life in a year and per thousand of live births in the same year and geographical region.

Early Neonatal Mortality Rate Early Neonatal Mortality Rate refers to number of neonatal deaths during the first seven days in a year and geographical region per thousand live births in the same year

Perinatal Mortality Rate Perinatal Neonatal Mortality Rate means and includes number of stillbirths after 28 weeks of pregnancy plus deaths within first week of delivery in a year and geographical region per births (live and still) in a year and geographical region.

Post Neonatal Mortality Rate Post Neonatal Mortality Rate refers to number of infant deaths after 28 days to one year of age per thousand live births in a given year.

Maternal Mortality Ratio Maternal Mortality Ratio is defined as number of women while pregnant or within 42 days of termination of pregnancy from any cause related to pregnancy and childbirth per 100,000 live births in a given year.

Maternal Mortality Rate Maternal Mortality Rate means number of maternal deaths per 100,000 women in reproductive ages 15–49.

Still Birth It indicates death of foetus after completing 28 weeks and till the time of birth.

Sex Ratio at Birth There is no one to one correspondence in the birth of a male and female. The registered births for a large number of countries over a wide range of periods indicate that sex ratio at births generally comes around 105 i.e. 105 male babies per 100 female babies.

Sex Ratio Sex Ratio indicates number of females per thousand males in a population.

Growth Rate The growth rate of a population refers to the number of individuals growing per year per 100 individuals of that population.

Dependency Ratio :Dependency Ratio means and includes proportion of children less than 15 years of age (child dependency) and elderly persons of more than 65 years of age (aged dependency) relative to the population of 'working ages' of 15–64 years.

Stable Population If the population of a given areas increase at constant rate, the population is said to be stable population

Stationary Population :If the size of a population in a given area is constant and its composition according to age and sex also remain constant over time, such a population is known as stationary population.

Acceptor Data Any person, male or female, who visits the programme to receive service and/or advice of family planning programme with the intention of using it can be termed or recorded as acceptor.

User Data User Data signifies the number of persons, male or female who are utilising services under a family planning programme at a given point of time. It is a prevalence concept.

Couple Protection Rate The Couple Protection Rate (CPR) is usually expressed as the percentage of women in the age group of 15–49 years of age from child birth/pregnancy in the year under consideration for specified region/location.

Effective Couple Protection Rate All contraceptive methods do not have equal effectiveness or success against the protection of pregnancy. For that to calculate, the number of couples are recoded with specific method by its user effectiveness and then worked out.

Contraceptive Prevalence Rate Contraceptive Prevalence Rate means the number of user of contraceptives in relation to number of eligible women. Eligible women here means the married women in reproductive age of 15–49 years of age.

Birth Interval :Birth Interval is used to measure the effect of on-going family planning programmes. This is worked out by calculating time elapse between two consecutive births.

Chapter 8
Data Profiles Relevant for the Management of Population Control Mechanism

Abstract Data embodies inherent messages which serve as critical input for policy framework, for its programme content or for use of its monitoring analysis. Data has, therefore, to be pure, properly validated and relevant. Untested data or data not properly verified are of no use and are not permissible for any use for any meaningful discourse on framing any policy or analysis. Depending on the purpose of study and analysis, data requirement could be met either from Primary sources or from Secondary sources. Usually, Census publications from the Government of India, which are secondary in nature from the perspective of population control managers, are acted upon for population related analysis. Similarly, SRS data or National Health and Family Welfare Survey data and the like are all important sources of Secondary data. The Primary data, on the other hand, are generated by the organisation through efforts of its own either by way of undertaking field visits and data collection e.g. Eligible Couple List. Similarly, Primary data are generated by way of collection of its own institutional data for monitoring and other purposes. The sole criteria for use of data for population related issues are of its purity and also of its relevance.

Keywords State level data · Demographic data · Migration related data · Vital statistics · TFR over the years · Couple protection rate · District level data · *Block wise and municipality wise data* · Eligible couple & child register · Couple protection rate etc

Data embodies inherent messages which serve as critical input for policy framework, for its programme content or for use of its monitoring analysis. Data has, therefore, to be pure, properly validated and relevant. Untested data or data not properly verified are of no use and are not permissible for any use for any meaningful discourse on framing any policy or analysis. Rather such untested data infects and derails the entire work process altogether. In the context of population related issues, where plural number of class, caste and religions are involved with multiple interests, data are inherently and potentially very sensitive and have to be obtained with due care and caution from sources for its acceptance without any reservation whatsoever. Any minor error is sure to be drummed up by the related interest group for avoidable hue and cry, connected as it is with emotive and relative sectional position and interest. The data to be stored and used has to be quality data in all aspects.

Depending on the purpose of study and analysis, data requirement could be met either from Primary sources or from Secondary sources. Usually, Census publications from the Government of India, which are secondary in nature from the perspective of population control managers, are acted upon for population related analysis. The size of population, decadal growth rate, density of population, Scheduled Castes, Scheduled Tribes and religion wise growth data etc. are some of the important data which are sourced from census publications. Similarly, SRS data or National Health and Family Welfare Survey data and the like are all important sources of Secondary data. The Primary data, on the other hand, are generated by the organisation through efforts of its own either by way of undertaking field visits and data collection e.g. Eligible Couple List. Similarly, Primary data are generated by way of collection of its own institutional data for monitoring and other purposes. The sole criteria for use of data for population related issues are of its purity and also of its relevance.

Further, as to the relevance of data, it all depends on the role of the organisation. The data requirement at the national level is different from those stationed at the State level. Similarly, data requirement at the district level is different from those posted at the Block Panchayat level. For the same reason, the data requirement for the Gram Panchayat level is different from those required for the Block panchayat level. The need for micro-level data is very important for population related issues since outcome of programme interventions of several linked departments are reflected therein. Thus, depending on the hierarchy and nature of the organisation, the data requirement would be distinct and specific. Accordingly, such level of organisation has to address the data structure for using it meaningfully and effectively. Because of diversity and varied level of socio-cultural status there is no scope to use any state level data as proxy indictor for the districts or other levels of the organisation. Such a practice of proxy indicator would dilute the very essence of purity of data meant for that level.

In the age of computer related data management system, there is a temptation to organise data structure in a big way for meeting the needs of plural and diverse stakeholders, thus making the data structure and retrieval system more complicated and professional dependent. Apart from time and cost associated with installing of such a computer enabling functioning system, the user-friendliness for day to day transactions at the micro-level gets unnecessarily disturbed. It is, therefore, to be seen that computer system and data structure should be as simple and relevant as possible for any organisational level to safeguard its user-friendliness.

The data need, from the perspective of population control and management issues, at the national level is phenomenal. It needs the country data as well as data for the states of India and UTs along with all the districts of India in a time format. Further, it needs a host of socio-cultural data for national and sub-national level for policy formulation and programme intervention. Additionally, it needs comparable data for the developed and other regions of the world for study and analysis for positioning India's commitment in the world of nations, the UN and the Sustainable Development Goals (SDGs). Normally speaking, as a general rule, the data needs are location specific of the organisation along with data of those located at one tier lower and one tier above its level. Thus for the state level organisation, the data need is for the state specific data, its disaggregated district level data and comparable national data. It would serve good purpose if the data of the best performing states could be captured for reference for way forward of a laggard state. Similarly, the district level unit would need to maintain district specific data of its own jurisdiction, Subdivision/Taluka/Municipality/Block Panchayat level data of its area along with comparable state specific data. It would be useful if the national status on those items could be captured for assessment of the district figure vis-a-vis its national position.

Since at the national level the data need and its generation are taken very seriously and professionally at the appropriate level, focus is concentrated here for the data need at the state level and below where neither the data are seriously and professionally captured nor made use of the available data. In real sense, the role of the state level and district level organisation is very important from population control and management point of view and, therefore, it is very crucial to have a structured data base at these levels. Thus, a set of minimum level of data structure that need to be maintained for the Family Welfare Set-up at the state level and down below are discussed hereunder.

8.1 State Level Data

8.1.1 *Demographic Data*

1. Trend in Census population: Size of population, Male and Female Population, Decadal Growth Rate & Sex Ratio—from 1951 to 2011
2. Rural and Urban Composition of Population, Census from 1951 to 2011
3. Density of population from 1951 to 2011
4. Population of Scheduled Castes and Scheduled Tribes, growth rate and their proportions to the total population over 1951–2011
5. Population of Muslims and other minorities, decadal growth rates and their proportions to the total population over 1951–2011
6. Child Population in the age-group 0–6 by sex and religion–Census 2001–2011
7. Literates and Literacy Rates by sex/SC/ST/Religion over 1951–2011
8. Distribution of Population by Age Groups 2001–2011(Census)
9. Projected Population Characteristics 2011–2026
10. Migration related data since 1971–2011 *and district wise break-up of all these Tables*

8.1.2 Vital Statistics

1. Crude Birth Rate, Death Rate, Natural Growth rate
2. TFR over the years
3. IMR over the years, separately for Rural and Urban years
4. Under-5 Child Mortality Rate

8.1.3 Family Planning/Welfare Data

1. Number of Married Couples (With Wife Aged between 15–44 Years), since 1991
2. Estimated eligible couples as per Eligible Couple List per 1000 population
3. 3 Reported coverage of Eligible Couples and its Percentage
4. 4.Couple Protection Rate

8.2 District Level Data

8.2.1 Demographic Data

1. Trend in Census population: Size of population, Male and Female Population, decadal increment, decadal growth rate & Sex Ratio–from 1951 to 2011
2. Rural and Urban Composition of Population, decadal increment, decadal growth rate in Census from 1951 to 2011
3. Density of population from 1951 to 2011
4. Population of Scheduled Castes and Scheduled Tribes, decadal increment, decadal growth rate and their proportions to the total population over 1951–2011
5. Population of Muslims and other minorities, decadal increment, decadal growth rates and their proportions to the total population over 1951–2011
6. Child Population in the age-group 0–6 by sex and religion—Census 2001–2011
7. Literates and Literacy Rates by sex/SC/ST/Religion over 1951–2011
8. Distribution of Population by Age Groups 2001–2011(Census)
9. Projected Population Characteristics 2011–2026
10. Migration related data since 1971–2011 *and Block wise and Municipality wise break-up of all these Tables*

8.2.2 Vital Statistics

Since District data on Vital Statistics are not usually available, State level data, as above, may be kept by way of reference. However, as and when such data are available from district specific survey/studies, those should be maintained for reference.

8.2.3 Family Planning/Welfare Data

1. Number of Married Couples (With Wife Aged between 15–44 Years), since 1991
2. Estimated eligible couples as per Eligible Couple List per 1000 population
3. Reported coverage of Eligible Couples and its Percentage
4. Couple Protection Rate

8.3 Sub-District Level Data

From population control and management point of view, the need for relevant and workable data base for down the district head quarter level, like those of Municipalities and levels of Panchayats (including similar levels of Health & Family units and Sub Centres), is most crucial These are the real areas where field level functionaries are entrusted with nuts and bolts of population control related issues. However, the data base in this location, unlike the data base structure of the district or the State, should be rather minimum; it has to be direct and specific to serve as a working reference tool. For the same reason, the preponderance of data base at this level would unnecessarily overload the functioning system and affect its efficiency. The simple kind of format of institutional data base at this level may be the following:

8.3.1 Demographic Data

1. Size of population, Male and Female Population, decadal increment and decadal growth rate from 2001 to 2011
2. Literates and Literacy growth rate from 2001 to 2011

8.3.2 Family Planning/Welfare Data

1. Estimated eligible couples as per Eligible Couple List per 1000 population
2. Estimated children from 0–6 years of age as per Eligible Couple & Child Register
3. Reported coverage of Eligible Couples under various services and its Percentage
4. Couple Protection Rate
5. Extent of Immunization coverage of children for various vaccinations etc.

Chapter 9
Family Welfare Structure in the Country and Issue-Based Management Support Structure

Abstract The moot point on organisational set-up is whether the country has put in place the right kind of organisational set-up for population control and family planning as envisaged in serial 20A in the Concurrent List. In fact, for execution of item 20A of the Concurrent List, there is no nodal ministry or department in the country who has been empowered to deal with all related issues concerning 'population control and family planning. The Ministry of Health and Family Welfare has only been empowered to deal with 'Family Welfare Matters', a sub-set function of 'population control and family planning' responsibility in the country. Additionally, in the context of India surpassing the huge size of population of China in the foreseeable year of 2030, the country needs to scale up a robust regime of population control to contain the runaway population. Further, the Sustainable Summit 2015 has enjoined a big responsibility to India to reach SDGs by 2030 which is only possible if sustainable population can be put in place at the first place. Moreover, the Paris Summit on climatic change also enjoins additional responsibility to control human-induced emission level at an internationally agreed level. All these belong to big-ticket reforms which call for revamping and restructuring the entire organisational set-ups of Family welfare to the level of "Population Control and Family Planning' for proactive, effective and outcome ensuring organisational set-up in the states of India to initiate the process for Sustainable population and enable us to reach SDGs by 2030.

Keywords Family welfare structure · Ministry of health and family welfare · Union business · List of business · Family welfare matters · National rural health mission · Departments of health and family welfare

The website of the Ministry of Health and Family Welfare (Swastha Aur Parivar Kalyan Mantralaya) mentions that, under Constitutional provision, the following broad sections of work are meant for the Department of Health and Family Welfare (Swastha Aur Parivar Kalyan Vibhag), namely:

I. Union Business
II. List of Business for Legislative and Executive purposes in respect of Union Territories
III. List of Business with which the Central Government deal in a legislative capacity only for the Union and in both legislative and executive capacity for all Union Territories.
IV. Miscellaneous Business
V. Family Welfare Matters

On perusal of the items of work, sections falling under I to IV relate to health issues of the Union Government or those of Union Territories. Only section V deals with Family Welfare details of which are mentioned hereunder:

- Policy and organisation of Family Welfare
- All matters relating to:
    (a) National Rural Health Mission
    (b) National Commission on Population
    (c) Reproductive and Child Health
- Inter-sectoral Coordination in accordance with National Population Policy
- Matters related to Janasankhya Sthiarta Kosh and Empowered Action Group
- Organisation and direction of education, training and research on all aspects of family welfare including higher training abroad
- Production and supply of Aids to Family Welfare
- Liaison with foreign countries and international bodies as regards relating to family welfare
- Family Welfare Schemes and Projects with external assistance
- International Institute of Population Science, Mumbai
- Development and Production of audio visual aids, extension education and information in relation to population and family welfare
- Promoting Public Private Partnership for the Family Welfare Programme
- All matters relating to following institutions:
    (a) Hindustan Latex Limited, Thiruvananthapuram
    (b) National Institute of Health and Family Welfare
- Implementation of Pre-conception and Pre-natal Diagnostic Technique (Prohibition of sex selection) Act, 1994 (57 of 1994)-Medical Termination of Pregnancy Act, 1971 (34 of 1971)

Against the background of job provision on Family Welfare, it would be relevant to look at the corresponding Organizational structure at the apex level of the Ministry at the Government of India and down the lines.

9.1 Apex Level

The Minister of Health and Family Welfare is the head of the Ministry of Health and Family Welfare on all items of works under "Constitutional Provisions", as mentioned in the website of the Ministry. In other words, the Minister is the head of the Family Welfare related function of the country for items as detailed under "V. Family Welfare Matters". Since Family Welfare Matters are not equivalent to population control and family planning, as would be evident from issues discussed in chapters falling under Part-A of this book, the mere Family Welfare Matters, as mentioned above in the allocation of business of the Ministry, is too inadequate to address the bigger issues of population control and family planning as enshrined in serial number 20A of the Concurrent List of the Constitution of India. In a sense, Family Welfare Matters" is a sub-set function of the bigger task of population control and family planning. There is in that sense no Minister in Charge of Population Control and family planning in our country to address to host of issues relevant to population control and family planning. The country is having only a Minister of Health and Family Welfare to address "Family Welfare Matters" related issues only.

Be, that as it may, the Family Welfare set-up in the country, as perceived by the government of the day, has come into being to discharge the overarching responsibility of Family Welfare. Such understanding has accordingly got reflected in the organisational architecture of the Secretariat, directorate and other set-ups down below. As per Ministry's website, the secretary of the Ministry continues to be in charge of both the Health and also the Family Welfare functions of the government. Incidentally, it is worthwhile to mention that up to September 2005, there had been a dedicated full time secretary on Family Welfare in the Ministry, which had since been discontinued while launching the National Rural Health Mission. Of the five Additional secretaries, one of them has been assigned the Family Welfare matters. Incidentally, as per distribution of works, the incumbent Additional secretary is required to look after some important components of Health matters of the Ministry as well. Among the twelve Joint Secretaries, two of them address Family Welfare matters along with Health related functions as well. In that sense, there is also no full time joint secretary in the Family Welfare segment as well.

The website of the Ministry classifies the following departments of the Ministry:

- Departments of Health and Family Welfare
- Department of Health Research

Now, the said website of the Ministry (as on 04.12.2015) further subdivides the Departments of Health and Family Welfare into the following departments:

- Blindness Control
- Bureau of Planning
- Cancer Control Programme
- CCA
- Central Design Bureau

- CHS
- Drugs Food Quality control division
- Emergency Medical Relief
- IC & IH
- Immunization

It would appear from above, that there does not exist any independent functional department of Family Welfare within the broad functional envelop of Health and Family Welfare department. This is unusual.

On the Organisational side, the said website of the Ministry (as on 04.12.2015) mentions about the Directorate of Health Services where the following set-ups are in place:

9.2 Directorate General Health Services

- CBHI
- CDSCO
- Central Health Education Bureau
- Leprosy Section
- MH
- MSO
- NCD Section
- Ophthalmology
- TB Section
- Registration of Public Testing Laboratories with MSO for testing the medicine
- Renewal of Registration Fresh registration of your firm Manufacturing Units with MSO
- Directorate General of Health Services
- International Health

There does not exist any separate Directorate of Family Welfare or any dedicated sub-directorate under Directorate General Health Services in the Ministry to pay focused attention on 'Family Welfare Matters'. This casual approach of no accountability in the organisational set-up is one of the reasons for sordid performance in the Family Welfare sector in the country.

The architectural design of the Organisational Structure at the State Head Quarter and down below in the districts and sub-district levels is primarily made by the concerned State. There exists in all states a secretariat of Family Welfare in the composite department of Health and Family Welfare along with a unit of Family Welfare in the Directorate of Health and Family Welfare. From the perspective of Family Welfare programmes implementation in the country, the infrastructure available in the districts and sub-districts in the states of India is very crucial and relevant. Over the years with Central Government support, the States of India have put in place Family welfare Unit both at the State Secretariat and Directorate level.

9.2 Directorate General Health Services

With 100 % funded Centrally Sponsored Scheme, the set-ups that have been put in place in the states of India include (a) Family Planning Cell at the State Secretariat, (b) State Family Welfare Bureau, (c) District Family Welfare Bureau, (d) Regional Family Planning Training Centres, (e) Establishment and Maintenance of Rural Family Planning Sub-centres, (f) Establishment and Maintenance of Urban Family Planning centres, (g) Establishment and Maintenance of Sterilisation Beds.

The website of the Ministry as on 04.12.2015 has also shown organisational sets-up of the states of India. On perusal of the same, it is found that there does not exist any uniform pattern of organisational set-ups among the states in India. Just as the demographic challenges are different from one State to another, so is the organisational set-up on family welfare. Further, such website does not suggest that EAG states have strengthened their Family Welfare organisational set-up to cope up with higher order demographic challenges. However, there is one distinct feature common to all the Family Welfare set-up in the states of India in that Family Welfare set-up does not enjoy any conspicuous identity and is subsumed by the Health system in the State. Virtually, it exists as an appendix of the health care system in the state. The brand image and importance is very important from functional point of view. This brand image of Family Welfare is non-existent in the States of India. This is the grey area of the organisational architecture in the Family Welfare in India.

Further, the core competence of the Ministry of Health and Family Welfare centres around preventive, curative, rehabilitative and promotive aspects of health and medical care in which Family welfare forms a part issue. As a result, the focus of action of the Ministry of Health & Family Welfare/Department of Health & Family Welfare is on the health of the nation. The preponderance of health related issues and the its intervention areas are too many and too complicated to keep the health professionals occupied with the health areas leaving only left-over time for family welfare related services. This is the real bane of the family welfare of the country.

Apart from formal government functionaries either in the Ministry of Health and Family welfare or in the departments of the state governments, a host of contractual and other functionaries have come to stay in the Family Welfare set-up in the state out of National Health Mission and the like. Additionally, there exits family welfare unit in the Railway hospitals, ESI and other bigger establishments, both public and private, addressing family welfare of targeted population. The vast network of private sector doctors and hospitals also provide family welfare related services. Moreover, a good number of NGOS and civil society organisations also address family welfare services in their adopted area. The area of structured coordination between the such service providers and the department of Health and Family Welfare is another grey area in the Family Welfare administration in the states of India.

The moot point on organisational set-up is whether the country has put in place the right kind of organisational set-up for population control and family planning as envisaged in serial 20A in the Concurrent List. Earlier, the issues relevant for population and family planning have been covered in Chaps. 2 and 3 where several issues relevant for population control and family planning have been listed and discussed. The issues are much broader than what are being addressed by the Ministry of Health and Family Welfare under 'Family welfare matters'. The demographic issues, socio-cultural issues, political and religious issues are not taken up as seriously as it ought to have been pursued as such items do matter in the outcome of family planning efforts but which, unfortunately, do not fall within the core functional area of the Ministry of Health and Family Welfare nor for that matter the issue of migration, immigration, influx etc. can at all be addressed by the Ministry of Health and Family Welfare. In fact, for execution of item 20A of the Concurrent List, there is no nodal ministry or department in the country who has been empowered to deal with 'population control and family planning', The Ministry of Health and Family Welfare has only been empowered to deal with 'Family Welfare Matters', a sub-set function of 'population control and family planning' responsibility in the country. Additionally, in the context of India surpassing the huge size of population of China in the foreseeable year of 2030, the country needs to scale up a robust regime of population control to contain the runaway population. Further, the Sustainable Summit, 2015 has enjoined a big responsibility to India to reach SDGs by 2030 which is only possible if sustainable population can be put in place at the first place. Moreover, the Paris Summit on climatic change also enjoins additional responsibility to control human-induced emission level at an internationally agreed level. All these belong to big-ticket reforms which call for revamping and restructuring the entire organisational set-ups of Family welfare to the level of "Population Control and Family Planning" for proactive, effective and outcome ensuring organisational set-up in the states of India to initiate the process for Sustainable population and enable us to reach SDGs by 2030.

Chapter 10
Decentralisation of Family Welfare Programmes in the Country

Abstract Decentralized governance is the very attribute of any modern state. It is in harmonious with the spirit and the concept of ownership and participation of the modern mind. The traditional system of governance by remote control from distant location is no substitute for a system of local governance and local participation. The local participation enables not only selection of locally relevant schemes with its bearing on living environment, it also promotes cost efficiency by subjecting openness to public scrutiny and by institutionalizing accountability to the stake holder community and other individuals. Further, involving local people may also result in investment in socially desirable services, particularly in drinking water, health care and primary education and in the process meets development deficit of the locality to a great extent. The 73rd and the 74th Amendments of the Constitution of India provides for decentralisation of Family Welfare Programmes in the country. Centralised family planning services is seldom effective in promoting family planning in the un-served and underserved locations in the countryside. There is great scope and dedicated space for rural and urban local bodies to share the national duty of population control and family planning in their respective areas.

Keywords Seventy third and the seventy forth amendments · Eleventh schedule · Twelfth schedule · Article 243G · Article 243W · Devolved functional responsibilities · Gram panchayat · Block panchayat · Zilla panchayat · Municipalities

Decentralized governance is the very attribute of any modern state. It is in harmonious with the spirit and the concept of ownership and participation of the modern mind. The traditional system of governance by remote control from distant location is no substitute for a system of local governance and local participation. The local participation enables not only selection of locally relevant schemes with its bearing on living environment, it also promotes cost efficiency by subjecting openness to public scrutiny and by institutionalizing accountability to the stake holder community and other individuals. Further, involving local people may also result in investment in socially desirable services, particularly in drinking water, health care and primary education and in the process meets development deficit of the locality to a great extent.

Decentralised governance may have different character and forms. From an operational angle, it might be divided into two ways (a) horizontal and (b) vertical. Horizontal decentralisation is characterised by dispersal of power among institutions at the same level whereas Vertical decentralisation is distinguished by delegation of some of the powers of the government to the lower tier of authorities including village institutions. Vertical decentralization of any government can itself take three forms:

It is the primary stage of decentralisation where limited power and authority are passed down. It is confined to exercising administrative discretion at the level of local offices of the ministries of the government. In this system although it does take place some dispersal of power, few decisions can be taken without reference to the government.

This falls in the intermediate stage of decentralisation where some defined authority and decision-making powers on specified items are passed on to local officials. But the government retains the right to overturn local decisions and can, at any time, take these powers back. Like deconcentration, it is confined to among the officials of the government. It does not cover local level institutions

It is the strongest form of decentralization. It covers local level institutions only. In this form of decentralisation, full decision-making powers are granted to local authorities and allowing them to take full responsibility without reference back to the government. Such bodies are empowered with financial power as well as the authority to design and execute local development projects and programmes.

The Constitution of India, while institutionalising local governance in the country under the Seventy Third and the Seventy Forth Amendments, adopted this devolution mode of vertical decentralisation as described above and accordingly enshrined '…the devolution of powers and responsibilities upon Panchayats, at the the appropriate level' under Article 243G and '…the devolution of powers and responsibilities upon Municipalities' under Article 243W. Further, to ensure that devolution of functional distribution to the tiers of the Panchayats or to the Municipalities is lawfully done, the Constitution of India has enshrined Eleventh Schedule and Twelfth Schedule, for the Panchayats and Municipalities respectively and requires in Article 243G and Article 243W that such devolution has to be passed by the Legislature of a State.

It would be useful to keep in the upfront the health and family welfare components of the Eleventh Schedule and Twelfth Schedule for objective analysis as mentioned in Table 10.1.

It would appear from above that the Eleventh and Twelfth Schedules have not enshrined identical functional items to empower the Panchayats or the Municipalities. While the Eleventh Schedule meant for the tiers of Panchayats provided Family Welfare as a separate functional item along with health care set-up like hospital, primary health care and dispensaries, the Twelfth Schedule meant for the municipalities mentions only the public health related function. Further, the Eleventh Schedule also contained additional item on women and child development

Table 10.1 Health and family welfare components of the eleventh schedule and twelfth schedule

Eleventh schedule (for the tier of panchayats)		Twelfth Schedule (for the municipalities)	
Sl no.	Name of the functional item	Sl no.	Name of the functional item
23.	Health and sanitation, including hospitals, primary health centres and dispensaries	6.	Public health, sanitation conservancy and solid waste management
24.	Family welfare		
25.	Women and child development		

Source Constitution of India

function for the Panchayts which also provide critical support to population stabilisation. In other words, the Constitution makers conceived proactive role of local self-governments for the health and family welfare sector in the rural areas and rather limited role for urban local self-governments in the same area. Possibly this imbalance would get corrected through a process of evolution over time. Be that as it may, there is utmost necessity to put in place decentralisation in the health and family welfare sector to obtain desired outcome. Unfortunately, however, very few state governments have empowered such units of local self-governments, as required under Article 243G or under Article 243W with distinct devolved functional responsibilities. Consequently, the institutional character on management and ownership on population related issues of such units of local self-governments has not grown up till date. This is a long void area and requires urgent resolution. The distribution of functional items for the State Government departments, thePanchayats (among the three tiers as well) and the Municipalities need to be worked out by the concerned State Legislature urgently for the distinct institutional responsibility.

As against this format of constitutional decentralisation through the local bodies, another form of delegating functional responsibilities at a local level through government set-up is in place in almost all the states of India, which very often passes for 'decentralisation' in health and family welfare areas. The induction of State Health & Family Welfare Samity, District Health & Family Welfare Samity and Block Health & Family Welfare Samity is to be mentioned in this regard. The Samity based parallel format of delegated authority is not in conformity with the spirit of devolution of functional responsibility as enshrined in the Constitution of India in Article 243G and 243W; rather it is a move to by-pass the devolution process itself. In fact, for decentralisation to have real meaning, such a set-up of functional organ has to be placed under the concerned local self-government only. The delegation of functional responsibilities to the Registered Health & Family Welfare Samity can never be a substitute for devolution of functional responsibilities to local self-governments.

Pending such Constitutional empowerment, the concerned state government has to work out a platform of working together of all its organs in any identified geographical location including the tiers of Panchayats (e.g. Gram Panchayat,

Block Panchayat or Zilla Panchayat) or Municipalities and the related health and family welfare unit of the state government. Needless to reiterate, the devolved functional responsibilities to the units of local self-governments would have been a better choice for collaborative arrangement for convergent action. Be that as it may, for ensuring good governance, a collaborative arrangement in the format of deputation of government functionaries to the office of such counterpart local self-government may ensure better programme outreach and of its quality. An example may clarify the position. The sub-centre set-up of the Health and Family welfare unit might be located within the jurisdictional G.P and the functional jurisdiction of such Sub-centre could as well be similar to the devolved responsibilities of the G.P. This will facilitate converge of services, optimise resources and ensure desired outcome at that level. The absence of such a collaborative arrangement, as it exists in many parts in our country, does not promote ownership, is not programme implementation friendly and spawns avoidable parallel service delivery points with different agenda to address local level prioritises and complexities.

Similar to the collaborative arrangement of GP level, the concerned state government need to work-out defined functional role for other tiers of Panchayat and municipalities via-a-vis the units of state government's health and family welfare set-up. This overdue reform area has been standing in the way of optimum outcome of programme efforts at any given area and proper decentralisation in health and family care in India.

Finally, the Civil societies and NGOs have embarked in a big way in the Family Welfare area being driven by commitment and also as programme partners under defined schemes. Under the National Policy for Voluntary Sector and also on the patronage of the erstwhile Planning Commission, such units have been making programme efforts to reach the under-served areas. The UN system also enlists them in the outreach and quality of delivery of services. In most cases, such NGOs work as parallel agencies without mainstreaming their activities with the programme of decentralised local self-governments. The need for shared responsibility under decentralised format is also an overdue area that needs resolution.

Chapter 11
Financing the Population Control and Family Planning Programme in the Country

Abstract The serial number 20A of the Concurrent List of the Constitution of India enshrines Population control and family planning meaning thereby that it is the joint responsibility of the Union Government and the State Government to administer it. Incidentally, neither the Union Government nor the State Government ever uses the word 'Population control and family planning' in its programme planning or in its budgeting to operationalise this functional responsibility of the Constitution. The lack of interest on the part of the governments has spawn a new version for 'Population control and family planning under the nomenclature of Family Welfare losing in the process the strong and focused intention of the constitutional provision of population control and family planning. Even in this case, many of the state governments feel that since it falls under the Concurrent List, the funding of this item be rather borne by the Union Government on the plea that the financial position of the state governments is not good enough to bear the related expenditure. The funding of the family welfare programme has thus become Union Government's responsibility. The ownership of the programme by the states in terms of sharing the cost of its implementation is not in place. The states have virtually played the role of implementing agents. The related role of the central government and the states of India in financing the family planning programmes over the areas, including its adequacy has been captured.

Keywords 20A of the concurrent list · Centrally sponsored scheme · Plan expenditure · Plan-wise expenditure · Plan and non-plan expenditure · NDC committee on population · Public expenditure · Percentage of FW expenditure

The serial number 20A of the Concurrent List of the Constitution of India enshrines Population control and family planning meaning thereby that it is the joint responsibility of the Union Government and the State Government to administer it. Incidentally, neither the Union Government nor the State Government ever uses the word 'Population control and family planning' in its programme planning or in its budgeting to operationalise this functional responsibility of the Constitution. It is for the same reason, this item remains outside the purview of any quantitative and qualitative monitoring on periodical basis, either at the level of the Union Government or at the level of State Governments. As a matter of fact, both the Union Government and the State Governments are averse to use the term Population control and family planning for a variety of reasons, chief being its possible strong backlash on electoral politics. It is for the same reason that the political and social leadership of the country does not take any interest on population control and family planning least it offends the sensibilities of voters and pose any negative impact on vote-bank politics. This is absolutely a NO NOZONE for the political parties and the social groups. The lack of interest on the part of people' representatives has spawn a new version for 'Population control and family planning under the nomenclature of Family Welfare losing in the process the strong and focused intention of the constitutional provision of population control and family planning. Even in this case, many of the state governments feel that since it falls under the Concurrent List, the funding of this item be rather borne by the Union Government on the plea that the financial position of the state governments is not good enough to bear the related expenditure. The funding of the family welfare programme has thus become Union Government's responsibility. The ownership of the programme by the states in terms of sharing the cost of its implementation is not in place. The states have virtually played the role of implementing agents.

The Family Welfare Programme thus emerged as a 100 % Centrally Sponsored Scheme. It is implemented only by the states of India utilising its infrastructure with full funding support of the Central Government as per norms approved by it. The programme has remained 'Plan expenditure' since its inception. The allocations of Family Welfare Programmes during various Five Years, as mentioned in the NDC Report, are given in Table 11.1.

Table 11.1 Plan-wise expenditure under the family welfare programme (Rs crores)

Plan period	Expenditure on public sector	Expenditure on family welfare[a]	Family welfare as % of total column 3/2 expenditure
1	2	3	4
First plan	1960.00	0.14	Negligible
Second plan	4672.00	2.15	0.05
Third plan	8576.50	24.86	0.30
Annual plan	6625.40	70.46	1.10
Fourth plan	15778.80	284.43	1.80
Fifth plan	394266.20	408.96	1.30
Sixth plan	110971.20	1425.73	1.30
Seventh plan	218729.60	3120.80	1.43
Eight plan	434100.00	6500.00	1.50

[a]This expenditure includes both plan and non-plan expenditure
Source NDC Report on population-page 77

The percentage allocation to this programme within the public sector plan outlay, as estimated by the NDC Committee on Population, ranged between 0.05 and 1.50 % during different plans up to Eighth Plan. The Committee was of the opinion that the programme experienced financial constraints because of increased cost to maintain already established infrastructure. The Committee noted substantial accumulated arrears to be paid to the states on this count, and had recommended that the Programme requires to be fully funded, not only to maintain existing infrastructure but also to start new initiative. It also observed that expenditure in family welfare sector is really an investment because reduction in the population growth rate would result into reduction in future allocations for social sectors e.g., education, employment, housing, health services etc. It had also recommended that the allocations might be raised to 2 % from the year 1993 to 94 and gradually increased to 3 % of the public sector expenditure by the end of the Eighth Plan. It had further recommended that the states should meet at least 10 % of their family planning expenditure, which are non-plan in nature from their state budget. The Committee was also of opinion that Railways, Posts and Telecommunications, Labour, Defence should give financial support for providing family welfare services to their government employees, and all departments, both of central and state should fully internalise family welfare expenditure within their respective budget.

From a different angle, public expenditures on Health and its components on Family Welfare have been captured from the website of the Ministry of H&FW from 2011–12 to 2014–15. Before dwelling on it, let us elucidate public expenditure:

Public Expenditure:

Public expenditure normally reflects business priority of the government. It is an indicator of the degree of concern of the Government of the day on the related priority area. This is more so in the context of the broad health care areas of the nation where there is pressing need to maintain at least 6 % of normal state budget for addressing manifold health issues of its citizens and also for improving its HDI status. In the context of alarming population growth scenario, there is also the necessity to invest sizable public fund for arresting its growth and more particularly on population control areas. The Ministry of Health and Family Welfare in its website has recorded public expenditure on Health and Family Welfare in India sourced from state budget documents, detailed demand for grants of MoHFW and concerned Central Ministries/Departments. This document gives totality in classification under plan and non-plan and provides actual expenditure, budget and revised estimates under different schemes/programmes in the health sector. However, the website has not mentioned related expenditure for years prior to 2011–12. Be that as it may, from different Tables, as shown in the website, public expenditure on Health and Family welfare by States and Union Territories from 2011 to 15, have been presented along with its corresponding percentage for the states of India at Table 11.2.

In the given background it would be interesting to see the size and level of population growth of the states of India and its related expenditure on Family welfare from Table 11.3.

Table 11.2 Public expenditure on family welfare from 2011 to 15

States	2011–12 Total health expenditure (actuals)	2011–12 Total FW expenditure (actuals)	Percentage of FW expenditure to total health expenditure in 2011–12	2012–13 Total health expenditure (actuals)	2012–13 Total FW expenditure (actuals)	Percentage of FW expenditure to total health expenditure in 2012–13
Andhra Pradesh	64,884,663	8,332,727	12.84	73,835,779	11,744,209	15.91
Arunachal Pradesh	3,121,327	111,338	3.57	2,825,459	1,342,355	47.51
Assam	14,752,476	1,702,373	11.54	15,386,767	1,852,485	12.04
Bihar	22,835,703	2,997,424	13.13	25,656,985	3,241,679	12.63
Chhattisgarh	11,831,987	1,186,353	10.03	13,146,516	1,287,399	9.79
Delhi	26,266,189	302,589	1.15	27,342,335	802,701	2.94
Goa	4,037,851	95,029	2.35	4,251,270	98,705	2.32
Gujarat	33,595,402	4,811,400	14.32	46,748,137	4,850,728	10.38
Haryana	13,456,244	1,037,062	7.71	17,214,127	1,240,591	7.21
Himachal Pradesh	9,009,981	1,011,349	11.22	10,969,570	1,236,346	11.27
Jammu and Kashmir	11,789,070	155,990	1.32	12,050,146	150,274	1.25
Jharkhand	10,457,334	690,826	6.61	10,101,541	806,257	7.98
Karnataka	33,610,048	3,688,635	10.97	40,421,373	4,283,243	10.60
Kerala	28,983,015	3,062,062	10.57	32,436,671	3,234,614	9.97
Madhya Pradesh	26,028,268	3,130,427	12.03	33,416,897	3,499,184	10.47
Maharashtra	54,396,126	5,175,647	9.51	64,325,731	5,857,165	9.11
Manipur	3,943,072	170,311	4.32	3,323,510	172,705	5.20

(continued)

Table 11.2 (continued)

States	2011–12 Total health expenditure (actuals)	2011–12 Total FW expenditure (actuals)	Percentage of FW expenditure to total health expenditure in 2011–12	2012–13 Total health expenditure (actuals)	2012–13 Total FW expenditure (actuals)	Percentage of FW expenditure to total health expenditure in 2012–13
Meghalaya	3,241,260	391,947	12.09	3,986,370	289,288	7.26
Mizoram	1,913,369	193,440	10.11	2,251,591	336,721	14.95
Nagaland	3,544,599	192,840	5.44	3,764,887	224,568	5.96
Odisha	13,888,629	956,548	6.89	18,102,834	2,138,300	11.81
Puducherry	3,340,432	57,858	1.73	3,003,082	65,018	2.17
Punjab	17,228,750	1,526,830	8.86	20,473,605	1,713,138	8.37
Rajasthan	34,305,827	7,574,715	22.08	39,394,875	7,563,865	19.20
Sikkim	2,259,270	132,472	5.86	2,448,382	167,012	6.82
Tamil Nadu	45,873,762	7,055,668	15.38	54,842,222	9,037,672	16.48
Tripura	3,413,372	200,622	5.88	3,175,109	211,347	6.66
Uttarakhand	8,033,287	845,472	10.52	9,663,077	875,385	9.06
Uttar Pradesh	68,264,865	13,987,847	20.49	87,980,918	24,506,546	27.85
West Bengal	40,750,985	4,758,063	11.68	40,955,252	4,622,210	11.29
Total	6.19E+08	76,435,864	12.35	723,495,018		0.00
Union Teritories (UTS)						
Andaman and Nicobar	1,589,456	–		1,976,685	–	
Chandigarh	1,880,114	–		2,250,955	–	
Dadra and Nagar Haveli	404,653	–		570,323	–	
Daman and Due	342,052	–		451,086	–	

(continued)

Table 11.2 (continued)

States	2011–12 Total health expenditure (actuals)	2011–12 Total FW expenditure (actuals)	Percentage of FW expenditure to total health expenditure in 2011–12	2012–13 Total health expenditure (actuals)	2012–13 Total FW expenditure (actuals)	Percentage of FW expenditure to total health expenditure in 2012–13
Lakshadip	346,575	–		415,887	–	
Total B	4,562,850	–		5,664,936		
MoHFW		97,894,696				
India	6.24E+08	174,330,560	27.95	729,159,954		

States	2013–14 Total health expenditure (budget estimate)	2013–14 Total FW expenditure (revised estimate)	Percentage of FW expenditure to total health expenditure in 2013–14	2014–15 Total health expenditure (revised estimate)	2014–15 Total FW expenditure (budget estimate)	Percentage of FW expenditure to total health expenditure in 2014–15
Andhra Pradesh	89,196,266	12,222,843	13.70	50,279,855	6,701,450	13.33
Arunachal Pradesh	3,834,166	134,871	3.52	3,970,893	257,955	6.50
Assam	26,935,577	2,586,793	9.60	35,569,395	2,419,445	6.80
Bihar	38,823,710	4,723,141	12.17	51,312,605	4,694,365	9.15
Chhattisgarh	20,100,473	1,744,000	8.68	29,655,712	1,988,570	6.71
Delhi	32,384,900	746,320	2.30	44,499,000	925,000	2.08
Goa	5,150,292	1,116,000	21.67	5,717,391	120,000	2.10
Gujarat	51,681,919	6,045,134	11.70	72,005,247	8,317,909	11.55
Haryana	22,700,381	1,403,300	6.18	30,040,731	1,641,916	5.47
Himachal Pradesh	13,033,953	1,553,182	11.92	13,611,945	1,508,955	11.09

(continued)

Table 11.2 (continued)

States	2013–14 Total health expenditure (budget estimate)	2013–14 Total FW expenditure (revised estimate)	Percentage of FW expenditure to total health expenditure in 2013–14	2014–15 Total health expenditure (revised estimate)	2014–15 Total FW expenditure (budget estimate)	Percentage of FW expenditure to total health expenditure in 2014–15
Jammu and Kashmir	18,709,411	260,005	1.39	22,252,485	315,590	1.42
Jharkhand	15,007,550	1,228,251	8.18	26,968,665	1,381,162	5.12
Karnataka	50,485,175	5,470,076	10.84	60,908,594	5,915,812	9.71
Kerala	36,289,750	4,245,894	11.70	45,589,803	5,454,796	11.96
Madhya Pradesh	40,528,308	4,820,922	11.90	61,646,567	5,940,800	9.64
Maharashtra	79,245,642	6,468,072	8.16	79,029,067	6,852,478	8.67
Manipur	4,179,261	202,336	4.84	2,926,441	202,336	6.91
Meghalaya	4,420,895	352,063	7.96	6,039,800	354,095	5.86
Mizoram	2,790,041	314,439	11.27	3,405,008	24,872	0.73
Nagaland	3,951,028	277,406	7.02	5,714,235	282,751	4.95
Odisha	22,598,509	2,550,096	11.28	39,176,721	2,765,399	7.06
Puducherry	3,647,627	76132	2.09	4,217,036	73,680	1.75
Punjab	28,846,843	2,413,687	8.37	28,743,895	2,097,596	7.30
Rajasthan	53,568,038	10,595,500	19.78	87,534,668	29,793,943	34.04
Sikkim	2,609,437	166,709	6.39	3,535,513	171,400	4.85
Tamil Nadu	63,989,723	2,051,971	3.21	69,197,199	11,717,830	16.93
Tripura	6,776,410	1,806,917	26.66	6,027,813	2,028,627	33.65
Uttarakhand	13,069,029	1,124,988	8.61	15,204,904	1,287,854	8.47
Uttar Pradesh	1.03E+08	26,619,163	25.81	1.45E+08	54,786,574	37.80

(continued)

Table 11.2 (continued)

States	2013–14 Total health expenditure (budget estimate)	2013–14 Total FW expenditure (revised estimate)	Percentage of FW expenditure to total health expenditure in 2013–14	2014–15 Total health expenditure (revised estimate)	2014–15 Total FW expenditure (budget estimate)	Percentage of FW expenditure to total health expenditure in 2014–15
West Bengal	57,620,727	5,210,597	9.04	58,758,316	5,277,947	8.98
Total	9.15E+08			1.11E+09		0.00
Union Teritories (UTS)						
Andaman and Nicobar	1,993,198	–		2,178,550		0.00
Chandigarh	2,490,505	–		2,510,308		0.00
Dadra and Nagar Haveli	634,590	–		949,336		0.00
Daman and Due	430,815	–		571,465		0.00
Lakshadip	542,484	–		544,874		0.00
Total B	6,091,592	–		7,054,533		0.00
India	9.21E+08			1.12E+09		0.00

Source Website Ministry of Health and Family Welfare, GoI

Table 11.3 Population scenario vis-a-vis Percentage of FW expenditure of total expenditure on Health

States/UTs	Population, 2011	Decadal growth of population, 2011	Percentage of public expenditure on FW, 2011–12	Percentage of public expenditure on FW, 2012–13	Percentage of public expenditure on FW, 2013–14	Percentage of public expenditure on FW, 2014–15
Andhra Pradesh	8,45,80,777	11.0	12.84	15.91	13.70	13.33
Arunachal Pradesh	13,83,727	26.0	3.57	47.51	3.52	6.50
Assam	3,12,05,576	17.1	11.54	12.04	9.60	6.80
Bihar	10,40,99,452	25.4	13.13	12.63	12.17	9.15
Chhattisgarh	2,55,45,198	22.6	10.13	9.79	8.68	6.71
Delhi	1,67,87,941	21.2	1.15	2.94	2.30	2.08
Goa	14,58,545	8.2	2.35	2.32	21.67	11.55
Gujarat	6,04,39,692	19.3	14.32	10.38	11.70	5.47
Haryana	2,53,51,462	19.9	9.71	7.21	6.18	11.09
Himachal Pradesh	68,64,602	12.9	11.22	11.27	11.92	1.42
Jammu and Kashmir	1,25,41,302	23.6	1.32	1.25	1.39	5.12
Jharkhand	3,29,88,134	22.4	6.61	7.98	8.18	59.71
Karnataka	6,10,95,297	15.6	10.97	10.60	10.84	9.71
Kerala	3,34,06,061	4.9	10.57	9.97	11.70	11.96
Madhya Pradesh	7,26,26,809	20.3	12.03	10.47	11.90	9.64
Maharashtra	11,23,74,333	16.0	9.51	9.11	8.16	8.67
Manipur	25,70,390	18.6	4.32	5.20	4.84	6.91
Meghalaya	29,66,889	27.9	12.09	7.26	7.96	5.86
Mizoram	10,97,206	23.5	10.11	14.95	11.27	0.73

(continued)

Table 11.3 (continued)

States/UTs	Population, 2011	Decadal growth of population, 2011	Percentage of public expenditure on FW, 2011–12	Percentage of public expenditure on FW, 2012–13	Percentage of public expenditure on FW, 2013–14	Percentage of public expenditure on FW, 2014–15
Nagaland	19,78,502	−0.6	5.44	5.96	7.02	4.95
Odisha	4,19,79,718	14.0	6.89	11.81	11.28	7.06
Puducherry	12,47,953	6.9	1.73	2.17	2.09	1.75
Punjab	2,77,43,338	13.9	8.86	8.37	8.37	7.30
Rajasthan	6,85,48,437	21.3	22.08	19.20	19.78	34.04
Sikkim	6,10,577	12.9	5.86	6.82	6.39	4.85
Tamil Nadu	7,21,47,030	15.6	15.38	16.48	3.21	16.93
Tripura	36,73,917	14.8	5.88	6.66	26.66	33.65
Uttarakhand	1,00,86,292	18.8	10.52	9.06	8.61	8.47
Uttar Pradesh	19,98,12,341	20.2	20.49	27.85	25.81	37.80
West Bengal	9,12,76,115	13.8	11.68	11.29	9.04	8.98
Union Territories (UTS)						
Andaman and Nicobar	3,80,581	6.9				
Chandigarh	10,55,450	17.2				
Dadra &Nagar Haveli	3,43,247	55.9				
Daman and Diu	2,43,247	53.8				
Lakshadip	64,473	6.3				
India	121,05,69,573	17.7				

Source Website of the MoFW, Government of India

Table 11.3 reveals the seriousness and concern of the states of India on Family Welfare given the vulnerability of population-stressed situation in their states. From Table 12.3, it would be seen that there is no correspondence between the size of population, its decadal growth and the percentage of Family Welfare expenditure. Since the volume of fund and percentage of Family Welfare expenditure is proportional to the social concern of the Government of the day, the data as above do not reflect that intensity.

Chapter 12
Monitoring Mechanism

Abstract Monitoring is defined as a process of measuring, recording, collecting, processing and communicating information to assist policy makers and project management to take right decision in a given situation. Monitoring is essentially a functioning tool to understand the flow character of a programme or a scheme right from the time of its commencement. It generates not only flow of data but deciphers also its inner meaning of the tied process signalling therein the irritating points that need to be resolved. In that mould, it acts as supervising the functioning process of a particular work plan at a given point of time. Further, when a policy related programme is monitored, it is possible that it may ask for revisiting the postulates and other fundamental premises and may even question the prudency to continue the policy any further point of time without any amendment or modifications. It is also possible that any pilot initiative in an emerging area, after receipt of encouraging monitoring input, might suggest expanding and replicating elsewhere. This chapter covers monitoring issues including what to be monitored, how to be monitored and who is to monitor. It also covers monitoring mechanism and decentralised monitoring structure.

Keywords Monitoring mechanism · Monitoring items · Monitoring indicators · Monitoring format · Monitoring authority · Monitoring meeting · Monitoring report · Concurrent evaluation

Monitoring is defined as a process of measuring, recording, collecting, processing and communicating information to assist policy makers and project management to take right decision in a given situation. Monitoring is essentially a functioning tool to understand the flow character of a programme or a scheme right from the time of its commencement. It generates not only flow of data but deciphers also its inner meaning of the tied process signalling therein the irritating points that need to be resolved. In that mould, it acts as supervising the functioning process of a particular work plan at a given point of time. Further, when a policy related programme is monitored, it is possible that it may ask for revisiting the postulates and other fundamental premises and may even question the prudency to continue the policy any further point of time without any amendment or modifications. It is also possible that any pilot initiative in an emerging area, after receipt of encouraging monitoring input, might suggest expanding and replicating elsewhere.

For productive monitoring four critical aspects are needed to be addressed:

(i) What is to be monitored

Monitoring items should be pointed, relevant and critical. Such items include physical and financial progress. In general perception, monitoring is usually understood as monitoring of expenditure and the focus of all activities is confined to financial monitoring. Monitoring per se includes time dimension of scheme expenditure as well. However, the popular approach of financial monitoring is one dimensional in that it does not give any idea of the emerging status of physical achievement. Financial monitoring without its concurrent physical status is not meaningful monitoring at all. As a matter of fact, physical monitoring is very important from project implementation point of view as it gives fair idea about linkage between quantum of money spent and the corresponding volume of physical output. Another way of looking at monitoring from project implementation point of view is Process monitoring and Outcome monitoring. A project is supposed to perform a set of tasks in a work schedule in a time format. Similarly, it is also supposed to show certain deliverables at different identified points of work programme. Monitoring assists the programme manager to keep track with the implementation process of several segments of such work programme. Such step by step monitoring helps to ensure projected vision of defined outcome in a given framework of time.

There are host of items for which monitoring technique is made use of. In the context of population control in India, there is scope for monitoring of influx and infiltration across the border states of India; it may as well be monitoring on adequacy and effectiveness of programme intervention in the demographically weaker states and districts in India; it could as well be monitoring the facts and issues that are critical for the relatively abnormal rate of growth of a particular religious faith or whether the socio-economic goals of the country have been following the right direction. There is good scope of monitoring the extent and quality of Civil Registration system of Births and Deaths or of the extent of coverage of the on-going marriage registration and prevention of under aged marriage and so on.

In so far as family planning is concerned, there is huge unmet need of monitoring on several important issues. Ironically enough, this is the most under-served areas of monitoring in India. The basis of family planning is Eligible Couple and Children Register which is hardly monitored in the states in India. Apart from being the primary referral document, it embodies the types of target-clients who have to be reached. Effective monitoring on this item alone would be a huge service to the cause of family planning programme in the country. As a follow-up of this ECCR monitoring, there could be FP service monitoring on Birth Spacing; monitoring on types of contraceptives canvassed (also male or female) and used in general but more particularly for those having more than two children in particular; whether counselling on permanent method has been initiated and the extent of its success etc. Similarly, based on ECCR, there could be monitoring on Ante-natal care of pregnant women, Breast-feeding, supplementary nutrition for under-weight children or on immunisation status of children etc.

Incidentally, the ushering of Target-free approach in family planning (FP) since April 1996 together with the policy decision of the Government of India to merge family planning and maternal and child health (MCH) services into Reproductive and Child health (RCH) services has made family planning programme less focused, non-visible and low priority in terms of importance, time, care, coverage and outreach of services. Medical and health related services occupying prime time usually got higher priority in supervision and monitoring. This continues even after the launch of the Rural Health Mission. The scope of intensive monitoring of family planning related services separately under Rural Health Mission and Urban Health Mission cannot be overemphasised. There is indeed a huge gap in the outreach of the family welfare services to the targeted eligible couples in our country.

Monitoring Indicators have now come into being for facilitating objectivity in monitoring exercise. Depending on the infrastructure set-up at any State, there is scope for designing distinct Monitoring Indicators/items for the level of Sub-centre, Primary Health Centre, District Heath Centre and State level Health Establishment and the like. It is for the same reason the design of the monitoring format would depend on what subjects/items will be monitored at that level. There is no uniform format for monitoring.

(ii) Who is to monitor

The primary responsibility of monitoring lies with the implementation authority. In case of government department it may be done by the department itself or delegated to its subordinate office in some cases. Be that as it may, there has to have one to one correspondence between the subjects of monitoring and the functional domain of such department or agency; otherwise the exercise would be simply mechanical and void of any life force. As a general rule, the monitoring authority would be the nodal department as per Rules of Business of the Government, either in the Union Government or in the State Governments. Regarding monitoring of influx and infiltration across the border states of India, or of adequacy and effectiveness of programme intervention in the demographically weaker states and districts in India or of monitoring the facts and issues that are critical for the relatively abnormal rate of growth of a particular religious faith or of the projected socio-economic goals of the country, the monitoring is needed to be undertaken by the department of the Ministry of Home Affairs. Similarly, monitoring of the extent and quality of Civil Registration system of Births and Deaths or of the extent of coverage of the ongoing marriage registration etc. need also to be done by the said Ministry and/or by the nodal department of the state government on a regular basis.

The family planning programme is the least monitored area in our country and even when it does take place, there is less professional approach in those exercises. One of the prerequisites for quality monitoring is decoding the messages embodied in the data structure and then re-activate the process to meet the identified deficit and systemise correction. Because of paramount importance of the Eligible Couple and Children Register, it needs to be monitored by the immediate supervisory authority in the Health & Family welfare set-up of the state government and also by the related Gram Panchayat in the rural areas/Ward Committee in the urban areas. Monitoring of Birth Spacing, contraceptive-use, Ante-natal care, Breast-feeding, supplementary nutrition and immunisation etc. need to be monitored by the immediate supervisory authority in the Health & Family welfare set-up of the state government.

The decentralised local bodies at various levels need to take up monitoring responsibility of functions as devolved on to them under Article 243G or Article 243W, as the case may be. The Gram Panchayat will monitor implementation of its own schemes and so by others. There is no scope of monitoring a local body's scheme by another tier of a local body. However, the DPC, as the apex Constitutional planning body in the district may take up monitoring in respect of all schemes or for schemes falling under any selective subject (e.g. drinking water, road linkage, waste disposal etc.) of all local bodies within its jurisdictional area, as it deems proper.

Another important area of monitoring relates to situation when a good number of monitoring authority participates on monitoring a single programme. An illustration will make it clear. Immunisation of children may be monitored at various levels of government, both at the Union Government or in the states. Normally the monitoring focus of such multiple monitoring are distinct from one another, though there may have some commonalities. In the case of the Union government, the focus is on the immunisation of the children of the country as such and the responsibility of providing all support including the required antigens to the states in India for 100 % coverage of immunisation. The state nodal department, on the other hand, is required to put in place infrastructure, distribution of required antigens and arranging service providers for ensuring full immunisation coverage to children in the state and it's monitoring on a regular basis. At the district, sub-district and field set-up, the monitoring focus of each of the organisations is different from one another. It varies from setting up of immunisation centres, informing such arrangement in the locality, distribution of IEC materials on immunisation, counselling of parents wherever needed, Home visit, mop-up for left-out cases and organisation of outreach camps where needed.

Another essential requisite for quality and effective monitoring is that monitoring team members must be well conversant with the schemes under monitoring review including its objects, sanctioned fund and its release, details of personnel on the job, timelines of completion etc. The domain knowledge of team members can only ensure meaningful interactions, locate the gaps or deficiencies and be in a position to suggest right prescription for course correction.

(iii) How is to monitor them

The structure of monitoring format is very crucial. It has to be designed in tune with the basic objectives of programme. Such structure of monitoring format has to be pretested, wherever possible, to ensure right flow of information and absence of any confusion in filling up the monitoring schedule/format by field functionaries. The field functionaries need to fill in the designed monitoring format as per time design and submit it to its monitoring authority. Normally, the descriptive version should not be encouraged for the sake of easier comprehension and its consolidation. Based on analysis of monitoring data, a status position is needed to be worked out. The monitoring status, with observations on course correction, if any, may then be communicated to all concerned for appropriate action. In case of necessity, it could be followed up by field inspection as well. A monitoring meeting may also be convened to sort out implementation issues for the desired way forward.

(iv) Monitoring of action taken report

Monitoring is for ensuring right course of programme implementation and, therefore, it is not a onetime affair. It is a continuous process till the scheme is completed. The prescriptions for correction based on a monitoring report, say MR-1, need to be comprehensively addressed before submitting monitoring report, say MR-2. Mere communication of findings without follow-up compliance report would make the entire exercise redundant. ATR is in fact the essence of monitoring.

Monitoring per se is not an end in itself. It is a means to ensure proper implementation of scheme by acting as an indivisible supervisor during the process of implementation. Moreover, it is not enough that monitoring technique should alone be applied: it is also required to be buttressed by the other technique known as concurrent evaluation. In fact monitoring and concurrent evaluation are two activities which go hand in hand. Adequate monitoring is a basic requisite for undertaking concurrent evaluation and it is impossible to evaluate a scheme unless it has been adequately monitored over time.

Part III
Sustainable Population in the Background of Sustainable Development Goals

Chapter 13
Sustainable Population in the Background of Sustainable Development Goals

Abstract The UN Sustainable Development Goals (SDGs) adopted on 25th September, 2015 aimed to end hunger, assure gender equity, and build a life of dignity for all over the world. It gave a clarion call of a new framework of "Transforming Our World: the 2030 Agenda for Sustainable Development". The SDGs composed of 17 goals and 169 targets to wipe out poverty, fight inequality and tackle climate change over the next 15 years. The SDGs are indeed visionary in character and an ideal framework in which all nations need to work out its actionable plans. Because of varying inter-country deficit on actionable areas of sovereign nations, SDGs have not gone into country-specific road-map for action. It is for this reason that SDGs did not mention any constraints in its overall approach to realise its visionary goals. However, constraints do exist in achieving SDGs just as it had existed under MDGs. Two of the top most goals under SDGs is to end poverty in all its forms everywhere (Goal 1) and end hunger, achieve food security and improved nutrition (Goal 2). SDGs also aim at promoting sustained, inclusive and sustainable economic growth, full and productive employment and decent work for all (Goal 8). An under-populated country can proceed to reach these goals easily; a country having stable population or optimum population can also move in that direction rather comfortably. But for a country, like India, where TFR is less than replacement level, where there is alarming high order density of population and high quotient of BPL population, where there exits shocking levels of unemployment, under-employment and adverse nutritional standard and relatively poor HDI, the quantum jump at on go to a SDGs level destination is rather a very difficult task. The huge size population and its incremental addition is responsible for this sordid state of affairs. It will be a gigantic task to implement the Goal 12 to ensure sustainable consumption and production patterns for a runaway population of more than 1.21 billion population for such over-populated country.

Keywords Sustainable Development Goals · Sustainable population · Our Common Future · Environmental sustainability · Bio capacity, ecological footprint · Global Footprint Network · Goal specific SDG targets

The UN Sustainable Development Goals (SDGs) adopted on 25th September, 2015 aimed to end hunger, assure gender equity, and build a life of dignity for all over the world. It gave a clarion call of a new framework of "Transforming Our World: the 2030 Agenda for Sustainable Development". The SDGs composed of 17 goals and 169 targets to wipe out poverty, fight inequality and tackle climate change over the next 15 years. Among the 17 goals to this agenda, focus was on complete poverty eradication, zero hunger, quality education, gender equality, sustainable cities and communities and clean water and sanitation. The 193-member General Assembly of the United Nations, headed by its Secretary General described the agenda as 'clarion call' to share prosperity, empower people's livelihood, ensure peace and heal our planet for the benefit of this and future generations. The secretary general further observed that the new agenda is a promise by leaders of all people everywhere. It is a universal, integrated and transformative vision for a better world. It is an agenda for the people, to end poverty in all its forms. He further mentioned that the 17 SDGs are our guide and to-do list for people and planet, and a blue print for success. This new agenda compels us to look beyond national boundaries and short term interests and act in solidarity for the long term.

For appreciating phases of our journey towards sustainable population, in the framework of SDGs, it would be useful to clarify some of the linked population concepts to comprehend in what way sustainable population is different from other traditional concepts and what precisely is its role and relevance for achieving Sustainable Development Goals.

- Under Population:

A country is said to be under-populated when the number of people is insufficient to take the fullest possible advantage of the natural and capital resources of that country. Normally such state of affairs existed in some bygone years of history when in a new country situation the resources were vast and there were not enough manpower to make use of available resources and register higher per capita income and better standard of living to its citizens. The size of population was always below the optimum population.

- Over population:

Overpopulation is a situation where the number of existing human population in a country exceeds the carrying capacity of such country. The size of population of such country is too large to make use of economic potential and optimum utilization of its natural capital resources. The existing resources, both natural and capital, in relation to the size of the population are too inadequate for gainful full employment of its citizens. As a result the per capita income is lower and the resultant standard of living. Poor quality of life is also reflected in life span of its citizen and in the morbidly and mortality profile of its citizens. The size of population of such country was always above optimum population. Solving overpopulation is essential in building a sustainable future.

- Optimum population:

The optimum population is the ideal population of a country where the per capita income would be the maximum with use of available resources or means of production. Any rise or fall in the size of the population above or below the optimum level will scale down income per head. Given the stock of natural resources, the technique of production and the stock of capital in a country, there is a definite size of population corresponding to the highest per capita income. Other things being equal, any deviation from this optimum-sized population will lead to a reduction in the per capita income. If the increase in population is followed by the increase in per capita income, the country is under-populated and it can afford to increase its population till it reaches the optimum level. On the contrary, if the increase in population leads to fall in per capita income, the country is over-populated and needs a decline in population till the per capita income is maximised. However, the optimum population is not a fixed concept. It is movable with the varying state of natural resources or means of production of such country. For instance, if there are improvements in the methods and techniques of production, the output per head will rise and the optimum point will shift upward.

- Stable Population:

The concept of Stable Population was first introduced in demography by Alfred J. Lotka as a particular case of Malthusian projections. Stable populations are theoretical models widely used by demographers to represent and understand the structure, growth and evolution of human populations. By definition, stable populations have age-specific fertility and mortality rates that remain constant over time. A stationary population is a simple example of a stable population with a zero growth rate, neither growing nor shrinking in size and is equivalent to a life table population. The age-specific fertility rates that generate this stationary life are not found in real life. The size of population in such situation will always be the same and the crude birth rate and death rates will never change and will always be equal. However, in a stationary population, the proportion in age-group is constant as in stable population, but unlike a growing or shrinking stable population, the numbers in each group are also constant over time. A stable population having constant birth rates and age distribution can emerge in a country situation and persist over time provided three conditions are satisfied, namely, (a) Age-specific fertility rates are constant (b) Age-specific death rates (i.e., the life table) are constant and (c) Age-specific net migration rates are zero.

Incidentally, The National Population Policy 2000 mentions that the long term objective of NPP 2000 is to achieve a stable population by 2045.

- Sustainable population

A sustainable population is one that can be maintained at that number of people indefinitely without adversely impacting the environment or the quality of life of the members of that population. Sustainable population growth means the rate at which the number of inhabitants of a given area can increase without overburdening the area's economic, social and natural resources. Many demographers use the term "carrying capacity" to refer to the number of inhabitants who can be supported in a particular area without degrading the physical and social environment at present or for future generations. When population growth exceeds an area's carrying capacity, growth is considered unsustainable because the existing resources, such as water and food, are not sufficient to sustain the growing population.

There does not exist any normative definition of sustainable population up to this point of time. Various international agencies go by their own understanding framework to define sustainable population. One strong view point based on Global Footprint Network data estimates that current population is three times the sustainable level. It shows that humanity uses the equivalent of 1.6 planet Earths to provide the renewable resources that are used and absorbed as waste. Based on the 7+ billion of population and also of an average European standard of living—which is about half the consumption level of an average American—the Earth could sustainably support only about 2 billion people. The underlying message is that there is one to one correspondence between standard of living, consequent use of natural resources and the earth capacity of liveable population.

Climate change has added a new dimension to any estimate of sustainable population and the planet's ability to support all 7+ billion of population. Climate scientists foresee lower crop yields of major grains such as wheat, rice, and maize. Rising sea levels may create hundreds of millions of climate refugees in years to come in addition to climate disruption followed by likely increasing levels of resource conflict and civil unrest. Adaptation to climate disruption will only be easier with a much smaller global population.

Thus, the world in general and the individual countries in particular need to be circumspect of the hard reality of changing resource base to uphold the existing size of population and factor in certain and irreversible climatic change of foreseeable nature and degrees. In addressing sustainable population, appropriate consideration need to be placed on demand pull forces that hit upon resource scenario in the post-climate changing scenario that the local area of the Earth can support.

Finally, as a signatory to the World summit of sustainable goals, each of Individual countries has to draw up a road-map for sustainable population by arresting and reversing the birth rate, immigration and infiltration, and the resource use. Sustainable population is the only requisite for sustainable development and the SDGs. The SDGs will remain illusory if sustainable population is not in place.

- Carrying capacity:

Carrying capacity in simple language means the maximum number of individuals that an area of land can support by their food requirements and without any detrimental effects. Ecologists however, define carrying capacity as the maximal population size of a given species that an area can support without reducing its ability to support the same species in the future. Because of variation of types and quantities of consumption, carrying capacity, for human habitations, varies markedly with culture and level of economic development. With a particular culture, the types of resource used and with a particular level of economic development the quantities of resources consumed vary signalling different carrying capacity of the locations. Carrying capacity may thus be defined as biophysical carrying capacity (the maximal population size that could be sustained biophysically under given technological capabilities) and social carrying capacities (the maxima that could be sustained under various social systems especially, the associated patterns of resource consumption).

"Population Matters in its website has dealt on Carrying capacity in a very lucid way:" the maximum number of individuals that can be supported sustainably by a given environment is known as its 'carrying capacity'. For most non-human species, the concept is quite simple. If carrying capacity is exceeded, the population declines because its environment can no longer support the excess numbers. In many situations this can happen very rapidly because excessive demand degrades or even devastates the environment and there is a sudden and catastrophic feedback effect. Such a feedback effect can not only eradicate those numbers of population in excess of the carrying capacity of an environment but under certain circumstances it can cause the near extinction of an entire species. A population can exceed the carrying capacity of its environment for a short while by using up the stored resources (i.e. natural capital) of its environment, but sooner or later the 'overshoot' will catch up. Once the capital is exhausted, population numbers inevitably fall because there are no longer enough resources available to support the number of individuals.

In the case of human populations, there is a large variation in per capita consumption levels between poor and affluent communities, so the basic definition of carrying capacity needs to be qualified and the given level of per capita consumption and waste generation needs to be taken into consideration. The carrying capacity of a given environment is much greater for people living at a subsistence level than it is for people with a typical Western European or North American lifestyle. It is also important to note that different geographic regions have a greater or smaller carrying capacity. Climate and local geography both play a crucial part.

- Sustainability:

Sustainability has been defined by Geir B. Ascheim in 1994 in the World Bank document as a requirement of our generation to manage the resource base such that the average quality of life that we ensure ourselves can be potentially shared by future generations. Development is sustainable if it is involved a non-decreasing average quality of life. A sustainable process is thus one that can be maintained without interruption. Sustainability is a necessary and sufficient condition for a population to be at or below any carrying capacity. Incidentally, Brundtland the author of Sustainable development concept way back in 1983 defined it as "development that meets the needs and aspirations of the present without compromising the ability of future generations to meet their own needs". Implicit in the desire for sustainability is the moral conviction that the current generation should pass on its inheritance of natural wealth, not unchanged, but undiminished in potential to support future generations. In other words, Living sustainably means balancing our consumption, our technology choices and our population numbers in order to live within the resources of the planet. It means maintaining a stable and healthy environment for both humanity and biodiversity.

- Sustainable development:

Brundtland, the pioneer of the Sustainable Development, defined it in 'Our Common Future' as that *development that meets the needs of the present without compromising the ability of future generations to meet their own needs. The concept has been addressed subsequently by scholars from various angles, examples of which are given hereunder*:

(i) 'Sustainable development (SD) is a process for meeting human development goals while maintaining the ability of natural systems to continue to provide the natural resources and ecosystem services upon which the economy and society depend. Sustainable development has in that sense three dimensions, domains or pillars, namely "economic, environmental and social" or "ecology, economy and equity" while some scholars include a fourth pillar, namely, culture, institutions or governance. The ecological sustainability of human settlements is part of the relationship between humans and their natural, social and built environments. Environmental sustainability concerns the natural environment and how it endures and remains diverse and productive while Ecological economics look upon it from the premise of natural capital of intergenerational equity and irreversibility of environmental change in that sustainability requires that human activity has to use up such amount of nature's resources that can be replenished naturally'.

(ii) The sustainability equation

Another way of looking upon sustainable development is from a mathematical equation. Paul R. Ehrlich, the eminent American biologist and educator, in his book, *The Population Bomb* summarised what is known as the Ehrlich or IPAT equation, $I = PAT$. I = impact on the environment or demand for resources, P = population size, A = affluence and T = technology.

The two most important conclusions deriving from this relationship are that Earth can support only a limited number of people in a sustainable manner and humanity has a clear choice between more people who have poorer lifestyles and fewer people who have a better quality of life.

(iii) Biocapacity and ecological footprint

Population Matters in its website has brought out two concepts relating to sustainable development—Biocapacity and ecological footprint developed by the Global Footprint Network and are quantified as global hectares (gha) which provide a common basis on which to compare the biological capability of the environment to provide food and other essential needs versus the demands placed by human communities on these ecological services. 'If the ecological footprint of a human population exceeds the biocapacity of its environment, the situation is unsustainable. Disturbingly, worldwide the total human ecological foot-print is 2.6 global hectares per capita (gha/cap) compared with a total worldwide bio-capacity of only 1.8 gha/cap (GFN Ecological Footprint Atlas 2009). This overshoot means that humanity is already using 1.4 times as many resources as are sustainably available. It means that we are in effect already 'living on the capital of the planet rather than its income'. The overshoot for high income countries is much more extreme than the overall average. Whereas low income countries have a typical footprint of 1.0 gha/cap, the average for high income countries is 6.1, of which the UK is typical at 6.12. Thus at the current we would already need 3.4 Planet Earths (i.e. 6.12/1.8) to support the total world population of 6.8 billion if everyone was to have typical UK living standards. Though the foot printing approach explicitly accounts for different levels of per capita consumption, it doesn't factor in the biocapacity needed for the preservation of other species—a clear moral problem that would concern many people and one with significant economic consequences for humanity. If capacity for other species is allowed for, we are in a situation of greater overshoot than the figures suggest'.

'All definitions of sustainable development look upon the world as a system—a system that connects space; and a system that connects time. When the world is looked as a system over space, air quality from one place to another or the harmful effect of pesticides from one country on to another, can be felt. And when the world is thought of as a system over time, the decisions of older generations over any subject areas continue to affect today; and the economic policies of today will have an impact on outcome of those policies tomorrow. Finally, when quality of life in the world is considered as a system, it boils down to outreach and access of ingredients of human development to all places in the nation and beyond. Thus the concept of sustainable development is rooted in this sort of systems thinking and the SDGs are the outcome of such systemic thinking. SDGs are a shining example of sustainable development'.

13.1 Background of Sustainable Development Goals—the End of Journey of Millennium Development Goals (MDGs)

The MDGs were agreed set of goals that international community set for itself for the first time at the UN Millennium Summit in September 2000 around a common 15-year agenda to tackle the indignity of poverty. It encompassed universally accepted human values and rights such as freedom from hunger, the right to basic education, the right to health and responsibility to future generations. The MDGs, as set for implementation and monitoring, were as follows:

Goal 1: Eradicate extreme poverty and hunger
Goal 2: Achieve universal primary education
Goal 3: Promote gender equality and empower women
Goal 4: Reduce child mortality
Goal 5: Improve maternal health
Goal 6: Combat HIV/AIDS, malaria and other diseases
Goal 7: Combat HIV/AIDS, malaria and other diseases
Goal 8: Develop a global partnership for development

Along with these eight goals, the MDGs had set itself targets for each of the goals totalling 21 targets, and also monitoring indicators within each of such targets aggregating 60 in all. The 15-year MDGs effort has produced the most successful anti-poverty movement in history as brought out by 'The Millennium Development Goals Report 2015'. A summary of such achievements, Goal wise, from the said Report is shown hereunder:

Goal 1: Eradicate extreme poverty and hunger

- Extreme poverty has declined significantly over the last two decades. In 1990, nearly half of the population in the developing world lived on less than $1.25 a day; that proportion dropped to 14 % in 2015.
- Globally, the number of people living in extreme poverty has declined by more than half, falling from 1.9 billion in 1990 to 836 million in 2015. Most progress has occurred since 2000.
- The number of people in the working middle class—living on more than $4 a day—has almost tripled between 1991 and 2015. This group now makes up half the workforce in the developing regions, up from just 18 % in 1991.
- The proportion of undernourished people in the developing regions has fallen by almost half since 1990, from 23.3 % in 1990–1992 to 12.9 % in 2014–2016.

Goal 2: Achieve universal primary education

- The primary school net enrolment rate in the developing regions has reached 91 % in 2015, up from 83 % in 2000.
- The number of out-of-school children of primary school age worldwide has fallen by almost half, to an estimated 57 million in 2015, down from 100 million in 2000.
- Sub-Saharan Africa has had the best record of improvement in primary education of any region since the MDGs were established. The region achieved a 20 % point increase in the net enrolment rate from 2000 to 2015, compared to a gain of 8 % points between 1990 and 2000.
- The literacy rate among youth aged 15–24 has increased globally from 83 to 91 % between 1990 and 2015. The gap between women and men has narrowed

Goal 3: Promote gender equality and empower women

- Many more girls are now in school compared to 15 years ago. The developing regions as a whole have achieved the target to eliminate gender disparity in primary, secondary and tertiary education.
- In Southern Asia, only 74 girls were enrolled in primary school for every 100 boys in 1990. Today, 103 girls are enrolled for every 100 boys.
- Women now make up 41 % of paid workers outside the agricultural sector, an increase from 35 % in 1990.
- Between 1991 and 2015, thse proportion of women in vulnerable employment as a share of total female employment has declined 13 % points. In contrast, vulnerable employment among men fell by 9 % points.
- Women have gained ground in parliamentary representation in nearly 90 % of the 174 countries with data over the past 20 years. The average proportion of women in parliament has nearly doubled during the same period. Yet still only one in five members are women.

Goal 4: Reduce child mortality

- The global under-five mortality rate has declined by more than half, dropping from 90 to 43 deaths per 1000 live births between 1990 and 2015.
- Despite population growth in the developing regions, the number of deaths of children under five has declined from 12.7 million in 1990 to almost 6 million in 2015 globally.
- Since the early 1990s, the rate of reduction of under-five mortality has more than tripled globally.
- In sub-Saharan Africa, the annual rate of reduction of under-five mortality was over five times faster during 2005–2013 than it was during 1990–1995.
- Measles vaccination helped prevent nearly 15.6 million deaths between 2000 and 2013. The number of globally reported measles cases declined by 67 % for the same period.
- About 84 % of children worldwide received at least one dose of measles-containing vaccine in 2013, up from 73 % in 2000.

Goal 5: Improve maternal health

- Since 1990, the maternal mortality ratio has declined by 45 % worldwide, and most of the reduction has occurred since 2000.
- In Southern Asia, the maternal mortality ratio declined by 64 % between 1990 and 2013, and in sub-Saharan Africa it fell by 49 %.
- More than 71 % of births were assisted by skilled health personnel globally in 2014, an increase from 59 % in 1990.
- In Northern Africa, the proportion of pregnant women who received four or more antenatal visits increased from 50 to 89 % between 1990 and 2014.
- Contraceptive prevalence among women aged 15 to 49, married or in a union, increased from 55 % in 1990 worldwide to 64 % in 2015.

Goal 6: Combat HIV/AIDS, malaria and other diseases

- New HIV infections fell by approximately 40 % between 2000 and 2013, from an estimated 3.5 million cases to 2.1 million.
- By June 2014, 13.6 million people living with HIV were receiving antiretroviral therapy (ART) globally, an immense increase from just 800,000 in 2003. ART averted 7.6 million deaths from AIDS between 1995 and 2013.
- Over 6.2 million malaria deaths have been averted between 2000 and 2015, primarily of children under five years of age in sub-Saharan Africa. The global malaria incidence rate has fallen by an estimated 37 % and the mortality rate by 58 %.
- More than 900 million insecticide-treated mosquito nets were delivered to malaria-endemic countries in sub-Saharan Africa between 2004 and 2014.
- Between 2000 and 2013, tuberculosis prevention, diagnosis and treatment interventions saved an estimated 37 million lives. The tuberculosis mortality rate fell by 45 % and the prevalence rate by 41 % between 1990 and 2013.

Goal 7: Combat HIV/AIDS, malaria and other diseases

- Ozone-depleting substances have been virtually eliminated since 1990, and the ozone layer is expected to recover by the middle of this century.
- Terrestrial and marine protected areas in many regions have increased substantially since 1990. In Latin America and the Caribbean, coverage of terrestrial protected areas rose from 8.8 to 23.4 % between 1990 and 2014.
- In 2015, 91 % of the global population is using an improved drinking water source, compared to 76 % in 1990.
- Of the 2.6 billion people who have gained access to improved drinking water since 1990, 1.9 billion gained access to piped drinking water on premises. Over half of the global population (58 %) now enjoys this higher level of service.
- Globally, 147 countries have met the drinking water target, 95 countries have met the sanitation target and 77 countries have met both.
- Worldwide, 2.1 billion people have gained access to improved sanitation. The proportion of people practicing open defecation has fallen almost by half since 1990.
- The proportion of urban population living in slums in the developing regions fell from approximately 39.4 % in 2000 to 29.7 % in 2014.

Goal 8: Develop a global partnership for development

- Official development assistance from developed countries increased by 66 % in real terms between 2000 and 2014, reaching $135.2 billion.
- In 2014, Denmark, Luxembourg, Norway, Sweden and the United Kingdom continued to exceed the United Nations official development assistance target of 0.7 % of gross national income.
- In 2014, 79 % of imports from developing to developed countries were admitted duty free, up from 65 % in 2000.

- The proportion of external debt service to export revenue in developing countries fell from 12 % in 2000 to 3 % in 2013.
- As of 2015, 95 % of the world's population is covered by a mobile-cellular signal.
- The number of mobile-cellular subscriptions has grown almost tenfold in the last 15 years, from 738 million in 2000 to over 7 billion in 2015.
- Internet penetration has grown from just over 6 % of the world's population in 2000 to 43 % in 2015. As a result, 3.2 billion people are linked to a global network of content and applications.

The MDG Report, 2015 captures the significant achievements of the targets worldwide, as mentioned above, but the progress has been uneven across regions and countries. This is more so to the poorer and to the disadvantages because of sex, age, disabilities, ethnicity and geographic considerations. The big gap still exists between the poorest and richer households, and between rural and urban areas. Further, climate change and environmental degradation still remains a critical challenge for the global community. Another gap that surfaced related to conflicts in the global scenario which account for a serious threat to human development. Millions of poor people still live in poverty and hunger without access to basic services. In fine, the MDG Report 2015 which documents the aspirational goals set out to achieve Millennium Declaration and highlights many success areas also mentions gaps that still remain. This has posed a new challenge at the end of MDGs' tenure.

- **Sustainable Development Goals (SDGs)**

At the meeting at the United Nations Headquarters in New York from 25 to 27 September 2015, the Heads of State and Government and High Representatives adopted a historic decision on a comprehensive, far-reaching and people-centred set of universal and transformative Goals and targets for the full implementation of the Sustainable Development Goals Agenda by 2030 for eradicating poverty in all its forms and dimensions, and achieving sustainable development in its three dimensions—economic, social and environmental—in a balanced and integrated manner. The new goals are to be built upon the achievements of the Millennium Development Goals and seek to address their unfinished business.

The Goals and targets that emerged after over two years of intensive public consultation and engagement with civil society and other stakeholders around the world paid particular attention to the voices of the poorest and most vulnerable. This consultation included valuable work done by the General Assembly Open Working Group on Sustainable Development Goals and by the United Nations, whose Secretary-General provided a synthesis report in December 2014. The 17- Sustainable Development Goals are as follows:

Goal 1. End poverty, end to in all its forms everywhere

Goal 2. End hunger, achieve food security and improved nutrition and promote sustainable agriculture Sustainable agriculture

Goal 3. Ensure healthy lives, at all ages and promote well-being for all at all ages

Goal 4. Ensure inclusive and equitable quality education and promote lifelong learning opportunities for all

Goal 5. Achieve gender equality and empower all women and girls

Goal 6. Ensure availability and sustainable management of water and sanitation for all

Goal 7. Ensure access to affordable, reliable, sustainable and modern energy for all

Goal 8. Promote sustained, inclusive and sustainable economic growth, full and productive employment, work for all and decent work for all

Goal 9. Build resilient infrastructure, promote inclusive and sustainable industrialization and foster innovation

Goal 10. Reduce inequality within and among countries Inequality

Goal 11. Make cities and human settlements inclusive, safe, resilient and sustainable

Goal 12. Ensure sustainable consumption and production patterns

Goal 13. Take urgent action to combat climate change and its impacts*

Goal 14. Conserve and sustainably use the oceans, seas and marine resources for sustainable development, Seas and marine resources for

Goal 15. Protect, restore and promote sustainable use of terrestrial ecosystems, sustainably manage forests, combat desertification, and halt and reverse land degradation and halt biodiversity loss

Goal 16. Promote peaceful and inclusive societies for sustainable development, provide access to justice for all and build effective, accountable and inclusive institutions at all levels

Goal 17. Strengthen the means of implementation and revitalize the global partnership for sustainable development

Before proceeding to Goal specific SDG targets, it would be worthwhile to touch upon additional background of the journey from MDGs to SDGs. Historically, the MDGs initiative is a path breaking event of monumental importance in that it set on motion a set on common social priorities worldwide. Further, it brought a paradigm shift on the role of global partner and global awareness, political accountability and public pressure to the *Report Card of development*, as Bill Gate liked to look at the MDGs Report.

The concept of the SDGs was born at the United Nations Conference on Sustainable Development, Rio +20, in 2012. The objective was to produce a set of universally applicable goals that balances the three dimensions of sustainable development: environmental, social, and economic.

SDGs go beyond MDGs in focusing on global system reforms to remove the barriers to sustainable development. While the first set of SDGs (No. 1–7) appears indeed as an extension of MDGs, the later set of goals is truly an extension of the agenda itself. The later parts have in turn two distinct characteristics: SDGs falling under numbers 8, 9 and 10 are reforms centred enablers of development covering areas like inclusion and jobs, infrastructure and industrialisation, and distribution. The final set of goals under SDGs No. 11–17 lays down the framework of governance covering areas like urbanization, consumption and production, climate change, resources and environment, peace and justice, and means of implementation and global partnership.

Considerable shift in approaches has been underway from MDGs to SDGs in that SDGs have now 5 Ps-people, planet, prosperity, peace and partnership.

Given the above perspective, the details of goals and targets of Sustainable Development Goals that were settled in the UN Summit on Sustainable Development are noted below:

Goal 1. End poverty in all its forms everywhere

1.1 By 2030, eradicate extreme poverty for all people everywhere, currently measured as people living on less than $1.25 a day

1.2 By 2030, reduce at least by half the proportion of men, women and children of all ages living in poverty in all its dimensions according to national definitions

1.3 Implement nationally appropriate social protection systems and measures for all, including floors, and by 2030 achieve substantial coverage of the poor and the vulnerable

1.4 By 2030, ensure that all men and women, in particular the poor and the vulnerable, have equal rights to economic resources, as well as access to basic services, ownership and control over land and other forms of property, inheritance, natural resources, appropriate new technology and financial services, including microfinance

1.5 By 2030, build the resilience of the poor and those in vulnerable situations and reduce their exposure and vulnerability to climate-related extreme events and other economic, social and environmental shocks and disasters

1.a Ensure significant mobilization of resources from a variety of sources, including through enhanced development cooperation, in order to provide adequate and predictable means for developing countries, in particular least developed countries, to implement programmes and policies to end poverty in all its dimensions

1.b Create sound policy frameworks at the national, regional and international levels, based on pro-poor and gender-sensitive development strategies, to support accelerated investment in poverty eradication actions

Goal 2. End hunger, achieve food security and improved nutrition and promote sustainable agriculture

2.1 By 2030, end hunger and ensure access by all people, in particular the poor and people in vulnerable situations, including infants, to safe, nutritious and sufficient food all year round

2.2 By 2030, end all forms of malnutrition, including achieving, by 2025, the internationally agreed targets on stunting and wasting in children under 5 years of age, and address the nutritional needs of adolescent girls, pregnant and lactating women and older persons

2.3 By 2030, double the agricultural productivity and incomes of small-scale food producers, in particular women, indigenous peoples, family farmers, pastoralists and fishers, including through secure and equal access to land, other productive resources and inputs, knowledge, financial services, markets and opportunities for value addition and non-farm employment

2.4 By 2030, ensure sustainable food production systems and implement resilient agricultural practices that increase productivity and production, that help maintain ecosystems, that strengthen capacity for adaptation to climate change, extreme weather, drought, flooding and other disasters and that progressively improve land and soil quality

2.5 By 2020, maintain the genetic diversity of seeds, cultivated plants and farmed and domesticated animals and their related wild species, including through soundly managed and diversified seed and plant banks at the national, regional and international levels, and promote access to and fair and equitable sharing of benefits arising from the utilization of genetic resources and associated traditional knowledge, as internationally agreed

2.a Increase investment, including through enhanced international cooperation, in rural infrastructure, agricultural research and extension services, technology development and plant and livestock gene banks in order to enhance agricultural productive capacity in developing countries, in particular least developed countries

2.b Correct and prevent trade restrictions and distortions in world agricultural markets, including through the parallel elimination of all forms of agricultural export subsidies and all export measures with equivalent effect, in accordance with the mandate of the Doha Development Round

2.c Adopt measures to ensure the proper functioning of food commodity markets and their derivatives and facilitate timely access to market information, including on food reserves, in order to help limit extreme food price volatility

Goal 3. Ensure healthy lives and promote well-being for all at all ages

3.1 By 2030, reduce the global maternal mortality ratio to less than 70 per 100000 live births

3.2 By 2030, end preventable deaths of newborns and children under 5 years of age, with all countries aiming to reduce neonatal mortality to at least as low as 12 per 1000 live births and under-5 mortality to at least as low as 25 per 1000 live births

3.3 By 2030, end the epidemics of AIDS, tuberculosis, malaria and neglected tropical diseases and combat hepatitis, water-borne diseases and other communicable diseases

3.4 By 2030, reduce by one third premature mortality from non-communicable diseases through prevention and treatment and promote mental health and well-being

3.5 Strengthen the prevention and treatment of substance abuse, including narcotic drug abuse and harmful use of alcohol

3.6 By 2020, halve the number of global deaths and injuries from road traffic accidents

3.7 By 2030, ensure universal access to sexual and reproductive health-care services, including for family planning, information and education, and the integration of reproductive health into national strategies and programmes

3.8 Achieve universal health coverage, including financial risk protection, access to quality essential health-care services and access to safe, effective, quality and affordable essential medicines and vaccines for all

3.9 By 2030, substantially reduce the number of deaths and illnesses from hazardous chemicals and air, water and soil pollution and contamination

3.a Strengthen the implementation of the World Health Organization Framework Convention on Tobacco Control in all countries, as appropriate

3.b Support the research and development of vaccines and medicines for the communicable and non-communicable diseases that primarily affect developing countries, provide access to affordable essential medicines and vaccines, in accordance with the Doha Declaration on the TRIPS Agreement and Public Health, which affirms the right of developing countries to use to the full the provisions in the Agreement on Trade-Related Aspects of Intellectual Property Rights regarding flexibilities to protect public health, and, in particular, provide access to medicines for all

3.c Substantially increase health financing and the recruitment, development, training and retention of the health workforce in developing countries, especially in least developed countries and small island developing States

3.d Strengthen the capacity of all countries, in particular developing countries, for early warning, risk reduction and management of national and global health risks

Goal 4. Ensure inclusive and equitable quality education and promote lifelong learning opportunities for all

4.1 By 2030, ensure that all girls and boys complete free, equitable and quality primary and secondary education leading to relevant and effective learning outcomes

4.2 By 2030, ensure that all girls and boys have access to quality early childhood development, care and pre-primary education so that they are ready for primary education

4.3 By 2030, ensure equal access for all women and men to affordable and quality technical, vocational and tertiary education, including university

4.4 By 2030, substantially increase the number of youth and adults who have relevant skills, including technical and vocational skills, for employment, decent jobs and entrepreneurship

4.5 By 2030, eliminate gender disparities in education and ensure equal access to all levels of education and vocational training for the vulnerable, including persons with disabilities, indigenous peoples and children in vulnerable situations

4.6 By 2030, ensure that all youth and a substantial proportion of adults, both men and women, achieve literacy and numeracy

4.7 By 2030, ensure that all learners acquire the knowledge and skills needed to promote sustainable development, including, among others, through education for sustainable development and sustainable lifestyles, human rights, gender equality, promotion of a culture of peace and non-violence, global citizenship and appreciation of cultural diversity and of culture's contribution to sustainable development

4.a Build and upgrade education facilities that are child, disability and gender sensitive and provide safe, non-violent, inclusive and effective learning environments for all

4.b By 2020, substantially expand globally the number of scholarships available to developing countries, in particular least developed countries, small island developing States and African countries, for enrolment in higher education, including vocational training and information and communications technology, technical, engineering and scientific programmes, in developed countries and other developing countries

4.c By 2030, substantially increase the supply of qualified teachers, including through international cooperation for teacher training in developing countries, especially least developed countries and small island developing States

Goal 5. Achieve gender equality and empower all women and girls

5.1 End all forms of discrimination against all women and girls everywhere

5.2 Eliminate all forms of violence against all women and girls in the public and private spheres, including trafficking and sexual and other types of exploitation

5.3 Eliminate all harmful practices, such as child, early and forced marriage and female genital mutilation

5.4 Recognize and value unpaid care and domestic work through the provision of public services, infrastructure and social protection policies and the promotion of shared responsibility within the household and the family as nationally appropriate

5.5 Ensure women's full and effective participation and equal opportunities for leadership at all levels of decision-making in political, economic and public life

5.6 Ensure universal access to sexual and reproductive health and reproductive rights as agreed in accordance with the Programme of Action of the

International Conference on Population and Development and the Beijing Platform for Action and the outcome documents of their review conferences

5.a Undertake reforms to give women equal rights to economic resources, as well as access to ownership and control over land and other forms of property, financial services, inheritance and natural resources, in accordance with national laws

5.b Enhance the use of enabling technology, in particular information and communications technology, to promote the empowerment of women

5.c Adopt and strengthen sound policies and enforceable legislation for the promotion of gender equality and the empowerment of all women and girls at all levels

Goal 6. Ensure availability and sustainable management of water and sanitation for all

6.1 By 2030, achieve universal and equitable access to safe and affordable drinking water for all

6.2 By 2030, achieve access to adequate and equitable sanitation and hygiene for all and end open defecation, paying special attention to the needs of women and girls and those in vulnerable situations

6.3 By 2030, improve water quality by reducing pollution, eliminating dumping and minimizing release of hazardous chemicals and materials, halving the proportion of untreated wastewater and substantially increasing recycling and safe reuse globally

6.4 By 2030, substantially increase water-use efficiency across all sectors and ensure sustainable withdrawals and supply of freshwater to address water scarcity and substantially reduce the number of people suffering from water scarcity

6.5 By 2030, implement integrated water resources management at all levels, including through transboundary cooperation as appropriate

6.6 By 2020, protect and restore water-related ecosystems, including mountains, forests, wetlands, rivers, aquifers and lakes

6.a By 2030, expand international cooperation and capacity-building support to developing countries in water—and sanitation-related activities and programmes, including water harvesting, desalination, water efficiency, wastewater treatment, recycling and reuse technologies

6.b Support and strengthen the participation of local communities in improving water and sanitation management

Goal 7. Ensure access to affordable, reliable, sustainable and modern energy for all

7.1 By 2030, ensure universal access to affordable, reliable and modern energy services

7.2 By 2030, increase substantially the share of renewable energy in the global energy mix

7.3 By 2030, double the global rate of improvement in energy efficiency

7.a By 2030, enhance international cooperation to facilitate access to clean energy research and technology, including renewable energy, energy efficiency and advanced and cleaner fossil-fuel technology, and promote investment in energy infrastructure and clean energy technology

7.b By 2030, expand infrastructure and upgrade technology for supplying modern and sustainable energy services for all in developing countries, in particular least developed countries, small island developing States, and land-locked developing countries, in accordance with their respective programmes of support

Goal 8. Promote sustained, inclusive and sustainable economic growth, full and productive employment and decent work for all

8.1 Sustain per capita economic growth in accordance with national circumstances and, in particular, at least 7 % gross domestic product growth per annum in the least developed countries

8.2 Achieve higher levels of economic productivity through diversification, technological upgrading and innovation, including through a focus on high-value added and labour-intensive sectors

8.3 Promote development-oriented policies that support productive activities, decent job creation, entrepreneurship, creativity and innovation, and encourage the formalization and growth of micro-, small- and medium-sized enterprises, including through access to financial services

8.4 Improve progressively, through 2030, global resource efficiency in consumption and production and endeavour to decouple economic growth from environmental degradation, in accordance with the 10-year framework of programmes on sustainable consumption and production, with developed countries taking the lead

8.5 By 2030, achieve full and productive employment and decent work for all women and men, including for young people and persons with disabilities, and equal pay for work of equal value

8.6 By 2020, substantially reduce the proportion of youth not in employment, education or training

8.7 Take immediate and effective measures to eradicate forced labour, end modern slavery and human trafficking and secure the prohibition and elimination of the worst forms of child labour, including recruitment and use of child soldiers, and by 2025 end child labour in all its forms

8.8 Protect labour rights and promote safe and secure working environments for all workers, including migrant workers, in particular women migrants, and those in precarious employment

8.9 By 2030, devise and implement policies to promote sustainable tourism that creates jobs and promotes local culture and products

8.10 Strengthen the capacity of domestic financial institutions to encourage and expand access to banking, insurance and financial services for all

8.a   Increase Aid for Trade support for developing countries, in particular least developed countries, including through the Enhanced Integrated Framework for Trade-Related Technical Assistance to Least Developed Countries

8.b   By 2020, develop and operationalize a global strategy for youth employment and implement the Global Jobs Pact of the International Labour Organization

Goal 9. Build resilient infrastructure, promote inclusive and sustainable industrialization and foster innovation

9.1   Develop quality, reliable, sustainable and resilient infrastructure, including regional and transborder infrastructure, to support economic development and human well-being, with a focus on affordable and equitable access for all

9.2   Promote inclusive and sustainable industrialization and, by 2030, significantly raise industry's share of employment and gross domestic product, in line with national circumstances, and double its share in least developed countries

9.3   Increase the access of small-scale industrial and other enterprises, in particular in developing countries, to financial services, including affordable credit, and their integration into value chains and markets

9.4   By 2030, upgrade infrastructure and retrofit industries to make them sustainable, with increased resource-use efficiency and greater adoption of clean and environmentally sound technologies and industrial processes, with all countries taking action in accordance with their respective capabilities

9.5   Enhance scientific research, upgrade the technological capabilities of industrial sectors in all countries, in particular developing countries, including, by 2030, encouraging innovation and substantially increasing the number of research and development workers per 1 million people and public and private research and development spending

9.a   Facilitate sustainable and resilient infrastructure development in developing countries through enhanced financial, technological and technical support to African countries, least developed countries, landlocked developing countries and small island developing States

9.b   Support domestic technology development, research and innovation in developing countries, including by ensuring a conducive policy environment for, inter alia, industrial diversification and value addition to commodities

9.c   Significantly increase access to information and communications technology and strive to provide universal and affordable access to the Internet in least developed countries by 2020

Goal 10. Reduce inequality within and among countries

10.1   By 2030, progressively achieve and sustain income growth of the bottom 40 % of the population at a rate higher than the national average

10.2   By 2030, empower and promote the social, economic and political inclusion of all, irrespective of age, sex, disability, race, ethnicity, origin, religion or economic or other status

10.3 Ensure equal opportunity and reduce inequalities of outcome, including by eliminating discriminatory laws, policies and practices and promoting appropriate legislation, policies and action in this regard
10.4 Adopt policies, especially fiscal, wage and social protection policies, and progressively achieve greater equality
10.5 Improve the regulation and monitoring of global financial markets and institutions and strengthen the implementation of such regulations
10.6 Ensure enhanced representation and voice for developing countries in decision-making in global international economic and financial institutions in order to deliver more effective, credible, accountable and legitimate institutions
10.7 Facilitate orderly, safe, regular and responsible migration and mobility of people, including through the implementation of planned and well-managed migration policies
10.a Implement the principle of special and differential treatment for developing countries, in particular least developed countries, in accordance with World Trade Organization agreements
10.b Encourage official development assistance and financial flows, including foreign direct investment, to States where the need is greatest, in particular least developed countries, African countries, small island developing States and landlocked developing countries, in accordance with their national plans and programmes
10.c By 2030, reduce to less than 3 % the transaction costs of migrant remittances and eliminate remittance corridors with costs higher than 5 %

Goal 11. Make cities and human settlements inclusive, safe, resilient and sustainable

11.1 By 2030, ensure access for all to adequate, safe and affordable housing and basic services and upgrade slums
11.2 By 2030, provide access to safe, affordable, accessible and sustainable transport systems for all, improving road safety, notably by expanding public transport, with special attention to the needs of those in vulnerable situations, women, children, persons with disabilities and older persons
11.3 By 2030, enhance inclusive and sustainable urbanization and capacity for participatory, integrated and sustainable human settlement planning and management in all countries
11.4 Strengthen efforts to protect and safeguard the world's cultural and natural heritage
11.5 By 2030, significantly reduce the number of deaths and the number of people affected and substantially decrease the direct economic losses relative to global gross domestic product caused by disasters, including water-related disasters, with a focus on protecting the poor and people in vulnerable situations

11.6 By 2030, reduce the adverse per capita environmental impact of cities, including by paying special attention to air quality and municipal and other waste management

11.7 By 2030, provide universal access to safe, inclusive and accessible, green and public spaces, in particular for women and children, older persons and persons with disabilities

11.a Support positive economic, social and environmental links between urban, peri-urban and rural areas by strengthening national and regional development planning

11.b By 2020, substantially increase the number of cities and human settlements adopting and implementing integrated policies and plans towards inclusion, resource efficiency, mitigation and adaptation to climate change, resilience to disasters, and develop and implement, in line with the Sendai Framework for Disaster Risk Reduction 2015–2030, holistic disaster risk management at all levels

11.c Support least developed countries, including through financial and technical assistance, in building sustainable and resilient buildings utilizing local materials

Goal 12. Ensure sustainable consumption and production patterns

12.1 Implement the 10-year framework of programmes on sustainable consumption and production, all countries taking action, with developed countries taking the lead, taking into account the development and capabilities of developing countries

12.2 By 2030, achieve the sustainable management and efficient use of natural resources

12.3 By 2030, halve per capita global food waste at the retail and consumer levels and reduce food losses along production and supply chains, including post-harvest losses

12.4 By 2020, achieve the environmentally sound management of chemicals and all wastes throughout their life cycle, in accordance with agreed international frameworks, and significantly reduce their release to air, water and soil in order to minimize their adverse impacts on human health and the environment

12.5 By 2030, substantially reduce waste generation through prevention, reduction, recycling and reuse

12.6 Encourage companies, especially large and transnational companies, to adopt sustainable practices and to integrate sustainability information into their reporting cycle

12.7 Promote public procurement practices that are sustainable, in accordance with national policies and priorities

12.8 By 2030, ensure that people everywhere have the relevant information and awareness for sustainable development and lifestyles in harmony with nature

12.a Support developing countries to strengthen their scientific and technological capacity to move towards more sustainable patterns of consumption and production

12.b Develop and implement tools to monitor sustainable development impacts for sustainable tourism that creates jobs and promotes local culture and products

12.c Rationalize inefficient fossil-fuel subsidies that encourage wasteful consumption by removing market distortions, in accordance with national circumstances, including by restructuring taxation and phasing out those harmful subsidies, where they exist, to reflect their environmental impacts, taking fully into account the specific needs and conditions of developing countries and minimizing the possible adverse impacts on their development in a manner that protects the poor and the affected communities

Goal 13. Take urgent action to combat climate change and its impacts[1]

13.1 Strengthen resilience and adaptive capacity to climate-related hazards and natural disasters in all countries

13.2 Integrate climate change measures into national policies, strategies and planning

13.3 Improve education, awareness-raising and human and institutional capacity on climate change mitigation, adaptation, impact reduction and early warning

13.a Implement the commitment undertaken by developed-country parties to the United Nations Framework Convention on Climate Change to a goal of mobilizing jointly $100 billion annually by 2020 from all sources to address the needs of developing countries in the context of meaningful mitigation actions and transparency on implementation and fully operationalize the Green Climate Fund through its capitalization as soon as possible

13.b Promote mechanisms for raising capacity for effective climate change-related planning and management in least developed countries and small island developing States, including focusing on women, youth and local and marginalized communities

Goal 14. Conserve and sustainably use the oceans, seas and marine resources for sustainable development

14.1 By 2025, prevent and significantly reduce marine pollution of all kinds, in particular from land-based activities, including marine debris and nutrient pollution

14.2 By 2020, sustainably manage and protect marine and coastal ecosystems to avoid significant adverse impacts, including by strengthening their resilience,

---

[1]Acknowledging that the United Nations Framework Convention on Climate Change is the primary international, intergovernmental forum for negotiating the global response to climate change.

13.1 Background of Sustainable Development Goals … 239

and take action for their restoration in order to achieve healthy and productive oceans

14.3 Minimize and address the impacts of ocean acidification, including through enhanced scientific cooperation at all levels

14.4 By 2020, effectively regulate harvesting and end overfishing, illegal, unreported and unregulated fishing and destructive fishing practices and implement science-based management plans, in order to restore fish stocks in the shortest time feasible, at least to levels that can produce maximum sustainable yield as determined by their biological characteristics

14.5 By 2020, conserve at least 10 % of coastal and marine areas, consistent with national and international law and based on the best available scientific information

14.6 By 2020, prohibit certain forms of fisheries subsidies which contribute to overcapacity and overfishing, eliminate subsidies that contribute to illegal, unreported and unregulated fishing and refrain from introducing new such subsidies, recognizing that appropriate and effective special and differential treatment for developing and least developed countries should be an integral part of the World Trade Organization fisheries subsidies negotiation

14.7 By 2030, increase the economic benefits to Small Island developing States and least developed countries from the sustainable use of marine resources, including through sustainable management of fisheries, aquaculture and tourism

14.a Increase scientific knowledge, develop research capacity and transfer marine technology, taking into account the Intergovernmental Oceanographic Commission Criteria and Guidelines on the Transfer of Marine Technology, in order to improve ocean health and to enhance the contribution of marine biodiversity to the development of developing countries, in particular small island developing States and least developed countries

14.b Provide access for small-scale artisanal fishers to marine resources and markets

14.c Enhance the conservation and sustainable use of oceans and their resources by implementing international law as reflected in UNCLOS, which provides the legal framework for the conservation and sustainable use of oceans and their resources, as recalled in paragraph 158 of The Future We Want

Goal 15. Protect, restore and promote sustainable use of terrestrial ecosystems, sustainably manage forests, combat desertification, and halt and reverse land degradation and Halt biodiversity loss

15.1 By 2020, ensure the conservation, restoration and sustainable use of terrestrial and inland freshwater ecosystems and their services, in particular forests, wetlands, mountains and drylands, in line with obligations under international agreements

15.2 By 2020, promote the implementation of sustainable management of all types of forests, halt deforestation, restore degraded forests and substantially increase afforestation and reforestation globally

15.3 By 2030, combat desertification, restore degraded land and soil, including land affected by desertification, drought and floods, and strive to achieve a land degradation-neutral world

15.4 By 2030, ensure the conservation of mountain ecosystems, including their biodiversity, in order to enhance their capacity to provide benefits that are essential for sustainable development

15.5 Take urgent and significant action to reduce the degradation of natural habitats, halt the loss of biodiversity and, by 2020, protect and prevent the extinction of threatened species

15.6 Promote fair and equitable sharing of the benefits arising from the utilization of genetic resources and promote appropriate access to such resources, as internationally agreed

15.7 Take urgent action to end poaching and trafficking of protected species of flora and fauna and address both demand and supply of illegal wildlife products

15.8 By 2020, introduce measures to prevent the introduction and significantly reduce the impact of invasive alien species on land and water ecosystems and control or eradicate the priority species

15.9 By 2020, integrate ecosystem and biodiversity values into national and local planning, development processes, poverty reduction strategies and accounts

15.a Mobilize and significantly increase financial resources from all sources to conserve and sustainably use biodiversity and ecosystems

15.b Mobilize significant resources from all sources and at all levels to finance sustainable forest management and provide adequate incentives to developing countries to advance such management, including for conservation and reforestation

15.c Enhance global support for efforts to combat poaching and trafficking of protected species, including by increasing the capacity of local communities to pursue sustainable livelihood opportunities

Goal 16. Promote peaceful and inclusive societies for sustainable development, provide access to justice for all and build effective, accountable and inclusive institutions at all levels

16.1 Significantly reduce all forms of violence and related death rates everywhere

16.2 End abuse, exploitation, trafficking and all forms of violence against and torture of children

16.3 Promote the rule of law at the national and international levels and ensure equal access to justice for all

16.4 By 2030, significantly reduce illicit financial and arms flows, strengthen the recovery and return of stolen assets and combat all forms of organized crime

16.5 Substantially reduce corruption and bribery in all their forms
16.6 Develop effective, accountable and transparent institutions at all levels
17.7 Ensure responsive, inclusive, participatory and representative decision-making at all levels
16.8 Broaden and strengthen the participation of developing countries in the institutions of global governance
16.9 By 2030, provide legal identity for all, including birth registration
16.10 Ensure public access to information and protect fundamental freedoms, in accordance with national legislation and international agreements
16.a Strengthen relevant national institutions, including through international cooperation, for building capacity at all levels, in particular in developing countries, to prevent violence and combat terrorism and crime
16.b Promote and enforce non-discriminatory laws and policies for sustainable development

Goal 17. Strengthen the means of implementation and revitalize the global partnership for sustainable development

Finance

17.1 Strengthen domestic resource mobilization, including through international support to developing countries, to improve domestic capacity for tax and other revenue collection
17.2 Developed countries to implement fully their official development assistance commitments, including the commitment by many developed countries to achieve the target of 0.7 % of ODA/GNI to developing countries and 0.15–0.20 % of ODA/GNI to least developed countries; ODA providers are encouraged to consider setting a target to provide at least 0.20 % of ODA/GNI to least developed countries
17.3 Mobilize additional financial resources for developing countries from multiple sources
17.4 Assist developing countries in attaining long-term debt sustainability through coordinated policies aimed at fostering debt financing, debt relief and debt restructuring, as appropriate, and address the external debt of highly indebted poor countries to reduce debt distress
17.5 Adopt and implement investment promotion regimes for least developed countries

Technology

17.6 Enhance North-South, South-South and triangular regional and international cooperation on and access to science, technology and innovation and enhance knowledge sharing on mutually agreed terms, including through improved coordination among existing mechanisms, in particular at the United Nations level, and through a global technology facilitation mechanism

17.7 Promote the development, transfer, dissemination and diffusion of environmentally sound technologies to developing countries on favourable terms, including on concessional and preferential terms, as mutually agreed

17.8 Fully operationalize the technology bank and science, technology and innovation capacity-building mechanism for least developed countries by 2017 and enhance the use of enabling technology, in particular information and communications technology

Capacity-building

17.9 17.9 Enhance international support for implementing effective and targeted capacity-building in developing countries to support national plans to implement all the sustainable development goals, including through North-South, South-South and triangular cooperation

Trade

17.10 Promote a universal, rules-based, open, non-discriminatory and equitable multilateral trading system under the World Trade Organization, including through the conclusion of negotiations under its Doha Development Agenda

17.11 Significantly increase the exports of developing countries, in particular with a view to doubling the least developed countries' share of global exports by 2020

17.12 Realize timely implementation of duty-free and quota-free market access on a lasting basis for all least developed countries, consistent with World Trade Organization decisions, including by ensuring that preferential rules of origin applicable to imports from least developed countries are transparent and simple, and contribute to facilitating market access

Systemic issues
Policy and institutional coherence

17.13 Enhance global macroeconomic stability, including through policy coordination and policy coherence

17.14 Enhance policy coherence for sustainable development

17.15 Respect each country's policy space and leadership to establish and implement policies for poverty eradication and sustainable development

Multi-stakeholder partnerships

17.16 Enhance the global partnership for sustainable development, complemented by multi-stakeholder partnerships that mobilize and share knowledge, expertise, technology and financial resources, to support the achievement of the sustainable development goals in all countries, in particular developing countries

17.17 Encourage and promote effective public, public-private and civil society partnerships, building on the experience and resourcing strategies of partnerships

Data, monitoring and accountability

17.18 By 2020, enhance capacity-building support to developing countries, including for least developed countries and small island developing States, to increase significantly the availability of high-quality, timely and reliable data disaggregated by income, gender, age, race, ethnicity, migratory status, disability, geographic location and other characteristics relevant in national contexts

17.19 By 2030, build on existing initiatives to develop measurements of progress on sustainable development that complement gross domestic product, and support statistical capacity-building in developing countries

13.2 Issues of Sustainable Population in the Background of Sustainable Development Goals

Sustainable Development Goals aim at ending poverty and hunger, in all their forms and dimensions, and to ensure that all human beings can fulfil their potential in dignity and equality and in a healthy environment. It also aims at protection of the planet from degradation, including through sustainable consumption and production, sustainably managing its natural resources and taking urgent action on climate change, so that it can support the needs of the present and future generations. Further, it aims at ensuring that all human beings can enjoy prosperous and fulfilling lives and that economic, social and technological progress occurs in harmony with nature and every country enjoys sustained, inclusive and sustainable economic growth and decent work for all.

The SDGs are indeed visionary in character and an ideal framework in which all nations need to work out its actionable plans. Because of varying inter-country deficit on actionable areas of sovereign nations, SDGs have not gone into country-specific road-map for action. It is for this reason that SDGs did not mention any constraints in its overall approach to realise its visionary goals.However, constraints do exist in achieving SDGs just as it had existed under MDGs. Two of the top most goals under SDGs is to end poverty in all its forms everywhere (Goal 1) and end hunger, achieve food security and improved nutrition (Goal 2). SDGs also aim at promoting sustained, inclusive and sustainable economic growth, full and productive employment and decent work for all (Goal 8). These need to be ensured to all its citizens today and also to those joining by 2030 irrespective of population load now or then. An under-populated country can proceed to reach these goals easily; a country having stable population or optimum population can also move in that direction rather comfortably. But for a country, like India, where TFR is less than replacement level, where there is alarming high order density of population and high quotient of BPL population, where there exits shocking levels of unemployment, under-employment and adverse nutritional standard and relatively poor HDI, the quantum jump at on go to a SDGs level destination is rather a very difficult task. The huge size population and its incremental addition is responsible for this sordid state of affairs. The impact on the level of consumption of huge size of population and its incremental addition is proportional to use-up of non-replacement level natural stock of resources in the given current technology. It will be a gigantic task to implement the Goal 12 to ensure sustainable consumption and production patterns for a runaway population of more than 1.21 billion population for such over-populated country.

Climate change has added a new dimension to any estimate of sustainable population and the planet's ability to support all 7+ billion of population. Climate scientists foresee lower crop yields of major grains such as wheat, rice, and maize. Rising sea levels may create hundreds of millions of climate refugees in years to come in addition to climate disruption followed by likely increasing levels of resource conflict and civil unrest. Adaptation to climate disruption will only be easier with a much smaller global population.

In fine, the world in general and the individual countries in particular, such as India, need to stabilise population size at a level consistent with sustainable population level in that demand pull forces do not hit upon resources at a rate the Earth of the local area can support. As a signatory to the World summit of sustainable goals, each of such individual countries has to draw up a road-map for arresting and reversing the birth rate, immigration and infiltration. Sustainable population is the only requisite for sustainable development and SDGs.

Part IV
Population Control Initiatives in India

Chapter 14
Health Policy in India

Abstract For the population control programme in India, the National Population Policy of India is destined to play the nodal role. However, since the formal National Population Policycame into being only at 2000, the policy guidelines on population control programme followed the prevailing Health Policy of the country. The first Health Policy of the country was published in 1983 and subsequently revised in 2002. The National Health Policy 2015 (draft) has now been under circulation for comments of all stakeholders for refinement. All these health policies looked upon population stabilisation issues from the prism of health sector angle. Accordingly, the salient features of such health policies have been captured hereunder to indicate how the country has been addressing the population control programme through health policies at the first instance.

Keywords Declaration at Alma Ata in 1978 · Inequities in health outcomes · Population control related issues · Population stabilization · The draft national health policy, 2015 · Gender mainstreaming · Safe abortion services · Reproductive tract illness

A policy is a system of principles to guide decisions and achieve rational outcomes. It also helps to decide on programs or spending priorities, from among different alternatives, on consideration of its likely impact. It is different from law in that while law can compel or prohibit behaviour, policy merely guides actions toward those that are likely to achieve a desired outcome. Policy also underscores a framework of political and administrative sanction to reach explicit goals.

For the population control programme in India, the National Population Policy of India is destined to play the nodal role. However, since the formal National Population Policy came into being only at 2000, the policy guidelines on population control programme followed the prevailing Health Policy of the country. Incidentally, the first Health Policy of the country was published in 1983 and subsequently revised in 2002. The National Health Policy 2015 (draft) has now been under circulation for comments of all stakeholders for refinement. All these health policies looked upon population stabilisation issues from the prism of health sector angle. Accordingly, the salient features of such health policies have been captured hereunder to indicate how the country has been addressing the population control programme through health policies at the first instance.

Health policy is usually defined as the decisions, plans, and actions that are undertaken to achieve specific health care goals within a society. According to the World Health Organization, an explicit health policy can achieve several things: it defines a vision for the future; it outlines priorities and the expected roles of different groups; and it builds consensus and informs people. The National Health Policy is usually a declaration of the determination of the Government of the day to address to the prevailing health challenges. It usually projects an intended health outcomes at an interval of time and charts out the pathway for infrastructural base for its destination goal for better health for all. Incidentally, the Joint WHO and UNICEF International Conference Declaration at Alma Ata in 1978 called on all governments to formulate national health policies according to their own circumstances to launch and sustain primary health care as part of their national health system.

14.1 National Health Policy, 1983

With this background in view the first National Health Policy (NHP) of the country was adopted in 1983. The NHP, 1983 was an attempt to synthesise recommendations of three important earlier committees, namely, the Bhore Committee of 1946, the Mudaliar Committee of 1962 and the Shrivastav Committee of 1975. The Bhore Committee, 1946, set up before India's independence, concentrated on preventive medicine and tried to link health with social justice. The Mudaliar Committee (1962) concentrated on medical policy on prevailing education and development of training infrastructure for static medical units. The Shrivastav Committee (1975) urged the training of a cadre of health assistants to serve as links between qualified medical practitioners and multipurpose workers.

The demographic scenario is usually kept in the upfront in any health policy for its appraisal later on. Demographic scenario of India, as revealed in the Census, 1981 has thus been kept here under to understand the extent of importance reflected in the Health Policy, 1983 (Table 14.1).

The salient features of the National Health Policy of 1983 were that it stressed the need for providing primary health care with special emphasis on prevention, promotion and rehabilitation. It mentioned time bound attention on nutrition, prevention of food adulteration, maintenance of quality drugs, improvement of water supply and sanitation, environmental protection, immunisation of children, maternal and child health services, school health programme, occupational health services and health care information services.

In the following Table 14.2 the Goals, Targets and Achievements of NHP 1983 have been shown.

14.1 National Health Policy, 1983

Table 14.1 Demographic scenario in 1981

Items as per census, 1981	India
Population as per census, 1981	683,329,097
Incremental population as in census, 1981	35,169,345
Decennial population growth as in census, 1981	24.7
Density of population as in census, 1981	216
Crude birth rate	33.9[a]
Infant mortality rate (SRS)	110

[a]*Source* National Commission on Population, Ministry of Health and Family Welfare, Government of India, May 2007

Table 14.2 Goals, targets and achievements of NHP 1983

Indicator	Goal by 2000	Achievement by 2000
IMR	60	70
PNMR	33	46
CDR	9	8.7
MMR	2	4
UFMR	10	9.4
Life expectancy Male Female	64 64	62.4 63.4
LBW	10 %	26 %
CBR	21	26.1
CPR	60 %	46.2 %
NRR	1	1.45
Growth rate	1.2	1.93
Family size	2.3	3.1

Source National Health Policy–Dr. J.P. Majra, Associate Professor, Department of Community Medicine, K.S. Hedge Medical Academy; website-similima.com

The results of the 1983 policy were mixed. There had taken place positive changes in incidence of diseases such as polio, malaria and leprosy. There had also taken place decreases in the Crude Birth Rate and Infant Mortality Rate. The infrastructural base in the form of more beds, centres and supporting manpower registered significantly. The NHP 1983, however, did not address health issues such as HIV/AIDS or life style diseases such as diabetes, cancer and cardiovascular diseases and also macro and micro-nutrient deficiencies, especially in women and children.

The National Health Policy was revised and published in 2002. Before capturing the salient features of National Health Policy, 2002, it would again be desirable to keep the highlights of the Census, 2001 in the upfront to understand the extent of importance reflected in the Health Policy, 2002 (Table 14.3).

Table 14.3 Demographic scenario in 2001

Items as per census, 2001	India
Population, 2001	1,028,737,436
Incremental population in census, 2001	182,316,397
Decennial population growth in census 2001	21.5
Density of population in census, 2001	325
Crude birth rate	25.4[a]
Infant mortality rate (SRS 2000)	66

[a]*Source* National Commission on Population, Ministry of Health and Family Welfare, Government of India, May 2007

The Ministry of Health and Family Welfare introduced the new health policy in 2002 and took care of new challenges and the new health goals.

14.2 National Health Policy, 2002

Based on the demographic and health scenario of the country, the Government of India formulates the National Health Policy of the country to address the then prevailing circumstances in the health sector and revises it with the onset of new situation and new challenges. Incidentally, the National Population Policy was formulated in 2000 and with the publication of the national health policy, 2002, it was intended that two policies would be very decisive to improve health standard in the country and neutralize the rapid growth of population. The cherished policy goal was based on the premise that population stabilization measures and general health initiatives, when effectively synchronized, will lead to maximize socio-economic wellbeing of the people. Though the two policies are separately formulated, the common points of both of them aim at prevention and control of communicable diseases, containment of HIV/AIDs infection, and universal immunization of children against all major preventable diseases, addressing the unmet needs for basic and reproductive health services and strengthening of infrastructures. The National Population Policy is being separately discussed in Chap. 15 while the objectives of the National Health Policy, 2002 are being discussed hereunder.

14.2.1 Objectives of the National Health Policy, 2002

The main objective of the Health Policy, 2002, as per the document of the government, is to achieve an acceptable standard of good health amongst the general population of the country. The approach would be to increase access to the decentralised public health system by establishing new infrastructure in deficit areas, and by upgrading existing institutions. Overriding importance would be

14.2 National Health Policy, 2002

given to ensuring a more equitable access to health services across the social and geographical expanse of the country. Emphasis will be given to increasing the aggregate public health investment through a substantially increased contribution by the Central Government. It is expected that this initiative will strengthen the capacity of public health administration at the State level to render effective service delivery. The contribution of the private sector in providing health services would be much enhanced, particularly for the population group which can afford to pay services. Primacy will be given to preventive and first line curative initiatives at the primary health level through increased share of sectoral allocation. Emphasis will be laid on rational use of drugs within the allopathic system. Increased access to tried and tested system of traditional medicine will be ensured. Within the broad objectives, NHP, 2002 will endeavour to achieve the time bound Goals as mentioned in the Box below:Goals to be achieved by 2002–2015

Eradicate polio and yaws	2005
Eliminate leprosy	2005
Eliminate kala azar	2010
Eliminate lymphatic filariasis	2015
Achieve zero level of growth of HIV/AIDS	2007
Reduce mortality by 50 % on account of TB, malaria and other vector and waterborne diseases	2010
Reduce prevalence of blindness to 0.5 %	2010
Reduce IMR to 30/1000 AND MMR to 100/Lakh	2010
Increase utilization of public health facilities from current Level of <20.0 >75	2010
Establish an integrated system of surveillance, National Health Accounts and Health Statistics	2005
Increase health expenditure by Government as a % of GDP from existing 0.9 to 20.0 %	2010
Increase share of Capital Grants to constitute at least 25 % of total health spending	2010
Increase State sector health spending from 5.5 to 7 % of the budget	2005
Further increase to 8 %	2010

Source National Health Policy, 2002

With the passage of thirteenth years of functional role of the National Health Policy, 2002, there has emerged contextual changes on health priorities necessitating new policy initiative. The new priorities, among other things, relate to the need to scale-up efforts to attain Millennium Development Goals with respect to maternal and child mortality. The country bears the ignominy of having 4 % of all female deaths in the age group of 15–49 year age group. Added to this is the inability of the prevailing health system to respond to prevention or access to treatment of a host of infectious diseases. The growing burden of non-communicable diseases is also significant. The second important change in the context is the robust growth of private health care industry much above the national growth rate. The incidence of high cost of health care services is an alarming third contextual factor linked as it is with major contributor to rural poverty. The fourth

Table 14.4 Demographic scenario in 2011

Items as per census, 2011	India
Population, 2011	1,21,05,69,573
Incremental population in census, 2011	18,19,59,458
Decennial population growth in census 2011	17.7
Density of population in census, 2011	382
Crude birth rate, 2012	21.6
Infant mortality rate (SRS 20)	44

and final contextual change is the enhanced fiscal capacity of the country to respond to this change.

The Government of India is now in the process of finalising the National Health Policy, 2015 and has accordingly put up a draft policy in the website for appropriate comments of the stake holders. The said draft policy has been captured hereunder. Here again the salient features of the Census, 2011 have been kept in the upfront (Table 14.4) to understand the extent of importance reflected in the Health Policy, 2015.

14.3 Draft National Health Policy 2015

The National Health Policy 2015, a draft in website circulation, seeks to address the urgent need to improve the performance of health systems in the country to move towards universal health coverage. It is a declaration of the determination of the Government to leverage economic growth to achieve health outcomes and an explicit acknowledgement that better health contributes immensely to improved productivity as well as to equity.

Earlier, the National Health Policy of 1983 and the National Health Policy of 2002 had served, in providing policy guidance and direction to formulate programmes for the health sector during the related Five-Year Plans. After 13 years of the last health policy, the context of the health scenario has changed in four major ways. Firstly, the Health Priorities have to conform to attain goals as set for the country under Millennium Development Goals with respect to maternal and child mortality. Maternal mortality now accounts for 0.55 % of all deaths and 4 % of all female deaths in the 15–49 year age group. The figures of maternal deaths are really very shocking and demands appropriate policy commitments to reduce the same. Additionally, the prevailing system has failed to respond to for many other health needs and infectious diseases, either in terms of prevention or access to treatment. There exists also a growing burden of non-communicable disease. The second important change in the context scenario is the emergence of a robust health care industry which grows at 15 % compound annual growth rate (CAGR) registering twice the rate of growth in all services and thrice the national economic growth rate. Thirdly, incidence of catastrophic expenditure due to health care costs is growing

14.3 Draft National Health Policy 2015

and is now being estimated to be one of the major contributors to poverty. The drain on family incomes due to health care costs can neutralize the gains of income increases through poverty reduction schemes. The fourth and final change in context is that economic growth has increased the fiscal capacity available. Therefore, the country needs a new health policy that is responsive to these contextual changes.

The primary aim of the National Health Policy, 2015, is to inform, clarify, strengthen and prioritize the role of the Government in shaping health systems in all its dimensions—investment in health, organization and financing of healthcare services, prevention of diseases and promotion of good health through cross sectoral action, access to technologies, developing human resources, encouraging medical pluralism, building the knowledge base required for better health, financial protection strategies and regulation and legislation for health.

The National Health Policy, 2015 has been drafted with situational analysis on 18 number of pressing items. However, three items having direct links with population control programme have been covered, as mentioned below:

- Achievement of Millennium Development Goals:
  India is set to reach the Millennium Development Goals (MDG) with respect to maternal and child survival. The MDG target for Maternal Mortality Ratio (MMR) is 140 per 100,000 live births. From a baseline of 560 in 1990, the nation had achieved 178 by 2010–12, and at this rate of decline is estimated to reach an MMR of 141 by 2015. In the case of under-5 mortality rate (U5MR), the MDG target is 42. From a baseline of 126 in 1990, in 2012 the nation has an U5MR of 52 and an extrapolation of this rate would bring it to 42 by 2015. However, in some states there is stagnation on these two indicators.
- Achievements in Population Stabilization:
  India has also shown consistent improvement in population stabilization, with a decrease in decadal growth rates, both as a percentage and in absolute numbers. Twelve of the 21 large States have achieved a TFR of at or below the replacement rate of 2.1 and three are likely to reach this soon. The challenge is now in the remaining six states of Bihar, Uttar Pradesh, Rajasthan, Madhya Pradesh, Jharkhand and Chhattisgarh but even here rates are declining. However these six States between them account for 42 % of the national population and 56 % of the annual population increase. In the remaining small States and Union Territories except Meghalaya, the Crude Birth Rate (CBR), is less than 21 per 1000. The national TFR has declined from 2.9 to 2.4. The persistent challenge on this front is the declining sex ratio.
- Inequities in Health Outcomes:
  Given these achievements, the draft Health Policy 2015 also takes note of high degree of health inequity in health outcomes and access to health care services as evidenced by indicators disaggregated for vulnerable groups. There are also urban-rural inequities and there are inequities across states. A number of districts, many in tribal areas, perform poorly even in states where overall averages are better and improving. Marginalized communities and poorer economic

Table 14.5 Disparities in health outcomes

Indicator	India			
	Total	Rural	Urban	% Differential
TFR (2012)	2.4	2.6	1.8	44 % difference
IMR (2012)	40	44	27	63 % difference
Indicator	States with good performances			States with great challenges
TFR (2012)	HP (1.7), Punjab (1.7), Tamil Nadu (1.7) and West Bengal (1.7)			Bihar (3.5), UP (3.3), Rajasthan (2.9), MP (2.9)
IMR (2010)	Kerala(12), Tamil Nadu (21), Delhi (24), Maharashtra (24)			Madhya Pradesh (54), Assam (54), Orissa (51), Rajasthan (47)
MMR (2010–12)	Kerala (66), Maharashtra (87), Tamil Nadu (90), Andhra Pradesh (110)			Assam (328), Uttar Pradesh/ Uttarakhand (292), Rajasthan (255), Odisha (235)

Source Draft National Health Policy, 2015

quintiles of the population continue to fare poorly. Outreach and service delivery for the urban poor, even for immunization services has been inadequate, as shown in Table 14.5.

The policy directions in the Draft National Health Policy, 2015 have been outlined at Chap. 4, in great details out of which those covering with population control related issues are captured hereunder.

"Population Stabilization including maintaining a gender balance has been and will continue to be one of the main components of national health policy. The changed situation however is that 21 States have already achieved replacement levels of fertility rates and in these states the strategic objectives now are better and safer contraceptive choices, with a further push back in age of marriage and improvement in spacing. In all 36 States however the fertility rates are declining rapidly and with improving levels of women's education, the demand for contraceptive services is established. Though declining, fertility rates continue to be unsustainably high in as many as nine States which account for over 35 % of the population. Here the policy imperative is to move away from camp based services with all its attendant problems of quality, safety and dignity of women, to a situation where these services are available on any day of the week or at least on a fixed day. Other policy imperatives are to increase the proportion of male sterilization from less than 5 % where it is currently, to at least 30 % and if possible much higher. The National Health Policy is explicit that coercive methods are not justified nor even effective to meet the goals. Improved access, education and empowerment would be the basis of successful population stabilization."

Further, the Draft National Policy also mentions about Women' Health and Gender Mainstreaming where it reiterates that Women's health issues and concerns go far beyond maternal health and the ability of the health sector has to address these issues and strengthened. Despite the introduction of new technologies, access to safe abortion services and for reproductive tract illness remains a major gap that must be seriously addressed.

Chapter 15
Population Policy in India

Abstract India was the first country in the world to launch the first national programme on family planning in 1952 emphasizing for reduction of birth rates 'to stabilize the population at a level consistent with the requirement of national economy'. Thereafter, the first statement on National Population Policy was made in 1976 while in 1977, a statement on Family Welfare Programmes was announced. Both the statements were laid on the Table of the House in Parliament, but were never discussed or adopted. The 1976 National Population Policy statement, however, could remain effective for a short period of just about one year only. The new Government that came in the office after the 1977 general elections, discarded major provisions made in the 1976 policy statement. The National Population Policy, 2000 is in that sense first approved population policy of the country so far, which has been comprehensively covered here.

Keywords National socio-demographic goals · Small family norm · Unmet needs for family welfare services · Under-served population · Increased participation of men · Social marketing schemes · Child marriage restraint act, 1976 · Pre-natal diagnostic techniques act, 1994

Population policy generally determines the principles, objectives and policies adopted by the State on identified population issues for the purpose of influencing the population growth status including all connected variables such as fertility, births, deaths, geographical distribution, immigration, population composition of sex, age structure including the proportion of the elderly along with general issues relating to health and education. Population policy forms a large umbrella policy covering all programs and activities directly and indirectly influencing population variables. Population policy also focuses on society's culture and values.

Population policy in India has a history since the Bhore committee of 1946. The Health Survey and Development Committee popularly referred to as the Bhore committee was constituted for making plans for post-war developments in the health fields and also for a comprehensive review of the field of population from the quantitative and qualitative points of view. The Bhore committee recommended that "The Population Problem should be the subject of continuous study". Accordingly the Census Act came into force in 1948 (Act No. 37 of 1948). The censuses of post independence era were conducted as per the provisions of this Act.

India was the first country in the world to launch the first national programme on family planning in 1952 emphasizing for reduction of birth rates 'to stabilize the population at a level consistent with the requirement of national economy'. Thereafter, the first statement on National Population Policy was made in 1976 while in 1977, a statement on Family Welfare Programmes was announced. Both the statements were laid on the Table of the House in Parliament, but were never discussed or adopted. The 1976 National Population Policy statement, however, could remain effective for a short period of just about one year only. The new Government that came in the office after the 1977 general elections, discarded major provisions made in the 1976 policy statement. The National Population Policy, 2000 is in that sense first approved population policy of the country so far. However, for chronological reason, it would be desirable to capture the salient features of the NPP statement, 1976 which were as follows:

(i) To raise the age of marriage to 21 for boys and 18 for girls;
(ii) To raise the monetary compensation for individual acceptance of family planning to Rs. 150 for sterilization with two living children, Rs. 100 with three children and Rs. 70 with four and more living children;
(iii) To introduce group incentives for the involvement of teaching and medical profession, Zila Parishads, Panchayat Samities, cooperative societies and also labour in the organised sector through their respective representative to national organization;
(iv) implementation of New multi-media national strategy for availing all media-newspapers, radio, TV, films etc., to move urban approach to rural oriented approach and to spread the knowledge of family planning and family limitation;
(v) To adopt small family norms for the employees of the Union Government and necessary changes to be made in their service conduct rules. The states may also follow the Central model;
(vi) Introduction of special measures to raise the level of female education in all states; and
(vii) Freezing the population base at the 1971 level for 25 years to determine central plan allocations to states and their representation in the Lok Sabha and earmark 8 % of Central Assistance to State Plans on the performance of family planning.

1976 National Population Policy statement, as mentioned earlier, could remain effective for a short period of just about one year only. The new Government that came in the office after the 1977 general elections, discarded major provisions made in the 1976 policy statement. In 1983, when the National Health Policy (which stated that replacement levels of total fertility rate 2 (TFR) should be achieved by the year 2000) was discussed and adopted in Parliament, the need for a separate

Population Policy for securing the small family norm through voluntary efforts together with the goal of population stabilization was commented upon. However, the formulation of National Policy on Population by itself was not an easy task in India linked as it is with diverse socio-religious fabric of our country. National Development Committee later seized on the matter and appointed an expert committee under the Chairmanship of Shri Karunakaran in 1991 to report on all possible issues connected for framing a policy on population. The said Report, known as the NDC Report on population, proposed the formulation of a National Population Policy to take a 'long term holistic view of development, population growth and environmental protection' and to suggest policies and guidelines (for) formulation of programmes" and 'a monitoring mechanism with short, medium and long term perspectives and growth'. It was also argued that the earlier policy statements of 1976 and 1977 were placed on the table of the Parliament but were never really discussed or adopted them. Specifically, it was recommended that 'a National Policy of Population should be formulated by the Government and adopted by Parliament'.

15.1 The National Population Policy, 2000

The Government of India appointed an Expert Group headed by Dr M.S. Swaminathan to prepare a draft of a population policy in 1993. The said draft, with a process of consultation with the state governments and interactions with other stake holders, and after securing the cabinet approval, shaped into the Population Policy of India in 2000.

The National Population Policy, 2000 begins with an introduction which reads as follows:

"The overriding objective of economic and social development is to improve the quality of lives that people lead, to enhance their well-being, and to provide them with opportunities and choices to become productive assets in society.

In 1952, India was the first country in the world to launch a national programme, emphasizing family planning to the extent necessary for reducing birth rates "to stabilize the population at a level consistent with the requirement of national economy"1. After 1952, sharp declines in death rates were, however, not accompanied by a similar drop in birth rates. The National Health Policy, 1983 stated that replacement levels of total fertility rate 2 (TFR) should be achieved by the year 2000.

On 11 May, 2000 India is projected to have 1 billion 3 (100 crore) people, i.e. 16 % of the world's population on 2.4 % of the globe's land area. If current trends continue, India may overtake China in 2045, to become the most populous country in the world. While global population has increased threefold during this century, from 2 to 6 billion, the population of India has increased nearly five times from 238 million (23 crore) to 1 billion in the same period. India's current annual increase in population of 15.5 million is large enough to neutralize efforts to conserve the resource endowment and environment".

Before addressing the contents of the National Population Policy, 2000, India's demographic achievements half a century after formulating the national family welfare programme, India were kept in the forefront of the Policy, namely,

- reduced crude birth rate (CBR) from 40.8 (1951) to 26.4 (1998, SRS);
- halved the infant mortality rate (IMR) from 146 per 1000 live births (1951) to 72 per 1000 live births (1998, SRS);
- quadrupled the couple protection rate (CPR) from 10.4 (1971) to 44 % (1999);
- reduced crude death rate (CDR) from 25 (1951) to 9.0 (1998, SRS);
- added 25 years to life expectancy from 37 to 62 years;
- achieved nearly universal awareness of the need for and methods of family planning, and
- reduced total fertility rate from 6.0 (1951) to 3.3 (1997, SRS).

The NPP, 2000 also began the policy statement with population projection which runs as follows:

India's population in 1991 and projections to 2016 are as follows in Table 15.1.

In this introductory part, the NPP Policy document captures a prelude statement on Policy as well in that stabilising population is an essential requirement for promoting sustainable development with more equitable distribution. However, it is as much a function of making reproductive health care accessible and affordable for all, as of increasing the provision and outreach of primary and secondary education, extending basic amenities including sanitation, safe drinking water and housing, besides empowering women and enhancing their employment opportunities, and providing transport and communications.

The National Population Policy, 2000 (NPP 2000) affirms the commitment of government towards voluntary and informed choice and consent of citizens while availing of reproductive health care services, and continuation of the target free approach in administering family planning services. The NPP 2000 provides a policy framework for advancing goals and prioritizing strategies during the next decade, to meet the reproductive and child health needs of the people of India, and to achieve net replacement levels (TFR) by 2010. It is based upon the need to simultaneously address issues of child survival, maternal health, and contraception, while increasing outreach and coverage of a comprehensive package of reproductive and child heath services by government, industry and the voluntary non-government sector, working in partnership.

Table 15.1 Population projections for India (million)	March 1991	March 2001	March 2011	March 2016
	846.3	1012.4	1178.9	1263.5

15.1.1 Objectives of the Population Policy, 2000

The immediate objective of the NPP 2000 is to address the unmet needs for contraception, health care infrastructure, and health personnel, and to provide integrated service delivery for basic reproductive and child health care. The medium term objective is to bring the TFR to replacement levels by 2010 through vigorous implementation of inter-sectoral operational strategies. The long term objective is to achieve a stable population by 2045, at a level consistent with the requirements of sustainable economic growth, social development, and environmental protection.

In pursuance of these objectives, focused National Socio-economic Goals to be achieved in each by 2010 were also set, as below:

National Socio-Demographic Goals for 2010

(1) Address the unmet needs for basic reproductive and child health services, supplies and infrastructure
(2) Make school education up to age 14 free and compulsory, and reduce drop outs at primary and secondary school levels to below 20 % for boys and girls.
(3) Reduce infant mortality rate to below 1000 live births
(4) Reduce maternal mortality ratio to below 100 per 100,000 live births
(5) Achieve universal immunization of children against all vaccine preventable diseases
(6) Promote delayed marriage for girls, not earlier than age 18 and preferably after 20 years of age.
(7) Achieve 80 % institutional deliveries and 100 % deliveries by trained persons
(8) Achieve universal access to information/counselling, and services for fertility regulation and contraception with a wide basket of choices
(9) Achieve 100 % registration of births, deaths, marriage and pregnancy
(10) Contain the spread of Acquired Immunodeficiency Syndrome (AIDS), and promote greater integration between the management of reproductive tract infections (RTI) and sexually transmitted infections (STI) and the National AIDS Control Organisation.
(11) Prevent and control communicable diseases
(12) Integrate Indian Systems of Medicine (ISM) in the provision of reproductive and child health services, and in reaching out to households.
(13) Promote vigorously the small family norm to achieve replacement levels of TFR
(14) Bring about convergence in implementation of related social sector programs so that family welfare becomes a people centred programme.

The NPP 2000 also contained a projection of population growth and stated that if the NPP 2000 is fully implemented, it is anticipated to have the size of population of 1107 million (110 crores) in 2010 instead of 1162 million (116 crores) and projected birth, infant mortality and total fertility rate, as indicated in Tables 15.2 and 15.3.

Table 15.2 Anticipated growth in population (million)

Year	If current trends continue		If TFR 2.1 is achieved by 2010	
	Total population	Increase in population	Total population	Increase in population
1991	846.3	–	846.3	–
1996	934.2	17.6	934.2	17.6
1997	949.9	15.7	949.0	14.8
2000	996.9	15.7	991.0	14.0
2002	1027.6	15.4	1013.0	11.0
2010	1162.3	16.8	1107.0	11.75

Table 15.3 Projection of crude birth rate, infant mortality rate and TFR, if the NPP 2000 is fully implemented

Year	Crude birth rate	Infant mortality rate	Total fertility rate
1997	27.2	71	3.3
1998	26.4	72	3.3
2002	23.0	50	2.6
2010	21.0	30	2.1

The NPP 2000 thereafter makes diagnostic observations on alarming population situation in India and states that population growth in India continues to be high on account of the following factors:

'The large size of the population in the reproductive age-group (estimated contribution 58 %). An addition of 417.2 million between 1991 and 2016 is anticipated despite substantial reductions in family size in several states, including those which have already achieved replacement levels of TFR. This momentum of increase in population will continue for some more years because high TFRs in the past have resulted in a large proportion of the population being currently in their reproductive years. It is imperative that the reproductive age group adopts without further delay or exception the "small family norm", for the reason that about 45 % of population increase is contributed by births above two children per family.

Higher fertility due to unmet need for contraception (estimated contribution 20 %). India has 168 million eligible couples, of which just 44 % are currently effectively protected. Urgent steps are currently required to make contraception more widely available, accessible, and affordable. Around 74 % of the population lives in rural areas, in about 5.5 lakh villages, many with poor communications and transport. Reproductive health and basic health infrastructure and services often do not reach the villages, and, accordingly, vast numbers of people cannot avail of these services.

High wanted fertility due to the high infant mortality rate (IMR) (estimated contribution about 20 %). Repeated child births are seen as an insurance against multiple infant (and child) deaths and accordingly, high infant mortality stymies all efforts at reducing TFR.

Over 50 % of girls marry below the age of 18, the minimum legal age of marriage, resulting in a typical reproductive pattern of "too early, too frequent, too many". Around 33 % births occur at intervals of less than 24 months, which also results in high IMR.'

The NPP, 2000 then states that 'We identify 12 strategic themes which must be simultaneously pursued in "standalone" or inter-sectoral programmes in order to achieve the National Socio-Demographic Goals for 2010. These are presented below:

(i) Decentralised Planning and Programme Implementation

The 73rd and 74th Constitutional Amendments Act, 1992, made health, family welfare, and education a responsibility of village panchayats. The panchayati raj institutions are an important means of furthering decentralised Planning and programme implementation in the context of the NPP 2000. However, in order to realize their potential, they need strengthening by further delegation of administrative and financial powers, including powers of resource mobilization. Further, since 33 % of elected panchayat seats are reserved for women, representative committees of the panchayats (headed by an elected woman panchayat member) should be formed to promote a gender sensitive, multi-sectoral agenda for population stabilisation that will "think, plan and act locally, and support nationally". These committees may identify area specific unmet needs for reproductive health services, and prepare need-based, demand driven, socio-demographic plans at the village level, aimed at identifying and providing responsive, people-centred and integrated, basic reproductive and child health care. panchayats demonstrating exemplary performance in the compulsory registration of births, deaths, marriages, and pregnancies, universalizing the small family norm, increasing safe deliveries, bringing about reductions in infant and maternal mortality, and promoting compulsory education up to age 14, will be nationally recognized and honoured.

(ii) Convergence of Service Delivery at Village Levels

Efforts at population stabilisation will be effective only if we direct an integrated package of essential services at village and household levels. Below district levels, current health infrastructure includes 2500 community health centres, 25,000 primary health centres (each covering a population of 30,000), and 1.36 lakh sub-centres (each covering a population of 5000 in the plains and 3000 in hilly regions) 4. Inadequacies in the existing health infrastructure have led to an unmet need of 28 % for contraception services, and obvious gaps in coverage and outreach. Health care centres are over-burdened and struggle to provide services with limited personnel and equipment. Absence of supportive supervision, lack of training in inter-personal communication, and lack of motivation to work in rural areas, together impede citizens' access to reproductive and child health services, and contribute to poor quality of services and an apparent insensitivity to client's needs. The last 50 years have demonstrated the unsuitability of these yardsticks for provision of health care infrastructure, particularly for remote, inaccessible, or sparsely populated regions in the country like hilly and forested areas, desert regions and tribal areas. We need to promote a more flexible approach, by extending basic reproductive and child health care through mobile clinics and counselling services. Further, recognizing that government alone cannot make up for the inadequacies in health care infrastructure and services, in order to resolve unmet needs and extend coverage, the involvement of the voluntary sector and the non-government sector in partnership with the government is essential.

Since the management, funding, and implementation of health and education programmes has been decentralised to panchayats, in order to reach household levels, a one-stop, integrated and coordinated service delivery should be provided at village levels, for basic reproductive and child health services. A vast increase in the number of trained birth attendants, at least two per village, is necessary to universalise coverage and outreach of ante-natal, natal and post-natal health care. An equipped maternity hut in each village should be set up to serve as a delivery room, with functioning midwifery kits, basic medication for essential obstetric aid, and indigenous medicines and supplies for maternal and new born care. A key feature of the integrated service delivery will be the registration at village levels, of births, deaths, marriage, and pregnancies. Each village should maintain a list of community midwives and trained birth attendants, village health guides, panchayatsewasahayaks, primary school teachers and a anganwadi workers who may be entrusted with various responsibilities in the implementation of integrated service delivery.

The panchayats should seek the help of community opinion makers to communicate the benefits of smaller, healthier families, the significance of educating girls, and promoting female participation in paid employment. They should also involve civil society in monitoring the availability, accessibility and affordability of services and supplies.

(iii) Empowering Women for Improved Health and Nutrition

The complex socio-cultural determinants of women's health and nutrition have cumulative effects over a lifetime. Discriminatory childcare leads to malnutrition and impaired physical development of the girl child. Under-nutrition and micronutrient deficiency in early adolescence goes beyond mere food entitlements to those nutrition related capabilities that become crucial to a woman's well-being, and through her, to the well-being of children. The positive effect of good health and nutrition on the labour productivity of the poor is well documented. To the extent that women are over-represented among the poor, interventions for improving women's health and nutrition are critical for poverty reduction.

Impaired health and nutrition is compounded by early childbearing and consequent risk of serious pregnancy related complications. Women's risk of premature death and disability is highest during their reproductive years. Malnutrition, frequent pregnancies, unsafe abortions, RTI and STI, all combine to keep the maternal mortality ratio in India among the highest globally.

Maternal mortality is not merely a health disadvantage; it is a matter of social injustice. Low social and economic status of girls and women limits their access to education, good nutrition, as well as money to pay for health care and family planning services.

The extent of maternal mortality is an indicator of disparity and inequity in access to appropriate health care and nutrition services throughout a lifetime, and particularly during pregnancy and child-birth, and is a crucial factor contributing to high maternal mortality.

Programmes for Safe Motherhood, Universal Immunisation, Child Survival and Oral Rehydration have been combined into an Integrated Reproductive and Child Health Programme, which also includes promoting management of STIs and RTIs. Women's health and nutrition problems can be largely prevented or mitigated through low cost interventions designed for low income settings.

The voluntary non-government sector and the private corporate sector should actively collaborate with the community and government through specific commitments in the areas of basic reproductive and child health care, basic education, and in securing higher levels of participation in the paid work force for women.

(iv) Child Health and Survival

Infant mortality is a sensitive indicator of human development. High mortality and morbidity among infants and children below 5 years occurs on account of inadequate care, asphyxia during birth, premature birth, low birth weight, acute respiratory infections, diarrhoea, vaccine preventable diseases, malnutrition and deficiencies of nutrients, including Vitamin A. Infant mortality rates have not significantly declined in recent years.

Our priority is to intensify neo-natal care. A National Technical Committee should be set up, consisting principally of consultants in obstetrics, paediatrics (neonatologists), family health, medical research and statistics from among academia, public health professionals, clinical practitioners and government. Its terms of reference should include prescribing perinatal audit norms, developing quality improvement activities with monitoring schedules and suggestions for facilitating provision of continuing medical and nursing education to all perinatal health care providers. Implementation at the grass-roots must benefit from current developments in the fields of perinatology and neonatology. The baby friendly hospital initiative (BFHI) should be extended to all hospitals and clinics, up to sub-centre levels. Additionally, besides promoting breast-feeding and complementary feeds, the BFHI should include updating of skills of trained birth attendants to improve new born care practices to reduce the risks of hypothermia and infection. Essential equipment for the new born must be provided at sub centre levels.

Child survival interventions i.e. universal immunisation, control of childhood diarrhoeas with oral rehydration therapies, management of acute respiratory infections, and massive doses of Vitamin A and food supplements have all helped to reduce infant and child mortality and morbidity. With intensified efforts, the eradication of polio is within reach. However, the decline in standards, outreach and quality of routine immunisation is a matter of concern. Significant improvements need to be made in the quality and coverage of the routine immunisation programme.

(v) Meeting the Unmet Needs for Family Welfare Services

In both rural and urban areas there continue to be unmet needs for contraceptives, supplies and equipment for integrated service delivery, mobility of health providers and patients, and comprehensive information. It is important to strengthen, energise and make accountable the cutting edge of health infrastructure at the village, sub-centre and primary health centre levels, to improve facilities for referral transportation, to encourage and strengthen local initiatives for ambulance services at village and block levels, to increase innovative social marketing schemes for affordable products and services and to improve advocacy in locally relevant and acceptable dialects.

(vi) Under-Served Population Groups

   (a) Urban Slums

Nearly 100 million people live in urban Slums, with little or no access to potable water, sanitation facilities, and health care services. This contributes to high infant and child mortality, which in turn perpetuate high TFR and maternal mortality. Basic and primary health care, including reproductive and child health care, needs to be provided. Coordination with municipal bodies for water, sanitation and waste disposal must be pursued, and targeted information, education and communication campaigns must spread awareness about the secondary and tertiary facilities available.

   (b) Tribal Communities, Hill Areas Population and Displaced and Migrant Population

In general, populations in remote and low density areas do not have adequate access to affordable health care services. Tribal populations often have high levels of morbidity arising from poor nutrition, particularly in situations where they are involuntarily displaced or resettled. Frequently, they have low levels of literacy, coupled with high infant, child, and maternal mortality. They remain under-served in the coverage of reproductive and child health services. These communities need special attention in terms of basic health, and reproductive and child health services. The special needs of tribal groups which need to be addressed include the provision of mobile clinics that will be responsive to seasonal variations in the availability of work and income. Information and counselling on infertility, and regular supply of standardised medication will be included.

   (c) Adolescents

Adolescents represent about a fifth of India's population. Adolescents represent about a fifth of India's population. The needs of adolescents, including protection from unwanted pregnancies and sexually transmitted diseases (STD), have not been specifically addressed in the past. Programmes should encourage delayed marriage and child-bearing, and education of adolescents about the risks of unprotected sex. Reproductive health services for adolescent girls and boys are especially significant in rural India, where adolescent marriage and pregnancy are widely prevalent. Their and sexually transmitted diseases (STD), have not been specifically addressed in the past. Programmes should encourage delayed special requirements comprise information, counselling, population education, and making contraceptive services accessible and affordable, providing food supplements and nutritional services through the ICDS, and enforcing the Child Marriage Restraint Act, 1976.

(d) Increased Participation of Men in Planned Parenthood

In the past, population programmes have tended to exclude menfolk. Gender inequalities in patriarchal societies ensure that men play a critical role in determining the education and employment of family members, age at marriage, besides access to and utilisation of health, nutrition, and family welfare services for women and children. The active involvement of men is called for in planning families, supporting contraceptive use, helping pregnant women stay healthy, arranging skilled care during delivery, avoiding delays in seeking care, helping after the baby is born and, finally, in being a responsible father. In short, the active cooperation and participation of men is vital for ensuring programme acceptance. Further, currently, over 97 % of sterilisations are tubectomies and this manifestation of gender imbalance needs to be corrected. The special needs of men include re-popularising vasectomies, in particular no scalpel vasectomy as a safe and simple procedure, and focusing on men in the information and education campaigns to promote the small family norm.

(vii) Diverse Health Care Providers

Give n the large unmet need for reproductive and child health services, and inadequacies in health care infrastructure it is imperative to increase the numbers and diversify the categories of health care providers. Ways of doing this include accrediting private medical practitioners and assigning them to defined beneficiary groups to provide these services; revival of the system of licensed medical practitioner who, after appropriate certification from the Indian Medical Association (IMA), could provide specified clinical services.

(viii) Collaboration With and Commitments from Non-Government Organisations and the Private Sector

A national effort to reach out to households cannot be sustained by government alone. We need to put in place a partnership of non-government voluntary organizations, the private corporate sector, government and the community. Triggered by rising incomes and institutional finance, private health care has grown significantly, with an impressive pool of expertise and management skills, and currently accounts for nearly 75 % of health care expenditures. However, despite their obvious potential, mobilising the private (profit and non-profit) sector to serve public health goals raises governance issues of contracting, accreditation, regulation, referral, besides the appropriate division of labour between the public and private health providers, all of which need to be addressed carefully. Where government interventions or capacities are insufficient and the participation of the private sector unviable, focused service delivery by NGOs may effectively complement government efforts.

(ix) Mainstreaming Indian Systems of Medicine and Homeopathy

India's community supported ancient but living traditions of indigenous systems of medicine has sustained the population for centuries, with effective cures and remedies for numerous conditions, including those relating to women and children, with minimal side effects. Utilisation of ISMH in basic reproductive and child health care will expand the pool of effective health care providers, optimise utilisation of locally based remedies and cures, and promote low-cost health care. Guidelines need to be evolved to regulate and ensure standardisation, efficacy and safety of ISMH drugs for wider entry into national markets.

Particular challenges include providing appropriate training, and raising awareness and skill development in reproductive and child health care to the institutionally qualified ISMH medical practitioners. The feasibility of utilising their services to fill in gaps in manpower at village levels, and at sub-centres and primary health centres may be explored. ISMH institutions, hospitals and dispensaries may be utilised for reproductive and child health care programmes. At village levels, the services of the ISMH "barefoot doctors", after appropriate training, may be utilised for advocacy and counselling, for distributing supplies and equipment, and as depot holders. ISMH practices may be applied at village maternity huts, and at household levels, for ante-natal, natal and post natal care, and for nurture of the new born.

(x) Contraceptive Technology and Research on Reproductive and Child Health

Government must constantly advance, encourage, and support medical, social science, demographic and behavioural science research on maternal, child and reproductive health care issues. This will improve medical techniques relevant to the country's needs, and strengthen programme and project design and implementation. Consultation and frequent dialogue by Government with the existing network of academic and research institutions in allopathy and ISMH, and with other relevant public and private research institutions engaged in social science, demography and behavioural research must continue. The International Institute of Population Sciences, and the population research centres which have been set up to pursue applied research in population related matters, need to be revitalised and strengthened.

Applied research relies upon constant monitoring of performance at the programme and project levels. The National Health and Family Welfare Survey provides data on key health and family welfare indicators every five years. Data from the first National Family Health Survey (NFHS-1), 1992–93, has been updated by NFHS-2, 1998–99, to be published shortly. Annual data is generated by the Sample

Registration Survey, which, inter alia, maps at state levels the birth, death and infant mortality rates. Absence of regular feedback has been a weakness in the family welfare programme. For this reason, the Department of Family Welfare is strengthening its management information systems (MIS) and has commenced during 1998, a system of ascertaining impacts and outcomes through district surveys and facility surveys. The district surveys cover 50 % districts every year, so that every 2 years there is an update on every district in the country. The facility surveys ascertain the availability of infrastructure and services up to primary health centre level, covering one district per month. The feedback from both these surveys enables remedial action at district and sub-district levels.

(xi) Providing for the Older Population

Improved life expectancy is leading to an increase in the absolute number and proportion of persons aged 60 years and above, and is anticipated to nearly double during 1996–2016, from 62.3 to 112.9 million 5. When viewed in the context of significant weakening of traditional support systems, the elderly are increasingly vulnerable, needing protection and care. Promoting old age health care and support will, over time, also serve to reduce the incentive to have large families. The Ministry of Social Justice and Empowerment has adopted in January 1999 a National Policy on Older Persons. It has become important to build in geriatric health concerns in the population policy. Ways of doing this include sensitising, training and equipping rural and urban health centres and hospitals for providing geriatric health care; encouraging NGOs to design and implement formal and informal schemes that make the elderly economically self-reliant; providing for and routinizing screening for cancer, osteoporosis, and cardiovascular conditions in primary health centres, community health centres, and urban health care centres at primary, secondary and tertiary levels; and exploring tax incentives to encourage grown-up children to look after their aged parents.

(xii) Information, Education and Communication

Information, education and communication (IEC) of family welfare messages must be clear, focused and disseminated everywhere, including the remote corners of the country, and in local dialects. This will ensure that the messages are effectively conveyed. These need to be strengthened and their outreach widened, with locally relevant and locally comprehensible media and messages. On the model of the total literacy campaigns which have successfully mobilised local populations, there is need to undertake a massive national campaign on population related issues, via artists, popular film stars, doctors, vaidyas, hakims, nurses, local midwives, women's organizations, and youth organizations.

15.1.2 Legislation, Public Support and New Structures

- Legislation

As a motivational measure, in order to enable state governments to fearlessly and effectively pursue the agenda for population stabilisation contained in the National Population Policy, 2000, one legislation is considered necessary. It is recommended that the 42nd Constitutional Amendment that freezes till 2001, the number of seats to the Lok Sabha and the Rajya Sabha based on the 1971 Census be extended up to 2026.

- Public Support

Demonstration of strong support to the small family norm, as well as personal example, by political, community, business, professional and religious leaders, media and film stars, sports personalities, and opinion makers, will enhance its acceptance throughout society. The government will actively enlist their support in concrete ways.

- New Structures

The NPP 2000 is to be largely implemented and managed at panchayat and nagarpalika levels, in coordination with the concerned state/Union Territory administrations. Accordingly, the specific situation in each state/UT must be kept in mind. This will require comprehensive and multisectoral coordination of planning and implementation between health and family welfare on the one hand, along with schemes for education, nutrition, women and child development, safe drinking water, sanitation, rural roads, communications, transportation, housing, forestry development, environmental protection, and urban development. Accordingly, the following structures are recommended:

(i) National Commission on Population

A National Commission on Population, presided over by the Prime Minister, will have the Chief Ministers of all states and UTs, and the Central Minister in charge of the Department of Family Welfare and other concerned Central Ministries and Departments, for example Department of Woman and Child Development, Department of Education, Department of Social Justice and Empowerment in the Ministry of HRD, Ministry of Rural Development, Ministry of Environment and Forest, and others as necessary, and reputed demographers, public health professionals, and NGOs as members. This Commission will oversee and review implementation of policy. The Commission Secretariat will be provided by the Department of Family Welfare.

(ii) State/UT Commissions on Population

Each state and UT may consider having a State/UT Commission on Population, presided over by the Chief Minister, on the analogy of the National Commission, to likewise oversee and review implementation of the NPP 2000 in the state/UT.

15.1 The National Population Policy, 2000

(iii) Coordination Cell in the Planning Commission

The Planning Commission will have a Coordination Cell for inter-sectoral coordination between Ministries for enhancing performance, particularly in States/UTs needing special attention on account of adverse demographic and human development indicators.

(iv) Technology Mission in the Department of Family Welfare

To enhance performance, particularly in states with currently below average socio-demographic indices that need focused attention, a Technology Mission in the Department of Family Welfare will be established to provide technology support in respect of design and monitoring of projects and programmes for reproductive and child health, as well as for IEC campaigns.

As to the funding, promotional and motivational measures for adoption of small family norm the NPP, 2000 states as below:

Funding:

The programmes, projects and schemes premised on the goals and objectives of the NPP 2000, and indeed all efforts at population stabilisation, will be adequately funded in view of their critical importance to national development. Preventive and promotive services such as ante-natal and post-natal care for women, immunisation for children, and contraception will continue to be subsidised for all those who need the services. Priority in allocation of funds will be given to improving health care infrastructure at the community and primary health centres, sub-centre and village levels. Critical gaps in manpower will be remedied through redeployment, particularly in under-served and inaccessible areas, and referral linkages will be improved. In order to implement immediately the Action Plan, it would be necessary to double the annual budget of the Department of Family Welfare to enable government to address the shortfall in unmet needs for health care infrastructure, services and supplies (in Appendix IV).

Even though the annual budget for population stabilisation activities assigned to the Department of Family Welfare has increased over the years, at least 50 % of the budgetary outlay is deployed towards non-plan activities (recurring expenditures for maintenance of health care infrastructure in the states and UTs, and towards salaries). To illustrate, of the annual budget of Rs. 2920 crores for 1999–2000, nearly Rs 1500 crores is allocated towards non-plan activities. Only the remaining 50 % becomes available for genuine plan activities, including procurement of supplies and equipment. For these reasons, since 1980 the Department of Family Welfare has been unable to revise norms of operational costs of health infrastructure, which in turn has impacted directly the quality of care and outreach of services provided.

Promotional and Motivational Measures for Adoption of the Small Family Norm:

The following promotional and motivational measures will be undertaken:

(i) Panchayats and ZilaParishads will be rewarded and honoured for exemplary performance in universalising the small family norm, achieving reductions in infant mortality and birth rates, and promoting literacy with completion of primary schooling.

(ii) The Balika Samridhi Yojana run by the Department of Women and Child Development, to promote survival and care of the girl child, will continue. A cash incentive of Rs. 500 is awarded at the birth of the girl child of birth order 1 or 2.

(iii) Maternity Benefit Scheme run by the Department of Rural Development will continue. A cash incentive of Rs. 500 is awarded to mothers who have their first child after 19 years of age, for birth of the first or second child only. Disbursement of the cash award will in future be linked to compliance with ante-natal check-up, institutional delivery by trained birth attendant, and registration of birth and BCG immunisation.

(iv) A Family Welfare-linked Health Insurance Plan will be established. Couples below the poverty line, who undergo sterilisation with not more than two living children, would become eligible (along with children) for health insurance (for hospitalisation) not exceeding Rs. 5000, and a personal accident insurance cover for the spouse undergoing sterilisation.

(v) Couples below the poverty line, who marry after the legal age of marriage, register the marriage, have their first child after the mother reaches the age of 21, accept the small family norm, and adopt a terminal method after the birth of the second child, will be rewarded.

(vi) A revolving fund will be set up for income-generating activities by village-level self help groups, who provide community-level health care services.

(vii) Crèches and child care centres will be opened in rural areas and urban Slums. This will facilitate and promote participation of women in paid employment.

(viii) A wider, affordable choice of contraceptives will be made accessible at diverse delivery points, with counselling services to enable acceptors to exercise voluntary and informed consent.

(ix) Facilities for safe abortion will be strengthened and expanded.

(x) Products and services will be made affordable through innovative social marketing schemes.

(xi) Local entrepreneurs at village levels will be provided soft loans and encouraged to run ambulance services to supplement the existing arrangements for referral transportation.

(xii) Increased vocational training schemes for girls, leading to self-employment will be encouraged.

(xiii) Strict enforcement of Child Marriage Restraint Act, 1976.

(xiv) Strict enforcement of the Pre-natal Diagnostic Techniques Act, 1994.

(xv) Soft loans to ensure mobility of the ANMs will be increased.

(xvi) The 42nd Constitutional Amendment has frozen the number of representatives in the Lok Sabha (on the basis of population) at 1971 Census levels. The freeze is currently valid until 2001, and has served as an incentive for

15.1 The National Population Policy, 2000

State Governments to fearlessly pursue the agenda for population stabilisation. This freeze needs to be extended until 2026.

The NPP, 2000 concludes with observations as below:

In the new millennium, nations are judged by the well-being of their peoples; by levels of health, nutrition and education; by the civil and political liberties enjoyed by their citizens; by the protection guaranteed to children and by provisions made for the vulnerable and the disadvantaged.

The vast numbers of the people of India can be its greatest asset if they are provided with the means to lead healthy and economically productive lives. Population stabilisation is a multisectoral endeavour requiring constant and effective dialogue among a diversity of stakeholders, and coordination at all levels of the government and society. Spread of literacy and education, increasing availability of affordable reproductive and child health services, convergence of service delivery at village levels, participation of women in the paid work force, together with a steady, equitable improvement in family incomes, will facilitate early achievement of the socio-demographic goals. Success will be achieved if the Action Plan contained in the NPP 2000 is pursued as a national movement.

15.2 Outcome of the National Population Policy, 2000

The outcome of the National Population Policy as reflected in the demographic scenario of the country after the publication of the Census Report, 2011 can be seen from the Table 15.4.

The National Population Policy published in February 2000 became practically invalid with the publication of Census Report, 2001. India has already exceeded the estimated population of NPP, 2000 in 2002 by about 14 million. Further, the NPP, 2000 has ceased to have any operational meaning any more with its expiry of policy-life in the year 2010, and more so, after the emergence of new demographic challenges following the Census Report, 2011. While the country is going to have a

Table 15.4 Demographic scenario in 2011

Items as per census, 2011	India
Population, 2011	1,210,569,573
Incremental population in census, 2011	190,376,151
Decennial population growth in census 2011	17.7
Density of population in census, 2011	382
Crude birth rate (SRS 2000) 2011	21.8
Total fertility rate (TFR) 2011	2.4
Infant mortality rate (SRS 2000) 2011	44

Sources Census publication and family planning statistics in India

new Health Policy 2015 shortly, with its component of population stabilisation, it is high time that the country revisits the policy area and formulates a new population policy based on updated vision and pressing need to have sustainable population in the background of SDGs as adopted on 25th September, 2015 by the Sustainable Development Summit.

Chapter 16
Family Welfare Approaches in the Five Year Plans

Abstract The Health Policy and the Population Policy of India overtime indicated the policy framework on family planning in our country. The approach of the government of India on population control, however, was reflected in the documents of the five year plans well before the formal adoption of either the Health Policy or the National Population Policy. India launched the National Family Planning Programme in 1951 with the objective of reducing the birth rate to the extent necessary to stabilize the population at a level consistent with the requirement of the National economy. Later, the National Family Planning programme was renamed as the National Family Welfare Programme recognized as national priority area and has been taken up as a 100% centrally sponsored programme in the five year plans. Subsequently, however, the National Family Welfare Programme has been submerged in the National Rural Health Mission. This evolution of the Family Welfare Programme in India over the five year plans has been captured here.

Keywords First Five Year Plan · Second Five Year Plan · Third Five Year Plan · Fourth Five Year Plan · Fifth Five Year Plan · Sixth Five Year Plan · Seventh Five Year Plan · Eight Five Year Plan · Ninth Five Year Plan · Tenth Five Year Plan · Eleventh Five Year Plan · Twelfth Five Year Plan

The Health Policy and the Population Policy of India overtime have been captured earlier to indicate the policy framework in which family planning or the population control programme in our country was supposed to be guided and directed. The approach of the government of India on population control, however, was reflected in the documents of the five year plans well before the formal adoption of either the Health Policy or the National Population Policy. India launched the National Family Planning Programme in 1951 with the objective of reducing the birth rate to the extent necessary to stabilize the population at a level consistent with the requirement of the National economy. Later, the National Family Planning programme was renamed as the National Family Welfare Programme recognized as national priority area and has been taken up as a 100 % centrally sponsored programme in the five year plans. Subsequently, however, the National Family Welfare Programme has been submerged in the National Rural Health Mission. This evolution of the Family Welfare Programme in India has primarily been sourced from the websites of the five year plans of the Planning Commission:

First Five Year Plan (1951–56):
The approach of the family planning in the first five year plan was based on prime considerations of the health and welfare of the family. Limiting the size of the Family and spacing the birth of the children was thought necessary to secure better health for the mother and better care of children. The document on the First Five Year Plan stated: "It is apparent that population control can be achieved only by the reduction of the birth-rate to the extent necessary to stabilise the population at a level consistent with the requirements of national economy. This can be secured only by the realisation of the need for family limitation on a wide scale by the people. The main appeal for family planning is based on considerations of health and welfare of the family. Family limitation or spacing of the children is necessary and desirable in order to secure better health for the mother and better care and upbringing of children. Measures directed to this end should, therefore, form part of the public health programme".

Creating a sufficiently strong motivation in favour of family planning in the minds of the people was attempted to put in place necessary framework for providing advice and service on acceptable, efficient, harmless and economic methods.

It was felt during that time that a programme like family planning and population control could not be properly implemented unless the Government of the day secures the following:

- obtain an accurate picture of the factors contributing to the rapid population increase in India;
- discover suitable techniques of family planning and devise methods by which knowledge of these techniques can be widely disseminated; and
- make advice, on family planning, an integral part of the service of Government hospitals and public health agencies.

The Plan outlay on Family Planning was Rs. 65 lakhs of the Ministry of Health for the family planning programme during the First Five Year Plan which included, among other things, the following:

(1) Medical officers working at Government hospitals and health centres were to be utilised for advice on methods of family planning including their suggestion for any chemical, mechanical or biological methods of contraception or sterilization as per procedure allowed by the Ministry of Health in U.K. in medical centres maintained by their local authorities.
(2) Field experiments on different methods of family planning for the purpose of determining their suitability, acceptability and effectiveness in different sections of the population were to be carried out.

(3) Efforts were to be made to develop suitable procedures to educate the people on family planning methods for the different economic and social sections of the population.
(4) Information network was to be set up for collecting and studying information about different methods of family planning (based on scientifically tested experience in India and abroad) and making such information available to professional workers.
(5) Research into the physiological and medical aspects of human fertility and its control were also to be taken up.

During the First Five Year Plan, 126 family planning clinics were set-up in urban areas and 21 in rural areas. Assistance in the shape of subsidies or grants was given to the states of India, local authorities, voluntary organisations and scientific institutions, family planning clinics and also to 19 research schemes relating to biological and demographic studies and problems.

Second Five Year Plan (1956–61):
The focus on family planning during the Second Five Year Plan was on regulating India's population from the standpoint of size and quality. Keeping the objectives set out in the first five year plan in mind, it was endeavoured to develop this programme further during the Second Plan.

The family planning programme in the Second Five Year Plan aimed at development on systematic lines for continuous study of population problems and set up an autonomous central board for family planning and population problems with objectives as given below:

- extension of family planning advice and service;
- establishment and maintenance of a sufficient number of centres for the training of personnel;
- development of a broad-based programme of education in family planning, which should include within its scope, sex education, marriage counselling and child guidance;
- research into biological and medical aspects of reproduction and of population problems;
- demographic research, including investigations of motivation in regard to family limitation as well as studies of methods of communication;
- inspection and supervision of the work done by different agencies, governmental and non-governmental, to which grants are made by the Central Board;
- evaluation and reporting of progress; and
- establishment of a well-equipped central organisation.

It also aimed to establish clinics, one for 50,000 populations, in all big cities and major towns. As regards small towns and rural areas, clinics were to be opened gradually in tandem with primary health units. The target for the Second Plan was to set up 300 urban and 2000 rural clinics. These clinics were tasked to create a general awareness of all issues pertaining to family welfare and also to provide services.

In the course of the Second Plan, the number of clinics increased to 549 in urban and 1100 in rural areas. In addition to these clinics, family planning services were provided at 1864 rural and 330 urban medical and health centres. A number of sterilisation centres were also established. The programme was guided by the Central and State Family Planning Boards. All States had set-up special units for family planning work. Considerable amount of research work had been undertaken at the Contraceptive Testing Unit at Bombay and elsewhere under the guidance of the Indian Council of Medical Research and at the All India Institute of Hygiene and Public Health, Calcutta. Demographic research centres were also set-up in Bombay, Calcutta, Delhi and Trivandrum. A number of valuable field investigations had been carried out, such as the India-Harvard-Ludhiana population study and the studies undertaken at Ramanagaram in Mysore, in the Lodi Colony in Delhi, at Najafgarh near Delhi, and at Singur near Calcutta. A broad based training programme had been developed which included centres for training of instructors, a rural training demonstration and experimental centre, development of training clinics into regional training centres, touring training teams and ad hoc training courses. Family planning had also been incorporated in the normal training programme of a number of teaching institutions for doctors and medical auxiliaries.

The Plan outlay for family planning during the Second Five Year Plan was fixed at Rs. 5 crores.

Third Five Year Plan (1961-66):
Apart from following the basic approach towards family planning as outlined in the first plan and followed-up since then in the Second, the Third Plan went a step ahead and incorporated a vision statement in the document's Chapter on Long-term Economic Development stating the necessity of having provisional estimates of increase in population over the next fifteen years and keep it at the very centre of planned development and link it up with programme efforts on stabilising the growth of population over a reasonable period. The Family Planning Programme also needed to get scaled-up status of adequate importance during the Third and subsequent Five Year Plans. This called for intensive education, provision of facilities and service-net work on a larger scale with close association with rural and urban social groups.

The family planning programme in the Third Plan provided for (a) education and motivation for family planning, (b) provision of services, (c) training, (d) supplies, (e) communication and motivation research, (f) demographic research, and (g) medical and biological research.

The programme involved a total plan outlay of Rs. 50 crores.

As per programmes drawn up for the Third Plan, the number of family planning clinics was supposed to increase from about 1800 at the end of the Second Plan to about 8200. Of the latter, about 6100 clinics would be in rural areas and 2100 in urban areas. Distribution of simple contraceptives and general advice were to be entrusted in a much larger measure to voluntary organisations, to para-medical personnel and to dais specially trained in family planning work. For meeting the requisite trained personnel especially women workers, it was emphasized to organise a large number of intensive short-term courses. In the urban areas, it was aimed at greater use of private medical practitioners in providing advice, distributing supplies and, to the extent possible, in undertaking sterilisation.

The need for indigenous manufacture of contraceptives was felt during this period to support large-scale family planning programme. Given the limited in-built capacity of hitherto local suppliers, it was considered necessary for the Government to participate in increasing production, take initiative in prescribing standards and specifications and also determining its prices. The provisions of supplies of contraceptives at free of cost for certain sections of the population, and at subsidised rates for others were also considered necessary. It was desired that detailed plans for the production of contraceptives, both by Government and by private firms, should be drawn up as a matter of high priority, keeping in view the objective that supplies would have to be made available, as rapidly as possible, on the needed scale.

An expanded programme of research undertaken in the Third Plan was the following:

- Development of studies of human genetics.
- Studies in the physiology of reproduction.
- Development of more effective local contraceptives, especially development of a suitable oral contraceptive.
- Follow-up of sterilisation cases, both male and female, to investigate possible after-effects in such cases.

An expert committee on oral contraceptives was also appointed to review periodically the developments in this field and to make recommendations. A committee to guide communication motivation and action-research in family planning was also constituted. Studies of the sociological problems involved in family planning were also undertaken.

During the second plan, infrastructural facilities for sterilisation operations were extended in several states and about 125,000 operations were carried out on the basis of voluntary choice. During the third plan, facilities for sterilisation were extended to district hospitals, sub-divisional hospitals and to such primary health centres as had the necessary facilities for surgical work. The number of sterilisation operations performed in the Third Plan period was 1.33 million.

A provision of Rs. 27 crores was actually made in the Third Plan. The expenditure incurred was Rs. 24.86 crores. During this period, family planning bureaux were organised at the State level and in 199 districts covering all States. At the end of the Third Plan, there were 3676 rural family welfare planning centres, 7081 rural sub-centres and 1381 urban family welfare planning centres. These centres provided supplies, services and advice on family planning. Twenty-eight centres were established for training in which 7641 personnel took regular courses and 34,484 short-term courses. Some progress was made in research, conducted in seven demographic centres and seven communication action research centres. Eight centres conducted studies on bio-medical aspects of family planning. For technical support, a Central Family Planning Institute was established at Delhi.

Fourth Five Year Plan (1969–74):
The Fourth Five Year Plan document accorded highest priority on Family Planning programme to respond to the mood and perception of public mind in general and also for the need of the country's economy in particular. Given the magnitude of the efforts involved for the estimated population of 527 million on October, 1968, restructuring of Family Planning organisation and stepping up of plan outlay assumed appropriate consideration.

The increase in population from 365 million in 1951 to 445 million in 1961 and 527 million in 1968 was the result of a sharp fall in mortality rate without any significant change in the fertility rate. The birth rate remained unchanged around 41 per thousand populations during the greater part of the past two decades up to 1965–66. Surveys of the Register General and the National Sample Survey Organisation indicated that the birth rate had come down to 39 per thousand populations for the country as a whole, the rate being somewhat higher in rural areas. In order to make economic development yield tangible benefits for the ordinary people, the birth rate needed to be brought down substantially as early as possible. It was aimed at reduction of birth rate from 39 per thousand to 25 per thousand population within the next 10–12 years. To address this target in the fourth plan, it was envisaged to draw up programme for creating facilities for the married couple about (i) group acceptance of the small sized family, (ii) personal knowledge about family planning methods; and (iii) ready availability of supplies and services.

The Indian Council of Medical Research approved mass utilisation of intra-uterine contraceptive device commonly known as the loop. Equal emphasis was also placed on sterilisation, condom and intra-uterine contraceptive device (IUCD). The initial response for the loop was found encouraging. A factory for producing IUCD was established at Kanpur with a daily production capacity of 30,009 loops, sufficient to take care of the country's needs. The facilities for IUCD insertions and sterilisations were provided not only free but also with some compensation to the individuals for out-of-pocket expenses, conveyance and loss of wages. These were available at static centres and mobile units. A central family planning corps was created to deploy female doctors in areas where there was a shortage. Of the conventional contraceptives, condoms constituted the most important item. A public sector factory at Trivandrum was set up with an initial capacity for producing 144 million pieces per annum and doubling and quadrupling the production, when necessary.

On the eve of the Fourth Plan, five Central Institutes and 43 State Family Planning Training Centres were in place. 4326 rural family welfare planning centres, 22,826 rural sub-centres and 1797 urban family welfare planning centres were also in operation. The progress in opening sub-centres was not very satisfactory due primarily to shortage of auxiliary nurse-mid-wives and absence of suitable accommodation for female workers in the rural areas. At the beginning of the Fourth Plan, 450 Family Planning annexes to Primary Health Centres were at different stages of construction (90 completed and 360 in progress) and so also of sub-centres (out of 2770 sub-centres taken up for construction, 1280 completed and 1490 were in progress). These incomplete structures were taken up for completion in the Fourth Plan.

Since April 1966 a separate Department of Family Planning has been constituted at the Centre. It co-ordinated family planning programmes at the Centre and in the States.

Family Planning remained as a 100 % Centrally sponsored programme and the entire expenditure were met by the Central Government. The Draft Plan outlay of Rs. 300 cores on Family Welfare were revised upwards subsequently at Rs. 315 crores to strengthen the programme and speed it up.

Fifth Five Year Plan(1974–79):
The Fifth Five Year Plan document, under its Demographic Profile section, mentioned that the National Population Policy lays down a target for birth rate of 25 per thousand and a population growth rate of 1.4 % by the end of the Sixth Plan period. The policy envisaged a series of fundamental measures including raising the minimum age for marriage, female education, spread of population values and the small family norm, strengthening of research in reproductive biology and contraception, incentives for individuals, groups and communities and permitting State Legislatures to enact legislation for compulsory sterilization. The targets laid down in the National Population Policy corresponded to those laid down in the Fifth Plan for achievement by the end of the Sixth Plan. The Fifth Five Year Plan also required that the family planning programmes had to be carried forward in an integrated manner along with Health, Maternity and Child Health Care and Nutrition services on the basis of the convergence strategy.

Two hundred additional post-partum centres were proposed to be opened up. Another unit of the Hindustan Latex Ltd. would have to be set up at Farakka to meet the increased demand of Nirodh. The India Population Project with Sl DA/I DA assistance have to be completed by the end of the Fifth Plan. Special multi-media motivation campaigns on pilot basis would have to be launched in Uttar Pradesh, Andhra Pradesh, and West Bengal. Maternity and child health programmes needed to be vigorously pursued and funds for this purpose were to be made available on the basis of performance. Research and evaluation facilities had to be strengthened. Funds for completion of incomplete buildings and for construction of essential buildings for Rural Family Welfare Planning Centres would need to be provided. 288 New Rural Family Welfare Planning Centres were to be opened in a phased manner.

A shock-jerking event of lasting significance took place during this plan period. During 1975–76 and also in 1976–77, phenomenal increase of sterilization cases took place by way of rigid enforcement of targets by field functionaries. This had resulted a setback of the family planning programme in the year 1977–78 and thereafter. The Government of the day then took a considered decision that family planning had to be an act of voluntary option and there was no scope for coercion whatsoever. It was to be implemented as an integral part of family welfare by way of extension and education. The National Family Planning Programme was also renamed as National Family Welfare Programme.

The Plan outlay for the Fifth Five Year Plan, which was initially provided at Rs. 516.00 crores was later on revised at Rs. 497.36 crores.

Sixth Five Year Plan (1980–85)

The Sixth Five-year Plan (1980–85) brought back family welfare at the centre stage of the development efforts. Earlier, the Working Group on Population Policy set up by the Planning Commission recommended the adoption of the long-term demographic goal of reducing the net reproduction rate (NRR) to one by 1996 for the country as a whole and by 2001 in all the States from the present level of 1.67 which required the followings:

- The average size of the family would have to be reduced from 4.2 children to 2.3 children.
- The birth rate per thousand population would have to be reduced from the level of 33 in 1978 to 21.
- The death rate per thousand population would have to be reduced from about 14 in 1978 to 9 and the infant mortality rate would be reduced from 129 to 60 or less.
- As against 22 per-cent t of the eligible couples protected with family planning at present, 60 % would need to be protected.
- The population of India will be around 900 million by the turn of century and will stabilise at 1200 million by the year 2050 AD.

The Sixth Plan document mentions that economic development is the best way out to bring about a fall in fertility rate. However, the country cannot wait for that eventuality. Nearly 40 % of the people belongs to the age group of 'below the age of 14 years' which means the existence of a high potential for rising growth of population in future. Limiting the growth of population was, therefore, one of the main objectives of the Sixth Plan. This was sought to be achieved through persuasion of people to adopt the small family norm voluntarily backed by appropriate programmes of supplies and services for contraception.

The Family Planning Programme was also aimed to be made a part of the total national effort for providing a better life to the people. Accordingly, along with eradication of unemployment and poverty reduction programme, special attention was to be paid to the education and employment of women and to liberate them from dependence and insecurity and improve their social status.

The Sixth Five Year Plan desired involvement of all Ministries/Departments and that Family Planning should be accepted as a national programme by all sections of the population. A national consensus on this subject was, therefore, to be developed. As a corollary to the national consensus, an integrated approach and co-ordination of activities were to be put in place in between public health, the Minimum Needs Programme and activities of different departments having a bearing on family planning and maternal and child care. Maternity and Child Health Care was particularly stressed because high morbidity and mortality rates among infants and mothers are generally believed to be responsible for the desire for more children.

The States would continue to get financial assistance from the Government of India on 100 % basis. The Outlays for the Sixth Plan were given in Table 16.1.

Besides continuation and strengthening of the existing activities, provisions were made in the Sixth Plan for completion of incomplete buildings and construction of 1100 new buildings for Rural Family Planning Centres, establishment of 51 Rural Family Planning Centres, 40,000 new sub-centres along with 10,000 female health supervisors, 800 urban family planning centres, 30 post-partum centres at district level and 300 post-partum centres at sub-divisional/taluka level hospitals and procurement of 700 additional vehicles. Certain geographical areas in 12 States and 46 districts, lagging in family planning, were identified for mounting of special health and family planning efforts under the 'Area Projects'. The capacity extension of Hindustan Latex Ltd. was also included in the Sixth Five Year Plan.

The Planning Commission also went through the performances of Family Welfare in the Sixth Plan which were as follows:

- Achievements fell short of the targets for the sterilisation programme. The performance in respect of IUD insertions and CC users reached a high level around 80 % and above.
- The effective couple protection achieved by March 1985 was of the order 32 %.

The Performance analysis also revealed that the national averages were substantially lower because of the relatively poor performance in the States of Uttar Pradesh, Bihar and Rajasthan which had a couple protection rate of less than 20 %. Madhya Pradesh and West Bengal had also a couple protection rates of 29 % as against the national average of 32 %. Further, the Family Welfare programme, since integrated with immunization and ante-natal care of the Maternal and Child Health (MCH) during Sixth Plan was far from satisfactory.

Table 16.1 Outlays for the Family Welfare Programme for the Sixth Five Year Plan (Rs. crores)

Sl. No.	Major items	Sixth Plan outlay
(1)	Services and supplies	687.70
(2)	Training	8.80
(3)	Research and evaluation	11.50
(4)	Mass media and education	32.00
(5)	Maternity and Child Health	250.30
(6)	Organisation	19.50
(7)	India population project	0.20
Total		1010.00

Sixth Plan outlay: family planning

Seventh Five Plan (1985–1990):
In the light of the progress made in the initial years of the Sixth Plan, the National Health Policy had set a target of a net reproduction rate of 1 by the year 2000 AD. The Family Welfare Programme envisaged the following goals for the year 1990:

- Effective couple protection rate 42 %
- Crude birth rate per thousand population 29.1 %
- Crude death rate per thousand population 10.4 %
- Infant mortality rate per 90 % thousand population
- Immunisation Universal coverage
- Ante-natal care 75 %

The Seventh Five Year Plan stipulated that to reach the above targets, particularly 42 % couple protection, 31 million sterilisations, 21.25 million IUD insertions and 14.5 million CC Users have to be achieved by the terminal year of the Seventh Plan. The targets to be reached regard to different methods in the Seventh Plan are given in Table 16.2.

The following strategies were to be acted upon for reaching the targets:

- Based on performances in the states, differential CPR targets would be fixed for the states to reach NRR = 1 by the year 2000.
- Targets for family planning, particularly sterilisation have to be achieved by special drives and camps on a sustained and continued basis.
- Educating and enlightening people on the benefits of late marriage would be emphasised.
- Inter-sectoral coordination and cooperation, and the involvement of non-governmental organisations and of informal leaders would be necessary for community participation for the voluntary acceptance of the Family Welfare. The training needs for them would be addressed.
- For the achievement of the "two child" norm, the child survival rate in our country would have to be improved upon and all the components of MCH programme adequately addressed.
- Vigorous steps would have to be taken to reduce maternal mortality. Since more than two-thirds of the women in the rural areas are still being attended by untrained Dais at childbirth, the Dais training programme needs to be augmented.

Table 16.2 Required acceptors of family planning methods (numbers in lakhs)

Period	Sterilisations	IUD	CC users and OP users	Increase in CPR (%)
1985–86	55.0	32.5	105	Form 32 %
1986–87	60.0	37.5	115	in April
1987–88	62.5	42.5	125	1985 to
1988–89	65.0	47.5	135	42 % in
1989–90	67.5	52.5	145	March 1990

Seventh Plan Outlays:

The Seventh Plan Outlay for the family welfare programme would be at Rs. 3256 crores with break-up for programme given in Table 16.3.

The achievements of the Family Welfare Programme of the country up to the end of the Seventh plan were as follows:

- Reduction in crude birth rate from 41.7 (1951–61) to 30.2 (SRS: 1990).
- Reduction in total fertility rate from 5.97 (1950–55) to 3.8 (SRS: 1990).
- Reduction in infant mortality rate from 146 (1970–71) to 80 (SRS: 1990).
- Increase in Couple Protection Rate from 10.4 % (1970–71) to 43.3 % (31.3.1990).
- Setting up of a large network of service delivery infrastructure.
- Over 118 million births were averted by the end of March, 1990.

Other achievements made during the Seventh Plan period were as follows:

- Based on the norm to have one sub-centre for every 5000 population in plain areas and for 3000 population in hilly and tribal areas, 1.30 lakhs sub-centres could be established in the country at the end of the Seventh Plan i.e. 31.3.1990.
- The Post-Partum programme was also progressively extended to the sub-district level hospitals. At the end of the Seventh Plan, 1012 sub-district level hospitals and 870 Health Posts were established in the country.
- Moreover, the Universal Immunization Programme started in 30 Districts in 1985–86 was extended to cover all the districts in the country by the end of the Seventh Plan.
- Additionally, a project for improving Primary Health Care in urban slums in the cities of Bombay and Madras with assistance from World Bank and Area Development Projects in selected districts of 15 major States were implemented with assistance from various donor Agencies.

Table 16.3 Seventh plan outlays Family Welfare Programme (Rs. crores)

Sl. No.	Programme	Outlays
1	Services and supplies	1356.92
2	Training	60.90
3	Information education and communication	105.0
4	Research and evaluation	25.00
5	ICMR	50.00
6	Maternity and child health	888.44
7	Organisation	125.00
8	Village health guides scheme	370.00
9	Area projects	275.00
	Total	3256.26

Annual Plans 1990–92

The basic approach on Family Welfare adopted during the Seventh Five Year Plan continued during Annual plans 1990–92. For effective community participation in Family Welfare, MahilaSwasthyaSanghs(MSS) was constituted at village level in 1990–91. MSS consisted of 15 persons, 10 representing the varied social segments in the community and five functionaries represented women's welfare activities at village level, such as, the Adult Education Instructor, Anganwari Worker, Primary School Teacher, MahilaMukhyaSevika and the Dai. Auxiliary Nurse Midwife (ANM) happened to be the Member-Convenor. Another new initiative undertaken during 1991–92 was the Child Survival and Safe Motherhood Project with amalgamation of Universal Immunization Programme and intensified MCH activities in high IMR States/Districts of the country.

Eight Five Plan (1992–1997):

High growth rate of the population continued to be one of the major problems facing the country. The 1991 Census recorded an annual addition of 18 million people to the country's population though rate of growth of population declined from 2.22 % in 1971–81 to 2.11 % in 1981–91. In this background, population control assumed an overriding importance in the Eighth Plan.

The family welfare approach and programme of the Eight Plan was also designed in the backdrop of the demographic goals and achievements of the Seventh Plan which were given in Table 16.4.

Further, while the Seventh Plan targets of achieving CPR of 42 % was achieved, this was not matched by a commensurate decline in the birth rate. The Family Welfare Programme remained a uni-sector programme of the Ministry of Health and Family Welfare and was not coordinated in effective manner with female literacy, age at marriage of the girls, status of women in the community, IMR, quality and outreach of health and family planning services and other socio-economic parameters. Additionally, the family welfare programme had also suffered on account of centralised planning and target setting.

Containing population growth was listed as one of the six major objectives of the Eighth Plan. Recognizing the fact that reduction in infant and child mortality is an essential pre-requisite for acceptance of small family norm, Eighth Five Year Plan attempted to integrate MCH and Family Planning as part of Family Welfare services at all levels. The National Development Council approved modified Gadgil Mukherjee Formula which for the first time gave equal weightage to performance in MCH Sector (IMR reduction) and FP Sector (CBR reduction) as a part basis for

Table 16.4 Goals and achievements of the Seventh Plan

	Seventh plan target	Performance status
Couple protection rate (C.P.R.)	42.0 %	44.1 (31.3.91)
Crude birth rate (CBR)	29.1	30.2 (1990)
Crude death rate (CDR)	10.4	9.6 (1990)
Infant mortality rate (IMR)	90	80 (1990)

computing central assistance to non-special category States. This initiative ensured that the inter linkages between Family Welfare Programme and Development was kept in focus in State Plans. The National Development Council also set up a Sub-Committee on Population to consider the problem of population stabilisation which had recommended that Family Welfare Programme should take cognizance of the area specific socioeconomic, demographic and health care availability differentials and allow requisite flexibility in programme planning and implementation.

Strategy for the Eighth Plan

One of the six most important objectives of the Eighth Plan was the target of reducing the birth rate from 29.9 per thousand in 1990 to 26 per thousand by 1997. The IMR was also to be brought down from 80 per thousand live births in 1990 to 70 by 1997.

The following strategies were adopted for achieving the goals of family welfare during the Eighth Plan.

- Convergence of services provided by various social services sectors, e.g., welfare, human resource development, nutrition, etc.
- Decentralised planning and implementation.
- As a natural corollary to decentralised planning and implementation, Panchayati Raj institutions like Gram Panchayat and ZilaParishads, etc. would be required to play significant role in planning, implementing and administering the programme.
- The younger couples being reproductively most active would be the focus of attention, with greater emphasis on spacing methods though the terminal methods would continue to remain as the important means of birth control. Medical Termination of Pregnancy (MTP) was also to be used as an important instrument in the entire scheme of family planning in the Eighth Plan.
- The targeted reduction in the birth rate would be the basis of designing, implementing and monitoring the programme against the current method of couple protection rate.
- The outreach and quality of family welfare services would be improved with completion and upgradation of the health services infrastructure.
- Child survival and safe motherhood initiatives will be vigorously pursued.
- The entire package of incentives and awards will be restructured to make it more purposeful.
- The involvement and commitment of practitioners of all systems of medicine in the Population Control Programme would be secured.
- The role of voluntary organisation in population control would be encouraged and integrated.
- Effective methods would be evolved to get the organised sector involved in the implementation of family welfare programme.
- Special efforts will he made to involve the community in the Family Planning Programme in all aspects of family welfare. Accordingly, Panchayats, youth clubs, village committees, Nehru YuvakKendras, women organisations etc.

would be partnered to accept the responsibility, ownership and the control of the programme along with grass-root level functionaries e.g. village dais, Village Health Guides (VHGs), Auxiliary Nurse Midwives (ANMs), Anganwadi workers, village extension workers and primary school teachers. The village level local functionary wills be the kingpin of these new initiatives.
- The village/neighbourhood tea shops, pan shops, public distribution system shops, pharmacies, cooperatives, etc., will be utilised for community based contraceptive sale and distribution.
- The social marketing programme will be extended to the oral pills as well.
- Information, Education and Communication, which are critical inputs will be further strengthened and expanded. Area specific IEC material will be developed and produced. The backbone of the IEC efforts will, however, remain the inter-personal communication for which the grass-root level female worker will have to be trained and effectively utilised.
- A new thrust in the research and development of methods aimed at regulation of fertility in the male, and of vaccines for fertility regulation, both in the male and female, will be given.
- For effective inter-sectoral coordination, a proper institutional set-up with representation of highest political and administrative authority would be put in place, as recommended also by the Committee on Population, constituted by the NDC.

Outlays:
The outlays for the Family Welfare Programme for the Eight Five Year Plan were fixed at Rs.6500 crores. Details are given in Table 16.5.

Performance During the Eighth Plan
During the Eighth Plan the Crude Birth Rate (CBR) and Infant Mortality Rate (IMR) declined to 27.4 and 72 as against the targets of 26 and 70 respectively. The

Table 16.5 Outlays for the Family Welfare Programme for the Eight Five Year Plan

Eighth plan Sl. No.	Family Welfare	Outlays in Rs-in crore)
1	Services and supplies	3086.00
2	Training	59.00
3	Information, education and communication	127.00
4	Research and evaluation	89.00
5	Maternity and child health	1982.00
6	Organisation	71.00
7	Village health guide scheme	140.00
8	Area projects	400.00
9	Other schemes	46.00
10	Provision for settlement of arrears payable to states	500.00
	Total	6500.00

Couple Protection Rate (CPR) increased to 45.4 % as against the target of 56 % during this period.

As to the performance of the Family Welfare Programme, between 1992 and 1996 the number of sterilisation remained unaltered. There had taken place an increase in IUD and OC use till 1995–96 and CC use till 1994–95. After the abolition of method-specific targets (Target Free Approach) in first April, 1996, the performance between 1995–96 and 1996–97 indicated that at the national level there had taken place a reduction in the acceptance of different methods of contraception. Comparison of the performance between 1995–96 and 1996–97 showed that there were substantial differences in performance between States. Tamil Nadu, Karnataka, Rajasthan had shown an improvement in acceptance of sterilisation as well as IUD but poorly performing States like Bihar and UP showed a further decline up to 50 % in performance.

Ninth Five Year Plan (1997–2002):

Reduction in the population growth rate has been recognised as one of the priority objectives during the Ninth Plan period. The Ninth Five Year Plan document records that the existing high population growth rate was due to: (1) the large size of the population in the reproductive age-group (estimated contribution 60 %); (2) higher fertility due to unmet need for contraception (estimated contribution 20 %); and (3) high wanted fertility due to prevailing high IMR (estimated contribution about 20 %). The Plan document also notes that while the population growth contributed by the large population in the reproductive age group will continue in the foreseeable future, the other two factors need to be effectively made use of for remedial action. Accordingly, the objectives during the Ninth Plan period was to reduce the population growth rate by meeting all the felt-needs for contraception and reducing the infant and maternal morbidity and mortality so that there is a reduction in the desired level of fertility.

The strategies adopted for addressing the objectives of Family Welfare during the Ninth Plan were as follows:

- to assess the needs for reproductive and child health at PHC level and undertake area-specific micro planning; and
- to provide need-based, demand-driven high quality, integrated reproductive and child health care.

In pursuance of them, the following programme designs were sought to be intervened:

- Bridging the gaps in essential infrastructure and manpower through a flexible approach and improving operational efficiency through investment in social, behavioural and operational research
- Providing additional assistance to poorly performing districts identified on the basis of the 1991 census to fill existing gaps in infrastructure and manpower.
- Ensuring uninterrupted supply of essential drugs, vaccines and contraceptives, adequate in quantity and appropriate in quality.

- Promoting male participation in the Planned Parenthood movement and increasing the level of acceptance of vasectomy.

Further, efforts were to be made to enhance the quality and coverage of family welfare services through:

- increasing participation of general medical practitioners working in voluntary, private, joint sectors and the active cooperation of practitioners of ISM(Indian System of Medicine) and H (Homeopathy);
- involvement of the Panchayati Raj Institutions for ensuring inter-sectoral coordination and community participation in planning, monitoring and management;
- involvement of the industries, organised and unorganised sectors, agriculture workers and labour representatives.

A review of the performance during the Ninth Plan revealed that the health systems in the states needed more time to adapt to decentralised planning and implementation of components of the RCH programme. As a result, Ninth Plan goals for CBR, couple protection rate, MMR and could not be achieved fully as would be evident from Table 16.6.

Tenth Five Year Plan (2002–2007)

The Tenth Five Year Plan document mentioned in its approach to family welfare that the on-going high population growth rate was due to:

- The large size of the population being in the reproductive age-group (estimated contribution 60 %);
- Higher fertility due to unmet need for contraception (estimated contribution 20 %); and
- High fertility due to prevailing high IMR (estimated contribution about 20 %).

It also stated that while population growth contributed by the large population size in the reproductive age group would continue in the foreseeable future, the remaining 40 % of growth could be substantially reduced by meeting the unmet needs for contraception and felt needs for maternal and child health to reduce IMR.

The Tenth Plan intended to operationalize efforts to assess and meet the unmet needs for contraception; achieve reduction in the high desired level of fertility through programmes for reduction in IMR and maternal mortality ratio

Table 16.6 Ninth Five Year Plan goals and achievements

Items	1997 Ninth Plan goal	2002 Ninth Plan achievements
Couple protection rate	51	45.7
Crude birth rate	24	25.0
Total fertility rate	2.9	3.0
Maternal mortality rate	300	301
Infant mortality rate	56	63

Source Family Welfare Statistics, 2011/Tenth Plan document

(MMR) andenable families to achieve their reproductive goals. If the reproductive goals of families are fully met, the country will be able to achieve the National Population Policy goal of replacement level of fertility by 2010. The medium and long term goals would be to continue this process to accelerate the pace of demographic transition and achieve population stabilisation by 2045.

Reductions in fertility, mortality and population growth rate were also the general objectives of the overall Tenth Five Year Plan. Three of the 11 monitorable targets for the Tenth Five Year Plan, as a whole, and beyond were:

- reduction in IMR to 45 per 1000 live births by 2007 and 28 per 1000 live births by 2012;
- reduction in maternal mortality ratio to 2 per 1000 live births by 2007 and 1 per 1000 live births by 2012; and
- reduction in decadal growth rate of the population between 2001 and 2011 to 16.2.

The Tenth Plan document lays faith that steep reduction in mortality and fertility, as envisaged, are technically feasible within the existing infrastructure and manpower as evidenced from performances in several States/districts. All efforts would be made to provide essential supplies, improve efficiency and ensure accountability to the sub-optimal performing states. This, in turn, was expected to result in substantial improvement in state and national indices and enable the country to achieve the goals set for the Tenth Plan.

Additionally, efforts would be met for restructuring the existing infrastructure for FW services, ensuring skill upgradation of the personnel, providing good quality integrated reproductive and child health services, improving the logistics of supply, operationalizing the referral system, involvement of the PRI in planning, monitoring and midcourse correction of the programme at local level, effective Inter-sectoral coordination between concerned sectors. Effective Information, Education, Communication & Motivation'.

The Family Welfare approach during the Tenth Plan was, in fact, in continuum of the shift that began earlier in the Ninth Plan with the Target Free Approach "The shift was from demographic targets to focussing on enabling couples to achieve their reproductive goals; from method specific contraceptive targets to meeting all the unmet needs for contraception to reduce unwanted pregnancies and also from numerous vertical programmes for family planning and maternal and child health to integrated health care for women and children. It was also distinguished as a shift from centrally defined targets to community need assessment and decentralised area specific micro-planning and implementation of program for health care for women and children, to reduce infant mortality and reduce high desired fertility; from quantitative coverage to emphasis on quality and content of care; from predominantly women centred programmes to meeting the health care needs of the family with emphasis on involvement of men in planned parenthood; from supply driven service delivery to need and demand driven service; and from improved logistics for ensuring adequate and timely supplies to meet the needs and choices and conveniences of the couples".

It was envisaged that if the reproductive goals of families could be fully met, the country would be able to achieve the National Population Policy goal of replacement level of fertility by 2010. The medium and long term goals of NPP, 2000 could be achieved by way of accelerating the pace of demographic transition and achieve population stabilisation by 2045. Early population stabilisation will enable the country to achieve its developmental goal of improving the economic status and quality of life of the citizens.

At Table 16.7, the name of the schemes undertaken, its outlay and actual expenditure have been shown:

The achievements in the Family Welfare sector, in terms of Indicators, during the Tenth Five Year Plan are now shown at Table 16.8.

The Family Welfare Statistics, 2011 published by the Ministry of Health and Family Welfare mentioned that Couple Protection Rate has increased from 45.7 % in 2001–02 to 46.2 % in 2006–07, a marginal increase of 0.5 %. This is a signal of the effectiveness of the NRHM-based family planning in India.

Eleventh Five Year Plan (2007–2012):

Ever since the NRHM assumed the role of family welfare programme in India in April, 2005, the family welfare lost its identity and became a part of a disease burden of the Health care system. The Eleventh Five Year plan posted a focused vision of health; however, it had no separate vision of family welfare or of population control and/or family planning. Nonetheless, the Plan document noted seven 'Time-Bound Goals for the Eleventh Five Year Plan' of which three goals belonged to family welfare, namely,

- Reducing Maternal Mortality Ratio (MMR) to 1 per 1000 live births.
- Reducing Infant Mortality Rate (IMR) to 28 per 1000 live births.
- Reducing Total Fertility Rate (TFR) to 2.1.

It is for the same reason that even the Eleventh Five year Plan document did not include any separate paragraph on approach towards family welfare nor had set it any hard demographic target for population stabilization. Since the make-over from family welfare to National Rural Health Mission has become complete, the Plan document has also mentioned expected outcomes from NRHM which are listed in Table 16.9.

The Eleventh Five Year Plan document summarized the Thrust areas to be pursued during the Eleventh Five Year Plan at paragraph 3.1.205 where, it mentioned, among other things, the following:

- Improving Health Equity under NRHM and NUHM
- Reducing Maternal mortality and improving Child Sex ratio through Gender Responsive Health care
- Reducing Infant and Child mortality through HBNC and IMNCI
- Training the TBAs to make them SBAs

Table 16.7 Outlay and expenditure of the Tenth Plan

Sl. No.	Name of the schemes	Outlay of outlay tenth plan (2002–2007)	Tenth plan (2002–2007) sum total of annual plan outlay	Tenth plan (2002–2007) actual expenditure
	Centrally sponsored schemes (CSS) of family welfare	24,169.20	28,011.97	23,854.74
1	Direction and Administration	1100.00	1176.66	999.93
2	Rural FW services (SCs)	9663.00	8881.29	7561.01
3	Urban FW services	580.00	638.17	539.48
4	Grants to state training institutions	480.00	500.37	411.08
5	Free distribution of contraceptives	940.00	760.22	627.97
7	Family welfare linked health insurance	150.00	105.10	10.63
8	Training	250.00	143.81	71.60
9	Procurement of supplies and materials	994.98	1141.30	335.14
10	Routine immunization	1557.88	1625.50	783.44
11	Pulse polio immunization	3110.00	3887.70	3999.56
12	IEC	539.50	569.87	542.42
13	Area projects	1750.00	1838.14	1250.60
14	Flexible pool for state PIPs	3041.84	6733.59	6713.10
	Central sector schemes (CS) of family welfare	1367.80	1611.53	1180.69
1	Social marketing area projects	20.00	35.00	0.00
2	Social marketing of contraceptives	660.00	790.04	599.701
3	FW Training and Res. Centre, Bombay	10.00	10.53	2.31
4	NIHFW, New Delhi	20.00	25.45	19.91
5	IIPS, Mumbai	10.00	9.57	8.09
6	Rural Health Training Centre, Najafgarh	45.00	12.42	1.56
7	Population Research Centres	45.00	39.13	30.01
8	CDRI, Lucknow	12.00	12.65	12.85
9	ICMR and IRR	100.00	150.00	162.44
10	Travel of experts/conference/meetings etc. (Melas)	57.00	17.00	47.84
11	International co-operation	9.00	8.44	6.73
12	NPSF/national commission on population	100.00	116.00	104.08

(continued)

Table 16.7 (continued)

Sl. No.	Name of the schemes	Outlay of outlay tenth plan (2002–2007)	Tenth plan (2002–2007) sum total of annual plan outlay	Tenth plan (2002–2007) actual expenditure
13	NGOs (PPP)	130.00	241.61	88.95
14	Other schemes	149.80	143.69	96.22
Transferred to states/weeded during Tenth Plan		589.00	417.50	291.12
1	District projects	51.00	105.00	40.95
2	Community incentive scheme	200.00	62.00	0.00
3	Transport	313.00	223.00	248.02
4	New initiatives	25.00	27.50	2.15
To NACO		0.00	200.00	265.99
Family welfare (Total)		26,126.00	30,241.00	25,592.54

Source Eleventh Five Year Plan

Table 16.8 Achievements during the Tenth Plan

Indicator	Goal for the Tenth Plan	Achievements
Decadal rate of population growth	16.2 %	15.9 % for 2001–2011 projected
IMR	45 per 1000 live births	58 per live births
MMR	2 per 1000	3.01 live births

Source Eleven Five Year Plan, planning commission

Table 16.9 NRHM goals during Eleventh Five Year Plan

Items	Goals
IMR	Reduction to 30/1000 live births by 2012
MMR	Reduction to 100/100,000 live births by 2012
TFR	Reduction to 2.1 by 2012

Source Eleventh Plan, planning commission

The Twelfth Five Year Plan, Volume III, showed at a Table at page (4) the monitorable Target, Target Baseline Level and Recent Status of the earlier Eleventh Plan, which are as follows (Table 16.10):

The Family Welfare Statistics, 2011 published by the Ministry of Health and Family Welfare mentioned, however, that Couple Protection Rate has decreased from 46.2 % in 2006–07 to 40.4 % in 2010–11. This is really alarming message for NRHM—based family planning in India.

Table 16.10 Eleventh Plan monitorable goals and achievements

Sl. No.	Eleventh plan monitorable target	Target baseline level	Recent status
1	Reducing maternal mortality ratio (MMR) to 100 per 100,000 live births	254 (SRS-2004-05)	212 (SRS-2007-09)
2	Reducing Infant mortality rate (IMR) to 28 per 1000 live births	57 (SRS-2006)	44 (SRS-2011)
3	Reducing total fertility rate (TFR) to 2.1.	2.8 (SRS 2006)	2.5 (SRS, 2010)

Source Twelfth Plan document, planning commission

Twelfth Five Year Plan (2012–2017):
The enormity of the population size, its implication on national economy and national planning efforts, and ways to contain the alarming situation remains unaddressed in the Twelfth Five Year Plan. The Twelfth Plan document at paragraph 20.3 summarised the existing weakness of the Health sector. Unfortunately, there was nothing to include the weakness of the Family Welfare segment when the Couple Protection Rate (CPR) at the end of Eleventh Plan was a mere 40.4 %. Further, the Strategy discussed at paragraph 20.19 of the same plan document does not have anything for the Family Welfare. The absence of focused approach on family welfare (or family planning) continued during the Twelfth Plan as well except by way of mentioning a casual comment like 'as compared to limiting methods, emphasis on family spacing methods like IUCD and male condoms has had a better impact in meeting the unmet needs of couples. A recent study has estimated that meeting unmet contraception needs could cut maternal deaths by one-third. There is, therefore, a need for much more attention to spacing methods such as, long term IUCD. IUCD insertion on fixed days by ANMs (under supervision of LHV for new ANMs) would be encouraged. Availability of MTP by Manual Vacuum Aspiration (MVA) technique and medical abortions will be ensured at fixed points where Mini-Laparotomy is planned to be provided. Services and contraceptive devices would be made easily accessible. This would be achieved through strategies including social marketing, contracting and engaging private providers. Postpartum contraception methods like insertion of IUD which are popular in countries like China, Mexico, and Egypt and male sterilisation would be promoted while ensuring adherence to internationally accepted safety standards'.

Be that as it may, the Twelfth Five Year Plan mentioned about the outcome of health sectors and that the National goals would be aggregate of State Goals. The State wise Targets of IMR and MMR under the Twelfth Plan have been mentioned at Table 16.11.

The IMR and MMR, as mentioned at Table 16.11 are designed to be implemented by promoting births by skilled birth attendants (SBAs) by means of rational posting of SBAs and equipping traditional birth attendants (TBAs) for safe deliveries. Further, meeting the unmet need for contraception would be addressed by giving focus on contraceptive "delivery points," doorstep delivery of contraceptives

Table 16.11 State wise targets of IMR and MMR in Twelfth Plan

Sl. No.	Name of the State/Union territories	Recent status		Twelfth plan	
		IMR	MMR	IMR	MMR
	India	44	212	25	100
1	Andhra Pradesh	43	134	25	61
2	Arunachal Pradesh	32	NA	19	–
3	Assam	55	390	32	177
4	Bihar	44	261	26	119
5	Chhattisgarh	48	269	28	122
6	Goa	11	NA	6	–
7	Gujarat	41	148	24	67
8	Haryana	44	153	26	65
9	Himachal Pradesh	38	NA	22	–
10	Jammu and Kashmir	41	NA	24	–
11	Jharkhand	39	261	23	109
12	Karnataka	35	178	15	80
13	Kerala	12	81	6	37
14	Madhya Pradesh	59	269	34	122
15	Manipur	11	NA	6	–
16	Maharashtra	25	104	15	47
17	Meghalaya	52	NA	30	–
18	Mizoram	34	NA	20	–
19	Nagaland	21	NA	12	–
20	Odisha	57	258	33	117
21	Punjab	30	172	16	78
22	Rajasthan	52	318	30	145
23	Sikkim	26	NA	15	–
24	Tamil Nadu	22	97	13	44
25	Tripura	29	NA	17	–
26	Uttar Pradesh	57	359	32	163
27	Uttarakhand	36	359	21	163
28	West Bengal	32	145	11	66
29	Andaman and Nicober Islands	23	NA	12	–
30	Delhi	28	NA	15	–
31	Chandigarh	20	NA	12	–
32	Dadra and Nagar Haweli	35	NA	20	–
33	Daman and Diu	22	NA	12	–
34	Lakshadweep	24	NA	14	–
35	Puducherry	19	NA	11	–

Source Twelfth Plan, planning commission

by the ASHA worker, promoting use of ML375 intra uterine contraceptive device (IUCD) as a short-term spacing method and enlisting more number of private providers for provision of services.

As to the reduction of Total Fertility Rate (TFR) to 2.1, the Twelfth Plan sounds eminently optimistic and states that India is on track for a TFR target of 2.1 by 2017 which is necessary to achieve net replacement level of unity and realise the cherished goal of the National Health Policy, 1983 and the National Population Policy, 2000.

The Twelfth Plan, like earlier plans, remained silent about sustainable population in India and the way forward to achieve it.

Chapter 17
Family Welfare Programmes in India

Abstract The Family welfare programmes of the country are funded by the Central Government under 100 % centrally sponsored scheme. The schemes which are so funded by the Union Government include the (a) Family Planning Cell at the State Secretariat, (b) State Family Welfare Bureau, (c) District Family Welfare Bureau, (d) Regional Family Planning Training Centres, (e) Establishment and Maintenance of Rural Family Planning Sub-centres, (f) Establishment and Maintenance of Urban Family Planning, (g) Establishment and Maintenance of Sterilisation Beds. The state governments implement the related Centrally Sponsored Schemes based on its yearly allocation. Normally, no component of state fund is included on these programmes, though there is no bar for any state to have its own budgeted schemes, beyond the central allocation, on these areas or to any incidental areas thereto. In a way, the schemes, as mentioned above, are the core schemes for the family planning in India and has been in place for the last few decades. In addition to the above initiatives, the central government takes the responsibility of all issues connected with contraceptive management of the country under the national family planning programme. The Rural Health Mission and Urban Health Mission have now become the focal points for family welfare intervention programmes in India.

Keywords National Family Welfare Programme Free Supply Scheme · Social Marketing Scheme · National Rural Health Mission (NRHM) · National Urban Health Mission · Accredited Social Health Activists · National Population Stabilization Fund · National Population Commission

The Family welfare programmes of the country are funded by the Central Government under 100 % centrally sponsored scheme. The schemes which are so funded by the Union Government include the following:

(a) Family Planning Cell at the State Secretariat,
(b) State Family Welfare Bureau
(c) District Family Welfare Bureau
(d) Regional Family Planning Training Centres
(e) Establishment and Maintenance of Rural Family Planning Sub-centres

(f) Establishment and Maintenance of Urban Family Planning
(g) Establishment and Maintenance of Sterilisation Beds

The state governments implement the related Centrally Sponsored Schemes based on its yearly allocation. Normally, no component of state fund is included on these programmes, though there is no bar for any state to have its own budgeted schemes, beyond the central allocation, on these areas or to any incidental areas thereto. In a way, the schemes, as mentioned above, are the core schemes for the family planning in India and has been in place for the last few decades. In addition to the above initiatives, the central government takes the responsibility of all issues connected with contraceptive management of the country under the national family planning programme.

- National Family Planning Programme

Family planning is the key to population control in any country. Depending on the state of contraceptive technology at any given time, the cultural practices in a given society, the access of such contraceptives and its after care services, where ever needed, and finally the reproductive behaviour of its eligible couples, the family planning programme character gets shaped. All the related parameters are not static but undergo process of changes, by and by. In the Indian context, there has taken place changes, now and then, in the area of contraceptive choices linked with it, among other things, the Government's policy on contraceptive of the day.

In this scenario, it may not be useful to trace the evolution of contraceptive choices and practices as such but be confined to discuss whether the Government did really apply its mind to make use of available contraceptive technology for meeting plural choices for unmet needs. For doing so, it would be fair to capture the on-going basket of contraceptive availability.

Contraceptive services under the National Family Welfare programme:
The contraceptive methods available currently in India may be broadly divided into two categories, (a) spacing methods and (b) permanent methods. There is another method (c) emergency contraceptive pill to be used in cases of emergency.

(a) Spacing Methods:
   Spacing methods fall in the category of reversible methods of contraception to be used by couples who wish to have children in future. These include the following:

   - Condoms:
     Condoms are meant for male partner and very protective against pregnancy. It is available without any cost from family welfare distribution network.
   - Oral contraceptive pills
     These are hormonal pills which have to be taken by a woman, preferably at a fixed time, daily. The strip also contains additional placebo/iron pills to be consumed during the hormonal pill free days.

- Copper-T

  Copper T is an effective spacing method meant for female partner. Earlier, Cu-T-200B was used in the in the National Family Welfare services. From 2003–04, advanced version of Intra Uterine Device i.e. Cu-T-380-A and from the year 2013–14 another advanced version of IUD-375 have been introduced in the programme.

(b) Permanent Methods:

  There are separate permanent methods for male and for female. For the male sterilization, vasectomy is conducted. The modern method of No Scalpel Vasectomy is now used in National Family Welfare Programme. For the female, female sterilization is conducted. The modern method of Laparoscopic and Minilap have now been in use in National Family Welfare Programme.

(c) Emergency Method:

  Emergency Contraceptive Pills (ECP or E- pills) has been a part of national contraceptive services from the year 2002–03. This contraceptive is to be used within 72 h of un-protected sex.

Distribution of contraceptives:

There are two modes of distribution of contraceptives in the country: (a) Free Supply Scheme and (b) Social Marketing Scheme.

(a) Free Supply Scheme:

Under Free Supply Scheme, contraceptives, namely Condoms, Oral Contraceptive Pills, Intra Uterine Device (Cu-T), Emergency Contraceptive Pills and Tubal Rings are procured and supplied free by the department of Health and Family Welfare of the central government to the States and UTs. The channel for supply of these contraceptives under Free Supply Scheme is Government network comprising Sub-Centres, Primary Health Centres, Community Health Centres and Government Hospitals, State AIDS Control Societies throughout the country.

Pregnancy Test Kits

These kits are being supplied free of cost through different service units.

(b) Social Marketing Scheme:

Under the Social Marketing Programme, both Condoms and Oral Pills are made available to the people at highly subsidized rates, through diverse outlets. The extent of subsidy ranges from 70 to 85 % depending upon the procurement price in a given year. Both these Contraceptives are sold through Social Marketing Organizations (SMOs). The SMOs are given Deluxe Nirodh condom at Rs 2.00 per packet of 5 pieces and this is sold @ Rs 3/- per packet of 5 pieces to the consumer. For women one cycle of Pills, which is required for one month, is given to the SMOs @ Re. 1.60/- and it is sold to the consumer @ Rs 3/- per strip (cycle) under the brand name-"Mala-D". Under the Social Marketing programme, currently Government brands and different SMO brands of condoms and OCPs are sold in the market. Based on the recommendation of the Working Group on Social Marketing of Contraceptives, SMOs have the flexibility to fix the price of branded condoms and OCPs within the range fixed by the Government.

Since December 1995, a non-steroidal weekly Oral Contraceptive Pill, Centchroman (Popularly known as Saheli&Novex), to prevent pregnancy is also being subsidized under the Social Marketing Programme. The weekly Oral pill is the result of indigenous research of CDRL, Lucknow. The pill is now available in the market at Rs 2.00 per tablet. The Government of India provides a subsidy of Rs 2.59 per tablet towards product and promotional subsidy.

In fine, the National Family Welfare services provide the following contraceptive methods at various levels of health system as shown in the following table:

Spacing methods	Limiting methods
IUCD 380 A and Cu IUCD 375	Female sterilization
Oral contraceptive pills	Laparoscopic and
Condoms	Minilap
Emergency contraceptive pills	Male sterilization (no scalpel Vasectomy)

Family planning method	Service provider	Service location
Spacing methods		
IUCD 380 A, IUCD 375	Trained & certified ANMs, LHVs, SNs and doctors	Sub centre & higher levels
Oral contraceptic pills (OCPs)	Trained ASHAs, ANMs, LHVs, SNs and doctors	Village level sub centre & higher levels
Condoms	Trained ASHAs, ANMs, LHVs, SNs and doctors	Village level Sub centre & higher levels
Emergency contraceptive pills		
Emergency contraceptive pills (ECPs)	Trained ASHAs, ANMs, LHVs, SNs and doctors	Village level Sub centre & higher levels
Limiting methods		
Minilap	Trained & certified MBBS doctor & specialised doctors	PHC & higher levels
Laparoscopic sterilization	Trained & certified MBBS doctor & specialised doctors	Usually CHC & higher levels
NSV: no scalpel vasectomy	Trained & certified MBBS doctor & specialised doctors	PHC & higher levels

With the expansion of concept of family welfare and more particularly with its integration to the health system of the country, a host of schemes have been put in place to address the broader area of family welfare with its impact on the fertility behaviour of the citizens. Historically, a good number of such initiatives were made and many of them have ceased to exist on the expiry of its term. Least it generates any confusion, only the current schemes/programmes, as available in the Annual Report of the ministry of Health and Family welfare, have been captured there.

National Health Mission (NHM):

'Broadly speaking, all health and family welfare programmes are now centred around the National Health Mission (NHM). The NHM has two Sub-Missions, the National Rural Health Mission (NRHM) and the National Urban Health Mission (NUHM). The main programmatic components include Health System Strengthening in rural and urban areas- Reproductive-Maternal-Neonatal-Child and Adolescent Health (RMNCH + A), and Communicable and Non-Communicable Diseases. The NHM envisages achievement of universal access to equitable, affordable & quality health care services that are accountable and responsive to people's needs.

(a) **National Rural Health Mission (NRHM):**

NRHM seeks to provide accessible, affordable and quality health care to the rural population, especially the vulnerable groups. Under the NRHM, the Empowered Action Group (EAG) States as well as North Eastern States, Jammu and Kashmir and Himachal Pradesh have been given special focus. The thrust of the mission is to carry out necessary architectural correction in the basic health care delivery system to ensure economic and social development and improving the quality of life of citizens. The Mission adopts a synergistic approach by relating health to determinants of good health viz. segments of nutrition, sanitation, hygiene and safe drinking water. It also aims at mainstreaming the Indian systems of medicine to facilitate health care. The Plan of Action includes increasing public expenditure on health, reducing regional imbalance in health infrastructure, pooling resources, integration of organizational structures, optimization of health manpower, decentralization and district management of health programmes, community participation and ownership of assets, induction of management and financial personnel into district health system, and operationalizing community health centres into functional hospitals meeting Indian Public Health Standards in each Block of the Country.

(b) **National Urban Health Mission (NUHM):**

The NUHM is designed to meet health needs of the urban poor, particularly the slum dwellers by making available to them essential primary health care services. NUHM would cover all state capitals, district headquarters and other cities/towns with a population of 50,000 and above (as per census 2011) in a phased manner. Cities and towns with population below 50,000 will be covered under NRHM. This is intended to be done by investing in high-calibre health professionals, appropriate technology through PPP, and health insurance for urban poor. NUHM was intended to focus on slums and other urban poor. At the State level, besides the State Health Mission and State Health Society and Directorate, the Mission programmes have to be managed by a State Urban Health Programme Committee. At the district level, similarly there would be a District Urban Health Committee and at the city level, a Health and Sanitation Planning Committee. At the ward slum level, there will be a Slum Cluster Health and Water and Sanitation Committee. For promoting public health and cleanliness in urban slums, it will also encompass experiences of civil society organizations (CSO) working in urban slum clusters.

(a) **Major initiatives under NRHM:**

It is worth mentioning various components under NRHM to have an idea of NRHM, more particularly about how it is addressing the family welfare needs of the country.

- ASHA:
  Community health volunteers called Accredited Social Health Activists (ASHAs) are in place to work as a link between the community and the public health system. ASHA is the first port of call for any health related demands of deprived sections of the population, especially women and children, who find it difficult to access health services in rural areas. ASHA Programme aims at bringing people back to Public Health System and increase in the utilization of their outpatient services, diagnostic facilities, institutional deliveries and in-patient care.
- Rogi KalyanSamiti/Hospital Management Society
  The society is the core of management structure. This committee is a registered society whose members act as trustees to manage the affairs of the hospital and is responsible for upkeep of the facilities and ensure provision of better facilities to the patients in the hospital. Financial assistance is provided to these Committees through untied fund to undertake activities for patient welfare. RogiKalyanSamitis (RKS) have been set up involving the community members in almost all District Hospitals (DHs), Sub-District Hospitals (SDHs), Community Health Centres (CHCs) and Primary Health Centres (PHCs).
- The Untied Grants to Sub-Centres (SCs)
  Untied fund assistance gives a new confidence to our ANMs in the field. The SCs are far better equipped now with Blood Pressure measuring equipment, Haemoglobin (Hb) measuring equipment, stethoscope, weighing machine, etc. This has facilitated provision of quality antenatal care and other health care services.
- The Village Health Sanitation and Nutrition Committee (VHSNC)
  It is an important tool of community empowerment and participation at the grassroots level. The VHSNC reflects the aspirations of the local community, especially the poor households and children. Untied grants of Rs 10,000 are provided annually to each VHSNC under NRHM, which are utilized through involvement of Panchayati Raj representatives and other community members. Capacity building of the VHSNC members with regards to their roles and responsibilities for maintaining the health status of the village is also addressed.

- Additional Human Resource Support:
  Health care service delivery requires intensive human resource inputs. There has been an enormous shortage of human resources in the public health care sector in the country. NRHM attempts to fill the gaps in human resources by providing additional health human resources to states including GDMOs, Specialists, ANMs, 9 Staff Nurses etc. on contractual basis. Apart from providing support for health human resource, NRHM also focuses on multi skilling of doctors at strategically located facilities identified by the states e.g. MBBS doctors are trained in Emergency Obstetric Care (EmOC), Life Saving Anaesthesia Skills (LSAS) and Laparoscopic Surgery. Similarly, due importance is given to capacity building of nursing staff and auxiliary workers such as ANMs. NRHM also supports co-location of AYUSH services in health facilities such as PHCs, CHCs and DHs.
- Janani SurakshaYojana (JSY)
  JRY aims to reduce maternal mortality among pregnant women by encouraging them to deliver in government health facilities. Under the scheme, cash assistance is provided to eligible pregnant women for giving birth in a government health facility.
- Janani Shishu Suraksha Karyakarm (JSSK):
  Launched on 1st June, 2011, JSSK entitles all pregnant women delivering in public health institutions to absolutely free and no expense delivery, including caesarean section. This marks a shift to an entitlement based approach. The free entitlements include free drugs and consumables, free diagnostics, free diet during stay in the health institutions, free provision of blood, free transport from home to health institution, between health institutions in case of referrals and drop back home and exemption from all kinds of user charges. Similar entitlements are available for all sick infants (up to 1 year of age) accessing public health institutions.
- National Mobile Medical Units (NMMUs):
  Support is provided to districts for MMUs under NRHM in the country. To increase visibility, awareness and accountability, all Mobile Medical Units have been repositioned as "National Mobile Medical Unit Service" with universal colour and design.
- National Ambulance Services:
  NRHM has supported free ambulance services to provide patients transport in every nook and corner of the country connected with a toll free number. Besides these, vehicles are empanelled to transport patients, particularly pregnant women and sick infants from home to public health facilities and back.
- Infrastructure:
  Up to 33 % of NRHM funds in High Focus States can be used for infrastructure development. In order to ensure that enhanced fund allocations to States/UTs and other institutions under the NRHM are fully coordinated, managed and utilized, Financial Management Group for NRHM (FMG-NRHM) has been set up.

- Mainstreaming of AYUSH:

  Mainstreaming of AYUSH has been taken up by allocating AYUSH facilities in PHCs, CHCs, DHs, health facilities above SC but below block level and health facilities other than CHC at or above block level but below district level.

- Mother and Child Tracking System (MCTS):

  MCTS is a name based tracking system, launched by the Government of India as an innovative application of information technology directed towards improving the health care service delivery system and strengthening the monitoring mechanism. MCTS is designed to capture information on and track all pregnant women and children (0–5 Years) so that they receive 'full' maternal and child health services and thereby contributes to the reduction in maternal, infant and child morbidity and mortality which is one of the goals of National Rural Health Mission. This tool also facilitates in generation of work plan for the field level health care service providers, to ensure timely and full range of services to them. MCTS employs mobile based SMS technology to alert health service providers and beneficiaries about the service delivery and for providing the due services on time. The system also facilitates with status note and actionable messages to policy makers, health managers and health administrators at different tiers of health care delivery system for necessary action on time.

New Initiatives:

- Rashtriya Bal Swasthya Karyakram (RBSK):

RBSK initiative was launched in February 2013 and provides for Child Health Screening and Early Intervention Services through early detection and management of 4 Ds i.e. Defects at birth, Diseases, Deficiencies, Development delays including disability.

- Rashtriya Kishor Swasthya Karyakram (RKSK):

RKSK is a new initiative, launched in January 2014 to reach out to 253 million adolescents in the country in their own spaces and introduces peer-led interventions at the community level, supported by augmentation of facility based services. This initiative broadens the focus of the adolescent health programme beyond reproductive and sexual health and brings in focus on life skills, nutrition, injuries and violence (including gender based violence), non-communicable diseases, mental health and substance misuse.

- Mother and Child Health Wings (MCH Wings):

100/50/30 bedded Maternal and Child Health (MCH) Wings have been sanctioned in public health facilities with high bed occupancy to cater to the increased demand for services.

- Free Drugs and Free Diagnostic Service:

Extremely high cost of drugs and diagnostics stand against accessible and affordable healthcare for all. An incentive to the extent of 5 % of the state's Resource Envelope under NRHM is in place for those states that implemented free essential drugs scheme for all patients accessing public health facilities. Under the National Health Mission—Free Drug Service Initiative and Free Diagnostics Service Initiative, substantial funding is being provided to state that implement these initiatives.

- National Iron + Initiative:

National Iron + Initiative is another new initiative to prevent and control iron deficiency to pregnant women and lactating mothers. It aims to provide IFA supplementation for children, adolescents and women in reproductive age group. Weekly Iron and Folic Acid Supplementation (WIFS) for adolescents is an important strategy under this initiative.

- Reproductive, Maternal, New born, Child and Adolescent Health Services (RMNCH + A):

RMNCH + A belongs to the category of a continuum of care approach and has been adopted under NRHM with the articulation of strategic approach to Reproductive Maternal, New born, Child and Adolescent health. Under RMNCH + A approach focus is brought on the adolescents at a critical life stage and set to establish linkages between child survival, maternal health and family planning efforts. It also aims to strengthen the referral linkages between community and facility based health services and between the various levels of health system itself.

- Delivery Points (DPs):

Health facilities that have a high demand for services and performance above a certain benchmark have been identified as "Delivery Points" with the objective of providing comprehensive reproductive, maternal, new born, child and adolescent health services (RMNCH + A) services at these facilities. Funds are allocated to strengthen these DPs in terms of infrastructure, human resource, drugs, equipment, etc.

- Universal Health Coverage (UHC):

Universal Health Coverage (UHC), is sought to be implemented through the forum of the National Health Mission where the existing public health quality care system of affordable nature of social protection is tied up with private sector supplementation to close the gap in the health care system. UHC pilot projects are designed to support at least one district of each state. The pilots are expected to demonstrate how access to care and social protection against the costs of care can be meaningfully expanded in the most cost effective manner, while at the same time reducing health inequity.

- Mother and Child Tracking Facilitation Centre (MCTFC):

MCTFC has been operationalized from National Institute of Health and Family Welfare (NIHFW). It is being operated by 80 Helpdesk Agents (HAs). It is set up to validate the data entered in CTS in addition to guiding and helping both the beneficiaries and service providers with up to date information on Mother and Child Care services through phone calls and Interactive Voice Response System (IVRS) on a regular basis. MCTFC will create awareness about Government mother and child health related programmes and seek feedback on services being provided. In addition, a module has been provided in the Mother and Child Tracking System (MCTS) portal so that States/UTs may utilise it to make calls for validation of MCTS data, getting feedback and raising awareness.

- Quality Assurance (QA):

The strategy is to shift focus from fragmented approach of different quality systems to one comprehensive approach of Quality Assurance. Based on best practices of existing quality system such as NABH, ISO, JCI and other scientific literature, comprehensive operational guidelines on Quality Assurance has been put in place wherein National Quality Assurance standards have been published. The road map for QA envisages development of a robust institutional mechanism within the states to make States self-sufficient and to have more sustainable system than existing systems. A total of 8 major areas of concerns-Service provision, Patient right, Inputs, Support Services, Clinical Services, Infection Control, Quality Management and Outcome have been identified. Standards have been developed for each area of concern and detailed checklists have been laid down to ensure conformance to these standards. All Public Health Facilities have been assessed, and Quality scored in a phased manner. Besides clinical care, due weightage are given to issues of patients' right, confidentiality, privacy, compliance to National Health Programme Guidelines, cleanliness at health facilities, etc. The facilities having a credible system of quality assurance (verified through district and state assessment) are assessed for National Level Certification. On successful attainment of the National certification, facilities are given incentives and QA certification.

- ASHA Certification:

A system for certification of ASHAs to enhance competency and professional credibility of ASHAs by knowledge and skill assessment has been put in place. The certification of ASHAs is done by National Institute of Open Schooling (NIOS).

- NGO Guidelines:

Guidelines for NGO involvement under NHM during Twelfth Five Year Plan have been issued which envisage greater state ownership for NGO led programmes and provide a broad framework to the States to partner with NGOs and facilitate their participation in capacity building, support for community processes service delivery, develop innovations through research and documentation, advocacy, and for supplementing capacities in key areas of the public health system to improve health care service delivery'.

(b) **Urban Health Mission (NUHM);**

'The National Urban Health Mission (NUHM) as a new sub-mission under the overarching National Health Mission (NHM) has come into being in the year 2013. NUHM envisages to meet health care needs of the urban population with particular care to the urban poor, by reaching out essential primary health care services and reducing their out of pocket expenses for treatment. This is intended to be achieved by strengthening the existing health care service delivery system, targeting the urban poor living in slums and partnering with various schemes relating to wider determinants of health like drinking water, sanitation, school education, etc. implemented by other Ministries like Housing & Urban Poverty Alleviation, Human Resource Development and Women & Child Development.

NUHM seeks to achieve its goal through:-

- Need based city specific urban health care system to meet the diverse health care needs of the urban poor and other vulnerable sections.
- Institutional mechanism and management systems to meet the health-related challenges of a rapidly growing urban population.
- Partnership with community and local bodies for a more proactive involvement in planning, implementation, and monitoring of health activities.
- Availability of resources for providing essential primary health care to urban poor.
- Partnerships with NGOs, for profit and not for profit health service providers and other stakeholders.

Under the Scheme, the following infrastructural coverage has been enlisted:

- One Urban Primary Centre (U-PHC) for every fifty to sixty thousand population
- One Urban Community Health Centre (U-CHC) for five to six U-PHC for big cities
- One Auxiliary Nursing Midwifery (ANM) for 10,000 population
- One Accredited Social Health Activist (ASHA) for 200–500 population

The estimated cost of NUHM for 5-year period is Rs 22,507 crore with Central Government share of Rs 16,955 crore. The centre-state funding pattern will be 75:25 for all the States except North-Eastern states including Sikkim and other special category states of Jammu and Kashmir, Himachal Pradesh and Uttarakhand, for whom the centre-state funding pattern will be 90:10.

NUHM intends to target slum dwellers and other marginalized groups like rickshaw pullers, street vendors, railway and bus station coolies, homeless people, street children, construction site workers and the like.

The Mission will be implemented in 779 cities and towns with more than 50,000 population and cover about 7.75 crore people. The interventions under the sub-mission is intended to result on

- Reduction of Infant Mortality Rate (IMR)
- Reduction of Maternal Mortality Rate (MMR)
- Universal access to reproductive health care
- Convergence of all health related interventions

The existing institutional mechanism and management system created and functioning under NRHM will be strengthened to meet the needs of NUHM. City wise Implementation Plan is to be prepared based on base-level survey and felt-need. The Programme Implementation Plans (PIPs) sent by the by the states are to be apprised and approved by the Ministry. Urban local bodies will be fully involved in implementation of the schemes'.

Other Important programmes on Population Stabilization:

(A) Jansankhya Sthirata Kosh:

'The Government of India had set up a National Population Stabilization Fund (NPSF) in the year 2004–05 with a one-time grant of Rs 100 crore in the form of a corpus fund. This is now known as Jansankhya Sthirata Kosh (JSK) . This is an autonomous body registered under the Societies Registration Act, 1860. JSK has to promote and undertake activities aimed at achieving population stabilisation at a level consistent with the needs of sustainable economic growth, social development and environment protection, by 2045. JSK can take all the policy related decisions. It can raise contributions from organisations and individuals that support population stabilisation. JSK implements two schemes, namely, (i) Santushti and (ii) Prerna.

The details of schemes are as follows:

(i) **Santushti Strategy**

Santushti is a strategy of Jansankhya Sthirata Kosh (JSK) for the highly populated states of India viz Bihar, Uttar Pradesh, Madhya Pradesh, Rajasthan, Jharkhand, Chhattisgarh and Odisha. Under this strategy, Jansankhya Sthirata Kosh, invites private sector gynaecologists and vasectomy surgeons to conduct sterilization operations in Public Private Partnership mode. According to this Scheme, an accredited private Nursing Home/Hospital can sign a tripartite MOU between the State Health Society as 1st party, accredited private health facility as 2nd party and JSK as the third party. Upon signing the MOU the private hospitals/nursing homes shall be entitled to incentive by JSK whenever it conducts 10 or more Tubectomy/Vasectomy cases in a month. The accreditation is done by the district and approved by the State Health Society.

(ii) **Prerna Strategy**

In order to help push up the age of marriage of girls and space the birth of children in the interest of health of young mothers and infants, Jansankhya Sthirata Kosh (National Population Stabilization Fund) has launched PRERNA, a

Responsible Parenthood Strategy in seven focus states namely Bihar, Uttar Pradesh, Madhya Pradesh, Chhattisgarh, Jharkhand, Odisha, and Rajasthan.

The strategy recognizes and awards couples who have broken the stereotype of early marriage, early childbirth and repeated child birth and have helped change the mind-sets of the community.

In order to become eligible for award under the scheme, the girl should have been married after 19 years of age and given birth to the first child after at least 2 years of marriage. The couple will get an award of Rs 10,000/- if it is a Boy child or Rs 12,000/- if it is a Girl child. If birth of the second child takes place after at least 3 years of the birth of first child and either parent voluntarily accept permanent method of family planning within one year of the birth of the second child, the couple will get an additional award of Rs 5000/- (Boy child)/Rs 7000/- (Girl child). The amount of award is given in the form of National Saving Certificate (NSC). The scheme is meant only for BPL families'.

(c) **Empowered Action Group on Population Stabilization:**

The Empowered Action Group (EAG) was set up in 2001 by the central government to facilitate preparation of area-specific programmes in eight States, namely, Bihar, Jharkhand, MP, Chhatisgarh, Orissa, Rajasthan, UP and Uttaranchal, which have lagged behind in containing population growth to manageable levels.

Reflecting upon the areas of concern, the measures needed to achieve systemic reforms and also emerge from the socio-demographic backwardness, it was required that the EAG needs to strengthen the systems of governance and monitoring by way of involving the community through local empowerment and convergence. In a federal structure like India, there is need to strengthen the Centre-State coordination before direct interventions can be made at district levels.

The data from rapid household surveys (DLHS-1) have indicated that the services are nearly non-existent in large parts of such EAG districts/states. For example, the couple protection rate (percentage of couples practicing any spacing/terminal method of family planning) was less than 30 per cent in 34 (out of 70) districts in Uttar Pradesh and 29 (out of 37) districts of Bihar. Similarly, in 21 districts of Bihar and 14 districts of Uttar Pradesh, less than 30 per cent pregnant women only were reached with ante-natal services.

The EAG was required to work with the participating States in formulating their action plan for improving service delivery. The plans will be prepared on the basis of the following guiding principles:

- Problems in the EAG States are less to do with the availability of funds than the issue of governance. Therefore, intervening action areas fall in the key aspects of human resource management, logistics management, mainstreaming of the ISM practitioners, integration of numerous health societies at State and district levels, regular release of funds to operational levels, joint planning/training for the field staff of the cognate departments, greater autonomy to the districts and

within districts, to hospitals and PRIs and have such actions to be integral parts of the plan.
- Within a State, incremental investments (that may be provided by the EAG) are to be utilized in bridging the intra-state demographic divide. A key objective in this regard would be to ensure, through a systemic re-structuring of manpower in association with physical improvement, that the district and sub-divisional hospitals in the backward districts in a State provide the full range of RCH services including 24-h availability of emergency obstetric services.

A major proportion of the funds available under the Rural Connectivity Scheme, Drinking Water Supply Scheme, the SJGSY Scheme and other Centrally Sponsored Schemes of the Department of Rural Development, are to be invested in the backward districts. Central assistance for these schemes will be integral'.

(c) Community Incentive Scheme:

Consequent to the adoption of National Population Policy, 2000 the Village Panchayats have become important stake holders in the health and family welfare services. One of the promotional and motivational incentives is to award the Gram Panchayats and Zillaparishads for exemplary performances in universalization of small family norm, achieving in reduction of IMR and Birth Rate and promoting literacy for primary schooling for achieving the goal of TFR by 2010.

The scheme was initially taken in 7 EAG states of Uttar Prdesh, Uttarakhand, Bihar Jharkhand, Chhattisgarh and Orissa covering 236 districts. Rajasthan was excluded since an incentive scheme on sterilization is already in place there.

(d) Sterilisation and IUD Insertion (Compensation Scheme):

The Central Government provides fund for loss of wages to acceptors of Sterilisation/IUD Insertion under the National Family Welfare programme to States/Union Territories for Tubectomy/Vasectomy/IUD insertion. This amount of compensation is for meeting the expenditure on drugs and dress as also on diet, transport and cash compensation to acceptors of sterilisation. The compensation package has now been raised at Rs 400.00 both for tubectomy and vasectomy. The amount of IUD insertion continues to be at Rs 20.00. Under this Compensation package while the State/UT could make adjustments among the components of compensation, a minimum amount of Rs 200.00 has to be paid to the acceptor of tubectomy and Rs 250.00 the acceptor of vasectomy. Further, the amount of compensation has to be paid at the time sterilisation only. Additionally, for quality of service a minimum amount of Rs 65.00 have to be spent for drugs and dressings for tubectomy, Rs 25.00 for vasectomy and Rs 20 for IUD. For EAG states, an additional amount of Rs 100.00 for tubectomy and Rs 200.00 could be given from EAG scheme.

(e) National Population Commission:

National Population Commission was formed on May 11, 2000 as India's headcount exceeded the one-billion mark at 12.56 h IST on that day when the

Government of India announced the setting up of a 100-member National Population Commission (NPC) with the then Prime Minister, Mr. AtalBihari Vajpayee, as its Chairman with the Deputy Chairman Planning Commission as vice chairman. Chief Ministers of all states, ministers of the related central ministries, secretaries of the concerned departments, eminent physicians, demographers and the representatives of the civil society were made members of the commission in a major bid to provide focused thrust to the stabilisation of population and stem its further growth.

The commission has the mandate

- to review, monitor and give direction for implementation of the National Population Policy with the view to achieve the goals set in the Population Policy
- promote synergy between health, educational environmental and developmental programmes so as to hasten population stabilization
- promote inter sectoral coordination in planning and implementation of the programmes through
- develop a vigorous peoples programme to support this national effort to the State plan'.

(f) National Institute of Health & Family Welfare

'National Institute of Health & Family Welfare (NIHFW) is an Apex Technical Institute, funded by Ministry of Health and Family Welfare for promotion of Health and Family Welfare programmes in the country through education, training, research, evaluation, consultancy and specialised services. NIHFW was established on 9th March, 1977 by merger of National Institute of Health Administration and Education (NIHAE) and National Institute of Family Planning (NIFP).

The National Institute of Health and Family Welfare is an apex technical institute of public health and family welfare in the country aiming to achieve wellbeing and vitality of the Indian people. The Institute has been addressing a wide range of issues on public health and family welfare management by its multi-disciplinary functions in the form of research, consultancy, education and training interventions for last more than three decades. In-service training of middle and senior level health professionals, and research have been the two major concerns of its mandate. It has been identified by the Ministry of Health and Family Welfare as the nodal institute for supervision, analysis and country report writing of 'Annual Sentinel Surveillance for HIV Infection', coordinating in-service training under National Rural Health Mission (NRHM) with the support of eighteen collaborating training institutions and to conduct ten weeks 'Professional Development Course in Management, Public Health and Health Sector Reforms for District Medical Officers' in collaboration with State Institutes of Health and Family Welfare and Central Training Institutions.'

(g) International Institute of Population Sciences:

International Institute for Population Sciences (IIPS) was established in 1956 under the joint sponsorship of Sir Dorabji Tata Trust, the government of India and the United Nations; it has established itself as the premier institute for training and research in Population Studies for developing countries in the Asia and Pacific region. Till July 1970, it was known as Demographic Training and Research Centre (DTRC) and thereafter it was known as The International Institute for Population Studies (IIPS) till 1985 when it was renamed International Institute for Population Sciences (IIPS) . Institute, now occupies key positions in the field of population and health in government of various countries, universities and research institutes as well as in reputed national and international organizations.

Chapter 18
The Story of Achievements

Abstract Unlike the compulsory birth planning system in China, India has opted out for the voluntary family planning in India. The status of achievement is bound to be different from a command system to a democratic format with laissez-faire approach. This is not all. The issue of population problems is seldom looked upon by the people of India at large as seriously and objectively as it deserves; it is the least monitored social concern in India. Just like varied perception of individuals, there are diverse stake holders from among economists, health professionals, social scientists, Women groups, religious leaders, political parties, etc. with plural opinions on to the enormity of the population problem in India. Such fractured social mind-set is not ideally conducive for family planning-friendly environment to bring out maximum possible outcome of the population control and the family planning initiatives in India.

Keywords Voluntary family planning · Demographic indicators · Projected and actual population in India · Family Planning Indicators · Lassiaze faire approach · Religious agenda · National Health Mission · Sustainable agenda

Unlike the compulsory birth planning system in China, India has opted out for the voluntary family planning in India. The status of achievement is bound to be different from a command system to a democratic format with laissez-faire approach. This is not all. The issue of population problems is seldom looked upon by the people of India at large as seriously and objectively as it deserves; it is the least monitored social concern in India. Just like varied perception of individuals, there are diverse stake holders from among economists, health professionals, social scientists, Women groups, religious leaders, political parties etc. with plural opinions on to the enormity of the population problem in India. Such fractured social mind-set is not ideally conducive for family planning-friendly environment to bring out maximum possible outcome of the population control and the family planning initiatives in India.

Be that as it may, India happens to be the first country in the world to have included population in the approach to the Five Year Plan even before the formal adoption of either the Health Policy or the National Population Policy. In the First Five Year Plan, India had launched the National Family Welfare Programme with the objective of reducing the birth rate to the extent necessary to stabilize the population at a level consistent with the requirement of the National economy. Since then the National Family planning Programme became an integral part of every Five Year Plan until 1977 when it was renamed as the National Family welfare Programme and formed a part of every edition of Five year Plan to improve the Family Welfare status of the country. At Chap. 16, the journey of achievements on each of the Five Year Plans has been captured. In this Chapter, the story of achievements has been sought to be presented through an overtime series of demographic, vital statistics and Family Planning Indicators which themselves are self-introductory.

Demographic Indicators:
The demographic indicators show phenomenal rise of the population of India with a robust size of decadal growth. The ever-expanding density of India is also a sharp pointer to the load pressure on the geographical space of India. Even the optimistic projected population could not be adhered to at the end of Census, 2011. The enormity of size of population of some of the selected States bears out its alarming character (Tables 18.1 and 18.2).

The country's headcount is almost equal to the combined population of the United States of America (USA), Indonesia, Brazil, Pakistan, Bangladesh and Japan—all put together. The combined population of UP and Maharashtra is bigger than that of the USA. Population of many Indian States is comparable with countries like United Kingdom (UK), Germany, Italy, Japan, Mexico, etc. as would appear from Table 18.3.

Vital Statistics Indicators:
Vital Statistics like Crude Birth Rate, Crude Death Rate, Infant Mortality Rate, Maternal Mortality Ratio, Total Fertility Rate reveal the state of achievements of broader population control and family welfare scenario. Efforts have been made to capture the same from 1951 to 2011. The latest SRS data for such indicators for the States of India have also been highlighted.

Table 18.1 Trend in census population from 1951–2011

Census year	Total population (in Lakh)			Decadal growth (%)	Density	Sex ratio
	Male	Female	Total			
1951	1855.3	1755.6	3610.9	13.31	117	946
1961	2262.9	2129.4	4392.3	21.64	142	941
1971	2840.5	2641.1	5481.6	24.80	177	930
1981	3533.7	3299.5	6833.3	24.66	216	934
1991	4339.6	4070.6	8464.2	23.87	267	926
2001	5322.2	4765.1	10,287.4	21.54	325	933
2011	6232.7	5815.9	12,108.6	17.7	382	943

Source census publications

18 The Story of Achievements

Table 18.2 Projected and actual population in India, States and Union Territories, 2011 (in '000)

India/state	Projected population, 2011	Actual population, 2011	Difference	Percentage difference
India	1,192,506.00	1,210,855.00	18,349.00	1.54
Andhra Pradesh	84,735.00	84,580.00	−155.00	−0.18
Arunachal Pradesh	1241.00	1383.00	142.00	11.44
Assam	30,568.00	31,205.00	637.00	2.08
Bihar	97,720.00	104,099.00	6379.00	6.53
Chhattisgarh	24,258.00	25,545.00	1287.00	5.31
Goa	1767	1458.00	−309.00	−17.49
Gujarat	59,020.00	60,439.00	1419.00	2.40
Haryana	25,439.00	25,351.00	−88.00	−0.35
Himachal Pradesh	6793.00	6864.00	71.00	1.05
Jammu and Kashmir	11,718	12,541.00	823.00	7.02
Jharkhand	31,472.00	32,988.00	1516.00	4.82
Karnataka	59,419.00	61,095.00	1676.00	2.82
Kerala	34,563.00	33,406.00	−1157.00	−3.35
Madhya Pradesh	72,200.00	72,626.00	426.00	0.59
Maharashtra	112,660.00	112,374.00	−286.00	−0.25
Manipur	2449.00	2856.00	417.00	16.62
Meghalaya	2621.00	2966.00	345.00	13.16
Mizoram	1004.00	1097.00	93.00	9.26
Nagaland	2249.00	1978.00	−271.00	−13.05
Odisha	40,750.00	41,974.00	1224.00	3.00
Punjab	22,678.00	27,743.00	65.00	0.23
Rajasthan	67,830.00	68,548.00	718	1.06
Sikkim	612.00	610.00	−2.00	−0.37
Tamil Nadu	67,444.00	72,147.00	4703.00	6.97
Tripura	3616.00	3673.00	57.00	1.58
Uttar Pradesh	200,764.00	199,812.00	−952.00	−0.47
Uttarakhand	9943.00	10,086.00	143.00	1.44
West Bengal	89,499.00	91,276.00	1777.00	1.99
Andaman &Nicobar Island	494.00	380.00	−114.00	−23.08
Chandigarh	1438.00	1055.00	−353.00	−26.63
Dadra and Nagar and Haveli	354.00	343.00	−11.00	−3.11
Daman and Diu	270.00	243.00	−27.00	−10.00
NCT of Delhi	18,451.00	16,787.00	−1664.00	−9.02
Lakshadweep	76.00	64.00	−12.00	−15.79
Puducherry	1391.00	1247.00	−144.00	−10.35

Source national health profile, 2013

Table 18.3 Population in states in India vs. countries in the world

States in India	Population in 2011	Country@	Population@
Uttar Pradesh	199.6	Brazil	195.4
Maharashtra	112.4	Japan	127.0
Bihar	103.8	Mexico	110.5
West Bengal	91.3	Philippines	93.6
Andhra Pradesh	84.7	Germany	82.1
Madhya Pradesh	72.6	Turkey	72.7
Tamil Nadu	72.1	Thailand	68.1
Rajasthan	68.6	France	62.8
Karnataka	61.1	United Kingdom	61.9
Gujarat	60.4	Italy	60.1
Orissa	41.9	Argentina	40.7
Kerala	33.4	Canada	33.9
Jharkhand	33.0	Morocco	32.4
Assam	31.2	Iraq	31.5
Punjab	27.7	Malaysia	27.9
Chhattisgarh	25.5	Saudi Arabia	26.2
Haryana	25.4	Australia	21.5

Source Family Welfare Statistics in India, 2011 (Revised); @*Source* state of world population 2010

Table 18.4 Crude birth rate, crude death rate, infant mortality rate, maternal mortality ratio, total fertility rate in India since 1951

Items	1951	1961	1971	1981	1991	2001	2011	2013
CBR	40.8	41.7	36.9	33.9	29.5	25.4	21.8	21.4
CDR	25.1	22.8	14.9	12.5	9.8	8.4	7.1	7.0
IMR	146	138	129	110	80	66	44	40
MMR	437	NA	NA	NA	398	301	178	NA
TFR	6.0	5.7	5.4	4.5	3.6	3.1	2.4	2.3

Source Chap. 3, estimates of fertility indicators-www.censusindia.gov.in/vital statistics/SRS, 2014/9th Plan-vol-2/Family Welfare statistics, 2011

The crude birth rate (defined as a ratio of the total number of births during given year and a given geographical area) at the all India level had declined from 40.8 in 1951 to 21.8 in 2011, registering a fall of about 53.43 %. During 1991–2013, the decline has been about 27.5 %, from 29.5 to 21.4 (Table 18.4).

The SRS, 2013 has shown the latest figures on vital statistics which reflects the extent of outreach of family welfare message and its adoption by the eligible couples. The CBR at national level was 21.4 in 2013 with 22.9 in rural to 17.3 in urban areas. Andhra Pradesh, Delhi, Himachal Pradesh, Jammu and Kashmir, Karnataka, Kerala, Maharashtra, Punjab, Tamil Nadu and West Bengal happened to be the major States having birth rate below 20 both in rural and urban areas. On the other hand, Bihar had the highest birth rate in rural areas (28.3) and Uttar Pradesh had the highest in urban areas (23.3) followed by Rajasthan (22.0). The lowest CBR was recorded in rural areas 15.0 of Kerala and in urban areas 10.9 of Himachal Pradesh (Table 18.5).

The total fertility rate (TFR), which measures average number of children born to a woman during her entire reproductive period, had declined from 6.0 in 1951 to 4.5 during 1981 and from 3.6 to 2.4 during 1991 to 2011. It has since declined at 2.3 in 2013, the corresponding TFR in urban areas and rural areas being 1.8 and 2.5 respectively.

Among the bigger States, TFR varies from 1.6 in West Bengal to 3.4 in Bihar. For rural areas, it varies from 1.7 in Himachal Pradesh, Punjab and Tamil Nadu to 3.5 in Bihar. For urban areas, such variation is from 1.2 in Himachal Pradesh and West Bengal to 2.5 in Bihar and Uttar Pradesh. At Table 18.6, levels of TFR by residence for India and bigger States, 2013 have been shown:

Total Marital Fertility Rate (TMFR) is the cumulative value of age specific marital fertility rates at the end of the reproductive period. It indicates the average number of children expected to be born per married woman during the entire span of her reproductive period. The TMFRs, as worked out in census India on the basis of ASMFRs (defined as the number of children born to married women in the said

Table 18.5 CBR in the states of India in 2013

India and bigger states	Total	Rural	Urban
India	**21.4**	**22.9**	**17.3**
Andhra Pradesh	17.4	17.7	16.7
Assam	22.4	23.5	15.4
Bihar	27.6	28.3	21.5
Chhattisgarh	24.4	25.8	17.9
Delhi	17.2	18.9	16.9
Gujarat	20.8	22.2	18.5
Haryana	21.3	22.4	19.0
Himachal Pradesh	16.0	16.5	10.9
Jammu and Kashmir	17.5	18.7	12.6
Jharkhand	24.6	25.9	18.5
Karnataka	18.3	19.1	16.7
Kerala	14.7	15.0	14.0
Madhya Pradesh	26.3	28.2	19.6
Maharashtra	16.5	17.2	15.4
Odisha	19.6	20.5	14.4
Punjab	15.7	16.3	14.7
Rajasthan	25.6	26.7	22.0
Tamil Nadu	15.6	15.7	15.5
Uttar Pradesh	27.2	28.1	23.3
West Bengal	16.0	17.7	11.4

Source Chap. 3, estimates of fertility indicators-www.censusindia.gov.in/vital statistics/SRS, 2013

Table 18.6 TFR (total fertility rate) by residence, India and bigger States, 2013

India and bigger states	Total	Rural	Urban
India	2.3	2.5	1.8
Andhra Pradesh	1.8	1.9	1.7
Assam	2.3	2.4	1.5
Bihar	3.4	3.5	2.5
Chhattisgarh	2.6	2.8	1.8
Delhi	1.7	1.8	1.7
Gujarat	2.3	2.5	2.0
Haryana	2.2	2.3	2.0
Himachal Pradesh	1.7	1.7	1.2
Jammu and Kashmir	1.9	2.0	1.3
Jharkhand	2.7	2.9	2.0
Karnataka	1.9	2.0	1.6
Kerala	1.8	1.9	1.8
Madhya Pradesh	2.9	3.1	2.0
Maharashtra	1.8	1.9	1.6
Odisha	2.1	2.2	1.5
Punjab	1.7	1.7	1.6
Rajasthan	2.8	3.0	2.3
Tamil Nadu	1.7	1.7	1.7
Uttar Pradesh	3.1	3.3	2.5
West Bengal	1.6	1.8	1.2

Source Chap. 3, estimates of fertility indicators-www.censusindia.gov.in/vital statistics/SRS, 2013

age group per 1000 women in the same age group), for the year 2013 are given below for India and bigger States separately for rural and urban areas. The TMFR for India is found to be 4.4 and varies from 4.2 in urban areas to 4.5 in rural areas. The TMFR is 5 and above in the States of Assam, Bihar, Jammu & Kashmir, Jharkhand and Uttar Pradesh (Table 18.7).

Family Planning Indicators

The great hype and sense of complicity in family planning achievements in India needs to be validated with reference to field data, which the government itself has collected. The Birth order, Birth interval, Medical attention at delivery and Methods of Family Planning, the extent of its coverage and the linked prevention of birth have been shown hereunder which themselves speak volumes of the real family planning achievements in India.

(i) Birth order

From family planning point of view, data on order of the live birth and interval between current and previous live births are very important. The SRS collects the data from 1990 onwards. These provide useful information on effectiveness on family planning initiatives on spacing of children and level of fertility. The

Table 18.7 Total marital fertility rates (TMFR) by residence in India and bigger states, 2013

India and bigger states	Total	Rural	Urban
India	4.4	4.5	4.2
Andhra Pradesh	3.3	3.2	3.4
Assam	5.0	5.1	4.1
Bihar	5.7	5.7	4.9
Chhattisgarh	4.5	4.5	3.9
Delhi	3.9	3.1	4.3
Gujarat	4.3	4.4	4.2
Haryana	3.9	3.9	3.8
Himachal Pradesh	4.4	4.3	5.6
Jammu and Kashmir	6.2	6.5	4.3
Jharkhand	5.2	5.2	5.4
Karnataka	3.9	3.9	3.9
Kerala	4.0	4.0	4.0
Madhya Pradesh	4.7	4.8	4.0
Maharashtra	2.9	2.7	3.7
Odisha	4.3	4.3	3.8
Punjab	4.3	3.9	5.2
Rajasthan	4.2	4.3	4.1
Tamil Nadu	4.2	4.0	4.6
Uttar Pradesh	6.0	6.0	6.4
West Bengal	3.5	3.6	3.0

Source Chap. 3, estimates of fertility indicators-www.censusindia.gov.in/vital statistics/SRS, 2013

estimated percentages on order of live birth and birth interval for India and bigger States have been shown at Tables 18.8 and 18.9 respectively.

It is found from Table 18.8 that 43.6 % of the current live births in India are first order births, and 32.4 % of total births are second order births. The third order births account for 13.5 % while the fourth and higher order births aggregate at 10.5 % of the total births. The third and fourth order births are broad indicators of unmet needs of contraception and inability to reach out to such eligible couples for required family planning counselling.

Among the bigger States, the percentage share of first order birth varies from 35.4 % in Bihar to 57.4 % in West Bengal. On the other hand, the percentage share of fourth and higher order births varies from 1.2 % in Andhra Pradesh to 19.6 % in Bihar.

(ii) Birth Interval

As mentioned earlier, birth spacing is an important initiative of the family planning services to control birth in a family and also protect the health of the mother and the child. Thus the birth spacing interval gives a good account of the effectiveness of family welfare services. At Table 18.9, the percentage distribution

Table 18.8 Percentage distribution of current live births by birth order, India and bigger states, 2013

India and bigger states	Birth order			
	Ist	2nd	3rd	4th and above
India	**43.6**	**32.4**	**13.5**	**10.5**
Andhra Pradesh	49.1	41.9	7.7	1.2
Assam	49.2	27.8	12.6	10.4
Bihar	35.4	27.5	17.5	19.6
Chhattisgarh	38.7	35.4	17.0	8.8
Delhi	48.3	32.0	12.1	7.6
Gujarat	47.0	30.8	13.2	9.0
Haryana	44.8	35.4	12.1	7.7
Himachal Pradesh	50.5	36.5	8.1	4.8
Jammu & Kashmir	43.2	31.8	14.1	11.0
Jharkhand	38.4	29.9	17.6	14.1
Karnataka	46.5	36.2	12.9	4.4
Kerala	48.7	38.3	10.7	2.3
Madhya Pradesh	41.0	33.2	14.5	11.3
Maharashtra	47.8	36.1	11.3	4.9
Odisha	45.1	34.4	12.4	8.1
Punjab	55.2	32.5	8.5	3.7
Rajasthan	38.5	30.2	15.6	15.7
Tamil Nadu	51.4	40.1	7.2	1.3
Uttar Pradesh	39.3	30.5	15.8	14.4
West Bengal	57.4	29.5	8.2	4.9

Source Chap. 3, estimates of fertility indicators-www.censusindia.gov.in/vital statistics/SRS, 2013

of second and higher order live births by interval between current and previous live birth has been shown for India and bigger States for the year 2013. At the All India level, 1.8 % of the live births occur within one year from the previous live birth. Such percentage varies from 0.2 in Kerala to 2.9 in Chhattisgarh. The percentage of births beyond three years of birth interval from the previous live birth for India is 40.7. It varies from 31.0 in Madhya Pradesh to 65.9 in Kerala. The birth interval data indicates scope for important in this area in quite a good number of states.

At Table 18.10, interval between current and previous live birth separately for rural and urban areas has been shown. At the National level, 1.9 % of live births have been reported within an interval of one year for rural and 1.2 % in urban areas. The data reveals that so far as spacing of children is considered, there is marginal difference between the rural and urban areas. It also indicates that about half of the births have spacing of 36 months and above in urban compared to about two fifth in rural areas. More than 70 % of births have birth interval of 24 and more months both in rural and urban areas.

18 The Story of Achievements

Table 18.9 Percentage distribution of second and higher order live births by interval, India and bigger States, 2013

India and bigger states	Interval between current and previous live birth (in months)			
	10–12	12–24	24–36	36+
India	**1.8**	**28.0**	**29.5**	**40.7**
Andhra Pradesh	0.9	33.3	31.0	34.8
Assam	1.5	17.8	22.9	57.8
Bihar	1.4	32.6	30.8	35.2
Chhattisgarh	2.9	26.0	33.5	37.7
Delhi	0.6	22.2	23.5	53.7
Gujarat	2.6	26.8	29.4	41.2
Haryana	1.8	29.9	31.0	37.4
Himachal Pradesh	1.9	30.9	27.3	39.9
Jammu and Kashmir	1.3	22.3	27.0	49.4
Jharkhand	2.5	24.1	29.4	44.0
Karnataka	0.6	31.3	33.7	34.3
Kerala	0.2	12.0	21.9	65.9
Madhya Pradesh	2.2	30.9	35.9	31.0
Maharashtra	1.1	28.8	30.2	40.0
Odisha	0.9	16.6	26.8	55.7
Punjab	1.7	26.8	28.6	42.9
Rajasthan	2.8	32.8	29.0	35.5
Tamil Nadu	0.8	26.6	30.4	42.2
Uttar Pradesh	2.2	29.2	28.2	40.4
West Bengal	1.1	17.8	24.5	56.5

Source Chap. 3, estimates of fertility indicators-www.censusindia.gov.in/vital statistics/SRS, 2013

(iii) Medical attention at delivery

The Family Planning professionals always campaign for institutional delivery for better services with various tied up incentives. The latest outcome of such campaign has been captured in the SRS data, 2013 which gives the percentage distribution of live births recorded in the year 2013 by type of medical attention received by the mother at the time of delivery at Government Hospital, Private Hospital, Qualified professional, Untrained functionary and others for India and bigger States separately for rural and urban areas. This indicates the extent of spread of the family planning infrastructure in the country and also the linked quality of the family welfare services. At the National level, 50.0 % births were attended by Government Hospitals and vary from 48.8 % in rural areas to 55.0 % in urban areas. Among the bigger States, it varies from 33.1 % in Jharkhand to 67.9 % in Rajasthan. About 24.4 % of birth occurred at Private Hospital. Medical attention by qualified professionals constitutes 12.7 % of total delivery whereas untrained and others constitute 12.9 %. More than three fourth of deliveries are occurring in

Table 18.10 Percentage distribution of second and higher order live births by interval and residence, India and bigger states, 2013

India and bigger states	Interval between current and previous live birth (in months)							
	Rural				Urban			
	10–12	12–24	24–36	36+	10–12	12–24	24–36	36+
India	**1.9**	**29.1**	**30.3**	**38.8**	**1.2**	**23.7**	**26.3**	**48.8**
Andhra Pradesh	1.2	34.3	31.4	33.2	0.2	30.9	30.1	38.8
Assam	1.5	18.4	23.3	56.8	0.9	10.7	18.3	70.1
Bihar	1.4	32.8	30.9	34.9	2.3	29.6	28.9	39.3
Chhattisgarh	2.9	26.2	33.7	37.2	2.5	24.2	31.8	41.6
Delhi	0.5	29.6	22.4	47.5	0.6	20.8	23.8	54.8
Gujarat	3.3	29.9	30.9	35.9	1.1	20.2	26.4	52.3
Haryana	1.6	31.6	32.4	34.3	2.0	25.1	27.1	45.7
Himachal Pradesh	2.1	31.6	27.8	38.5	0.0	20.3	19.7	60.1
Jammu and Kashmir	1.4	23.1	27.3	48.2	0.3	16.8	24.9	58.0
Jharkhand	2.7	24.7	29.5	43.2	1.3	19.6	28.8	50.4
Karnataka	0.6	33.1	36.5	29.8	0.7	27.2	27.2	44.8
Kerala	0.2	12.3	21.7	65.8	0.2	10.7	22.5	66.6
Madhya Pradesh	2.4	32.3	37.4	27.9	1.3	22.6	27.0	49.0
Maharashtra	1.2	32.7	33.8	32.3	0.8	22.8	24.6	51.7
Odisha	1.0	16.6	27.2	55.2	0.2	16.5	22.5	60.8
Punjab	2.0	28.6	29.1	40.3	1.2	23.1	27.6	48.1
Rajasthan	3.0	33.8	30.4	32.8	1.7	28.1	22.9	47.2
Tamil Nadu	1.0	29.8	32.4	36.9	0.5	22.0	27.7	49.8
Uttar Pradesh	2.2	29.8	28.2	39.8	2.1	25.7	28.0	44.1
West Bengal	1.3	17.9	25.4	55.4	0.5	17.5	20.3	61.7

Source Chap. 3, estimates of fertility indicators-www.censusindia.gov.in/vital statistics/SRS, 2013

institutions and conducted by the qualified professional as per report of SRSs shown at Table 18.11.

(iv) Methods of Family Planning, the extent of coverage

Various methods of Family Planning have been canvassed over the years by the Family Planning Professionals for temporary or permanent methods. At Table 18.12, it has been shown, year wise, estimated number of births that were averted by the methods of Family Planning for the country as a whole as worked out by the related government department. The extent of fully protected couples by all methods of family planning since 1980–81 has also been shown at Table 18.13. A separate estimation made by non-government professionals for avertion of births through family welfare methods has also been shown at Table 18.14.

To sum up the achievements, the Population control and Family Planning Programme in India has achieved a modest success in that it could avert 442.747 million additional births as per government estimate as on 2010–11 since 1956,

18 The Story of Achievements

Table 18.11 Per cent distribution of live births by type of Medical Attention received by the mother at delivery by residence, India and bigger States, 2013

	Govt. Hospital			Private Hospital			Qualified professional			Untrained functionary and others		
	Total	Rural	Urban	Total	Rural	Urban	Total	Rural	Urban	Total	Rural	Urban
India	**50.0**	**48.8**	**55.0**	**24.4**	**20.9**	**37.1**	**12.7**	**14.4**	**6.1**	**12.9**	**15.9**	**1.7**
Andhra Pradesh	50.0	52.4	44.0	43.6	38.9	55.7	5.7	7.8	0.2	0.7	1.0	0.1
Assam	48.5	47.0	62.6	25.2	24.6	30.6	12.5	13.3	5.9	13.7	15.1	0.9
Bihar	39.7	38.9	49.1	20.6	19.2	37.5	12.0	12.3	9.3	27.6	29.7	4.1
Chhattisgarh	43.4	41.3	57.4	23.1	22.7	25.9	20.8	21.8	13.8	12.7	14.1	2.9
Delhi	61.7	61.7	61.6	29.4	27.9	29.7	6.1	6.3	6.1	2.8	4.1	2.5
Gujarat	42.7	43.0	42.1	45.9	40.3	56.6	9.3	13.5	1.1	2.2	3.2	0.2
Haryana	39.0	40.0	36.5	38.3	34.8	47.6	19.7	21.4	15.2	2.9	3.7	0.6
Himachal Pradesh	55.8	54.2	79.9	15.0	15.0	14.2	20.5	21.6	4.5	8.7	9.2	1.4
Jammu and Kashmir	65.1	62.3	81.4	16.5	16.7	15.0	8.6	9.6	2.6	9.8	11.4	1.0
Jharkhand	33.1	30.3	49.6	18.1	15.0	37.0	17.5	18.4	12.1	31.3	36.2	1.4
Karnataka	64.4	66.7	59.6	28.5	24.0	37.7	3.9	5.4	1.0	3.2	3.9	1.7
Kerala	45.9	48.6	37.2	53.6	51.0	62.1	0.3	0.2	0.5	0.2	0.2	0.1
Madhya Pradesh	52.5	49.5	68.3	23.0	22.4	25.9	12.3	14.0	3.5	12.2	14.0	2.3
Maharashtra	54.5	52.6	57.4	40.0	39.0	41.6	3.4	5.1	0.7	2.1	3.2	0.3
Odisha	58.5	57.3	69.5	13.8	13.0	21.5	11.2	11.7	6.4	16.5	18.1	2.6
Punjab	36.8	36.1	38.1	47.5	44.0	54.0	13.9	17.6	7.2	1.7	2.3	0.7
Rajasthan	67.9	66.9	71.9	15.1	14.2	18.7	13.3	14.4	8.7	3.7	4.5	0.7
Tamil Nadu	62.1	65.1	58.0	31.3	24.2	41.0	6.3	10.3	0.7	0.3	0.4	0.2
Uttar Pradesh	43.3	42.8	45.7	14.8	11.9	30.3	22.1	22.6	19.3	19.8	22.7	4.6
West Bengal	63.6	63.2	65.2	13.6	10.7	25.7	8.3	9.2	4.8	14.5	16.9	4.3

Source Chap. 3, estimates of fertility indicators-www.censusindia.gov.in/vital statistics/SRS, 2013

Table 18.12 Number of births (in million) averted since 1956 by various methods of family planning

Year	Sterilisations	I.U.D.	CC	OP	Other methods (col. 4 + 5)	Total (all methods)	Cumulative total
1	2	3	4	5	6	7	8
1956–60	0.036	0.000			0.000	0.036	0.036
1961	0.034	0.000			0.000	0.034	0.070
1962	0.055	0.000			0.000	0.055	0.125
1963	0.086	0.000	0.009		0.009	0.095	0.219
1964	0.119	0.000	0.039	0.000	0.039	0.158	0.377
1965–66	0.179	0.024	0.055	0.000	0.055	0.258	0.635
1966–67	0.316	0.171	0.064	0.000	0.064	0.552	1.187
1967–68	0.510	0.280	0.054	0.000	0.054	0.844	2.031
1968–69	0.866	0.323	0.069	0.000	0.069	1.258	3.289
1969–70	1.159	0.330	0.127	0.000	0.127	1.666	4.905
1970–71	1.400	0.330	0.188	0.000	0.188	1.917	6.823
1971–72	1.585	0.318	0.239	0.000	0.239	2.141	8.964
1972–73	1.957	0.300	0.274	0.000	0.274	2.532	11.496
1973–74	2.437	0.260	0.296	0.000	0.296	2.993	14.488
1974–75	2.463	0.232	0.335	0.000	0.335	3.030	17.518
1975–76	2.584	0.201	0.321	0.002	0.323	3.107	20.625
1976–77	3.068	0.227	0.409	0.009	0.418	3.713	24.338
1977–78	4.401	0.231	0.408	0.015	0.423	5.055	29.393
1978–79	4.329	0.202	0.374	0.018	0.393	4.924	34.317
1979–80	4.286	0.221	0.381	0.019	0.408	4.907	39.224
1980–81	4.300	0.245	0.368	0.020	0.387	4.933	44.157
1981–82	4.369	0.262	0.452	0.023	0.475	5.106	49.263
1982–83	4.587	0.299	0.553	0.031	0.585	5.471	54.734
1983–84	5.013	0.400	0.724	0.074	0.798	6.210	60.944
1984–85	5.533	0.620	0.913	0.202	1.115	7.269	68.212
1985–86	5.975	0.831	1.012	0.303	1.315	8.121	76.733
1986–87	6.484	1.076	1.102	0.342	1.444	9.003	85.336
1987–88	7.005	1.330	1.184	0.438	1.622	9.956	95.293
1988–89	7.467	1.559	1.347	0.499	1.846	10.873	106.165
1989–90	7.824	1.777	1.491	0.582	2.074	11.674	117.840
1990–91	8.039	1.940	1.659	0.667	2.326	12.305	130.144
1991–92	8.189	2.081	1.684	0.739	2.423	12.693	142.837
1992–93	8.311	3.041	1.642	0.760	2.402	12.753	155.590
1993–94	8.466	2.110	1.806	0.772	2.578	13.154	168.745
1994–95	8.640	2.355	2.017	1.031	3.047	14.042	182.786
1995–96	8.798	2.609	2.042	1.143	3.185	14.593	197.379
1996–97	8.888	2.746	2.004	1.190	3.194	14.828	212.207

(continued)

Table 18.12 (continued)

Year	Sterilisations	I.U.D.	CC	OP	Other methods (col. 4 + 5)	Total (all methods)	Cumulative total
1997–98	8.874	2.667	1.985	1.284	3.269	14.810	227.017
1998–99	8.894	2.694	1.967	1.515	3.482	15.070	242.087
1999–2000	8.927	2.709	2.044	1.656	3.700	15.336	257.423
2000–01	9.006	2.737	2.106	1.791	3.897	15.640	273.064
2001–02	9.112	2.741	2.099	1.834	3.933	15.787	288.850
2002–03	9.223	2.773	2.203	2.059	4.262	16.258	305.108
2003–04	9.341	2.769	2.658	2.213	4.871	16.981	322.089
2004–05	9.459	2.760	2.767	2.256	5.018	17.237	339.326
2005–06	9.563	2.762	2.823	2.141	4.965	17.289	356.615
2006–07	9.618	2.760	3.054	2.219	5.273	17.651	374.266
2007–08	9.644	2.734	3.103	2.322	5.425	17.803	392.069
2008–09	9.730	2.717	2.885	2.407	5.291	17.738	409.807
2009–10	9.820	2.634	2.144	2.007	4.151	16.605	426.412
2010–11	9.901	2.612	1.921	1.900	3.821	16.335	442.747

Source Family Welfare of India, Statistics 2011, Revised in 30.11.2011

Table 18.13 Couple currently and effectively protected in India by various methods of family planning since 1980–81 (000's)

Year	Couples currently protected due to all methods	%	Couples effectively protected due to all methods	%
1980–81	28,365	24.4	26,444	22.8
1985–86	50,109	38.7	45,163	34.9
1990–91	71,933	49.6	64,071	44.1
1991–92	72,088	48.6	64,655	43.6
1992–93	73,948	48.7	65,946	43.5
1993–94	79,561	51.3	70,361	45.4
1994–95	81,737	51.6	72,552	45.8
1995–96	84,368	52.2	75,094	46.8
1996–97	83,946	51.0	74,746	46.5
1997–98	84,213	50.8	75,226	45.4
1998–99	81,888	48.6	74,221	44.0
1999–2000	83,790	51.9	79,016	46.2
2000–01	89,109	51.3	79,352	45.6
2001–02	90,366	51.1	80,812	45.7
2002–03	96,473	53.7	84,510	47.6

(continued)

Table 18.13 (continued)

Year	Couples currently protected due to all methods	%	Couples effectively protected due to all methods	%
2003–04	98,724	54.1	86,140	47.2
2004–05	98,332	53.1	85,808	46.3
2005–06	101,455	54.0	87,702	46.7
2006–07	102,166	53.6	88,144	46.2
2007–08	103,938	53.7	89,884	46.5
2008–09	94,131	48.1	83,967	42.9
2009–10	91,494	46.1	82,497	41.6
2010–11	91,121	46.6	82,502	40.4

Source Family Welfare of India, Statistics 2011, Revised in 30.11.2011

Table 18.14 Couples effectively protected and births averted through Family Planning Efforts

Year	Couples effectively protected through				Births averted	
	Sterilisation (%)	IUD (%)	Other methods (%)	All methods (%)	Number (million)	Proportional to potential births (%)
1971–72	9.9	1.4	1.1	12.4	2.56	11.16
1972–73	12.3	1.2	1.2	14.7	3.08	13.23
1973–74	12.4	1.1	1.4	14.9	3.10	13.41
1974–75	12.6	1.0	1.2	14.8	3.17	13.32
1975–76	14.2	1.0	1.7	16.9	3.78	15.21
1976–77	20.7	1.1	1.7	23.5	5.59	21.15
1977–78	20.1	0.9	1.5	22.5	5.33	20.25
1978–79	19.9	0.9	1.6	22.4	5.42	20.16
1979–80	19.9	1.0	1.4	22.3	5.51	20.07
1980–81	20.1	1.0	1.7	22.8	5.86	20.52
1981–82	20.7	1.1	2.9	24.7	6.68	22.23
1982–83	22.0	1.4	2.5	25.9	7.24	23.31
1983–84	23.7	2.2	3.7	29.6	8.85	26.64
1984–85	24.9	2.9	4.4	32.2	10.04	28.98
1985–86	26.5	3.7	4.7	34.9	11.29	31.41
1986–87	27.9	4.5	5.1	37.5	12.69	33.75
1987–88	29.0	5.2	5.7	39.9	14.02	35.91
1988–89	29.8	5.9	6.2	41.9	15.08	35.71
1989–90	39.1	6.3	6.9	43.3	15.91	38.97
1990–91	30.3	6.7	7.2	44.2	16.51	39.78
1991–92	30.3	6.3	6.9	43.5	16.14	39.15
1992–93	30.3	6.3	6.9	43.5	16.23	39.15
1993–94	30.3	6.8	8.3	45.4	17.62	40.86

(continued)

Table 18.14 (continued)

Year	Couples effectively protected through				Births averted	
	Sterilisation (%)	IUD (%)	Other methods (%)	All methods (%)	Number (million)	Proportional to potential births (%)
1994–95	30.2	7.2	8.5	45.9	18.18	41.31
1995–96	30.2	7.8	8.5	46.5	18.56	41.85
1996–97	29.6	7.4	8.3	45.3	17.74	40.77
1997–98	29.3	7.3	8.8	45.4	17.83	40.86
1998–99	29.1	7.4	7.5	44.0	16.90	39.60
1999–2000	29.0	7.3	9.9	46.2	18.45	41.58
2000–01	28.9	7.2	9.6	45.7	18.22	41.13
2001–02	28.7	7.1	9.9	45.7	18.26	41.13
2002–03	28.5	7.0	11.5	47.0	19.24	42.30
2003–04	28.0	7.0	11.9	46.9	19.13	42.21
2004–05	28.2	6.8	11.3	46.3	18.63	41.67
2005–06	27.9	6.7	12.1	46.7	18.98	42.03
2006–07	27.6	6.5	12.1	46.2	18.66	41.58
2007–08	27.5	6.4	12.5	46.5	19.01	41.85
2008–09	27.4	6.1	9.4	42.9	16.66	38.61
2009–10	27.2	6.0	8.3	41.6	15.76	37.44
2010–11	26.7	5.7	8.0	40.4	15.05	36.36

Source 40 years of Planned Family Planning Efforts in India (as Estimated) by AlokRanjanChaurasia 'Shyam' Institute, Bhopal, India Ravendra Singh National Academy of Statistical Administration, New Delhi, India

while non-government family planning expert, AlokRanjanChaurasia of 'Shyam' Institute, Bhopal and Ravendra Singh of National Academy of Statistical Administration, New Delhi, India have put the said avertion at 15.05 million in forty years from 1971 till 2011. No estimates on birth avertion are, however, available on public domain on what could have been achieved had the family planning been properly implemented in the states of India and the extent of missed opportunity. Incidentally, way back in 2000, the National population Commission published the Report of the Working Group on Strategies to address Unmet Needs which mentioned that 'India has around 29 million women having unmet needs for family planning'. It has also referred to the NFHS I (1992–93) and NFHS II(1998–99) data which reported that the unmet needs for contraception were 19.5 and 15.8 % respectively. The situation has not improved since then. The Annual Report of the Ministry of Health and Family Welfare, 2014–15 has mentioned that as per DLHS III (2007–08), the unmet needs of contraception stood at 21.3 %. Further, the Family Welfare Statistics, published by the Ministry of Health and Family Welfare in 2011, as shown in Table 18.13, captured the Couple Protection Rate from all methods since 1980–81 which reveals the enormity of unprotected couples, size wise and percentage wise up to 2010–11. These indirectly point to the level of missed opportunity to avertion of birth, though no quantitative estimate of possible birth avertion was available. The slow moving improvements of CBRs and the inability in reaching TFR at 2.1 even in 2016, after a lapse of 65 years of family planning initiatives, are contra-indications of the real effectiveness of the family planning in the country. The demographic indicators of decadal growth rate and the net periodical addition of population point to our helplessness in containing runaway population. The most worrying aspect is lack of our sensitivity to the impact of burgeoning number of population on our economy and body-politic. The country appears to be least concerned to accept the increasing load of population as a problem in improving the quality of life, the size of BPL population, the nutrition profile, the employment scenario, the natural stock of resources and in upscaling the human development profile. Further, the lassiaze faire approach of family planning has become very conducive to promote religious agenda to increase numbers with no social impunity. Additionally, the international borders of India look like welcome posts for infiltrators, and unabated flow of them add the number of population of India merrily on regular basis as this important aspect of population control is hardly addressed on ground of political expediency and vote-bank politics. Unfortunately enough, the prevention of infiltration as a part of population control programme has not been in place in India and the Ministry of Home Affairs has not been tied up as a co-partner for population control. Ironically enough, the Ministry of Health and Family Welfare now looks upon population problem merely as health care management problem within the bounds of National Health Mission. Moreover, the extent of carrying capacity of population in our country does never feature as an agenda for policy debate either in the government or in other social platforms. The country treats the tricking sound of additional number of population in the Population Watch as a disturbing noise and never bother to revisit and reinvent the course correction mechanism. India has, ipso facto, turned out to be an unique model of demographic transition with limitless boundary. For the same reason, India does not envisage any sustainable population nor does it have any sustainable agenda in its policy front.

Chapter 19
India's Perceived Challenges for Securing Sustainable Population in the Context of Sustainable Development Goals

Abstract The destination point of UN Sustainable Development Goals (SDGs) adopted on 25th September, 2015 is to put in place a new world free from hunger, gender inequity, and human indignity. The summit gave a clarion call of a new framework of "Transforming Our World: the 2030 Agenda for Sustainable Development". One of the collateral objectives is to put in place the state of sustainable population. Unless sustainable population is secured, it is well-nigh impossible to have sustainable development. With 121, 05, 69, 573 number of population as per census, 2011, India happens to be second most populous country in the world sustaining 17.5 % of the world population on 2.4 %of the world's surface area. It is on way to overtake China on carrying maximum load of population in the world. In the circumstances, India is in need of framing a Sustainable Population Policy and also its Strategy. A Parliamentary Committee is needed to address all mundane issues connected with sustainable population and draw up a Report for intensive discussions in the Parliament on sustainable population policy and strategy. The uppermost consideration should always be the realisation that there are physical limits to the earth-space of India and its life support systems. Technology may give temporary respite in some areas with perceptible fortune for short duration; the limits to population growth are permanent. India is a case of limitless population growth. It is a classic case of over population. India has now no choice but to pursue a regime of hard options for a period of time as there is no quick fix for sustainable population.

Keywords The 2030 agenda for sustainable development · UN projection · Sustainable population concept · Civilization crisis · Demographic disaster · Sustainable population policy for india

The destination point of UN Sustainable Development Goals (SDGs) adopted on 25th September, 2015 is to put in place a new world free from hunger, gender inequity, and human indignity. The summit gave a clarion call of a new framework of "Transforming Our World: the 2030 Agenda for Sustainable Development". One of the collateral objectives, which in turn happened to transform itself as a necessary condition of sustainable development, is to put in place the state of sustainable population in a given country. Unless sustainable population is secured, it is well-nigh impossible to have sustainable development. With 121, 05, 69, 573 number of population as per census, 2011, India happens to be second most populous country in the world sustaining 17.5 % of the world population on 2.4 % of the world's surface area. It is on way to overtake China on carrying maximum load of population in the world. In the circumstances, it is very important to have an objective assessment as to whether India will be able to secure sustainable population and meet the essential requirement for sustainable development.

The carrying capacity of any sustainable population depends, among other conditions, the geographical size of a country. In fact, availability of geographical space may be considered as the primary requirement for maintaining a particular size of population. It would thus be relevant to see the relative geographical area of India vis-à-vis size of the earth and of other major countries of the world. In Table 19.1, the relative geographical position of India has been shown:

The dynamics of population load changes fast and the respective Population clock of the country aims to capture such dynamic growth of population. At Table 19.2 the relative position of India and other major populous countries of the world, as revealed in the Population Clock, has been captured as on 25th November, 2015. It is a fact that the figures on population clock may not be as exact as the census figures; nonetheless it is a dependable indicator of the emerging scenario. Thus the population load of India, which represented 17.5 % of the world population in 2011, moved up words at 17.6 % as per Population clock figure on 25th November, 2015, as would be evident from the following Table 19.2

Population and Development in volume 41 Number 3, September 2015, in section Documents, at page 557 mentioned the UN Population Division's series of

Table 19.1 The First 10 largest countries of the world

Country/Territory	Rank	Area (km^2)	% of total	Continent/Region
Earth	–	148,940,000	–	Solar System
Russia	1	17,098,242	11.3	Asia/Europe
Canada	2	9,984,670	6.7	America/North
United States	3	9,629,091	6.5	America/North
P.R. China	4	9,598,094	6.4	Asia/Easter
Brazil	5	8,514,877	5.7	America/South
Australia	6	7,692,024	5.2	Australia
India	7	3,287,263	2.3	Asia/South-Central
Argentina	8	2,780,400	2	America/South
Kazakhstan	9	2,780,400	1.8	Asia/South Central
Algeria	10	2,381,741	1.6	Africa/Northern

Source www.nationsonline.org/oneworld/countryby_area.htm

Table 19.2 Dependence of population

Rank	Country (or dependent territory)	Population	Date	% of world population	Source
1	China	1,373,310,000	21 November, 2015	18.9	Official population clock
2	India	1,280380,000	21 November, 2015	17.6	Official population clock
3	United States	322,277,000	21 November, 2015	4.42	Official population clock
4	Indonesia	255,461,700	1 July, 2015	3.51	Official projection
5	Brazil	205,223,000	21 November, 2015	2.82	Official population clock
6	Pakistan	188,925,000	1 July, 2015	2.59	UN projection
7	Nigeria	182,202,000	1 July, 2015	2.5	UN projection
8	Bangladesh	159,404,000	21 November, 2015	2.19	Official population clock
9	Russia	146,443,030	21 November, 2015	2.01	Official population clock
10	Japan	126,890,000	21 November, 2015	1.74	Official population clock

Source Wikipedia

population projections for twenty-first century, termed the 2015 Revision which was released in July 2015. The 2015 Revision's medium-variant projections for the ten largest countries as of 2015 (together making up 58 % of the world's population) are shown at Table 19.2

The data speak volumes of the population load of India and its alarming scenario. Geographically speaking, India is projected to carry the highest number of population in 2030 with a mere 2.3 (or 2.4 as per Census India's estimate) per cent of space of the earth. This itself raise a question as to whether India will be able to absorb this size of population without any damage to the fundamentals of the Indian economy. The data further raises serious question on how far the on-growing huge size of population over and above 2011 population would be sustainable to meet the parameters of sustainable development goals in 2030 Table 19.3. The huge size of population, current status of human development and its needed up-scaling for tomorrow, including the emerging additional size of population, its impact on natural resources along with climatic change factors, are areas of big concern for India. Normal economic sense does suggest that such a scenario will bring about untold stress on all aspects of economic and social life. Further, the incremental per year population addition from this higher base of number would be fearsome, to say the least. The UN projection of 1705 million in 2050 and 1660 million in 2100 are figures that signal the onset of a period of population disaster Table 19.4.

Table 19.3 Population in 2015 and projected population (medium- variant) in 2030, 2050, and 2100 of the ten largest countries in 2015(millions)

Country	2015	2030	2050	2100
China	1376	1416	1348	1004
India	1311	1526	1705	1660
United States	322	356	389	450
Indonesia	258	295	322	314
Brazil	208	229	238	200
Pakistan	189	245	310	364
Nigeria	182	263	399	752
Bangladesh	161	186	202	170
Russian Federation	143	139	129	117
Mexico	127	148	164	148

Source United Nations Department of Economic and Social Affairs/Population Division, World Population Prospects: The 2015 Revision. Key Findings and Advance Tables 52
Based on the related data in Table 19.1 and Table 19.2, a Table has been prepared to indicate the relative population size of India in the year 2030, the terminal year of SDGs

Table 19.4 The size of geographical area vis-à-vis the size of population and its projection on 2030

Country/Territory	% of geographical area of the earth	Population estimation in 2015	Projected population in 2030
China	6.4	1376	1416
India	2.3	1311	1526
United States	6.5	322	356
Indonesia	1,3	258	295
Brazil	5.7	208	229
Pakistan	0.53	189	245
Nigeria	0.62	182	263
Bangladesh	0.10	161	186
Russian Federation	11.3	143	139
Mexico	1.3	127	148

Source www.nationsonline.org/oneworld/country_by_area.ht

In the given context, the choice before the country is to initiate robust population control measures, as permissible in a democratic country and embark on damage limitation exercise on demographic front on areas of concern. For that matter, two very important areas, namely, the nature of CBR and the status of TFR in India deserve particular attention. However, the All-India aggregate population scenario on these areas might not be very helpful for course correction in view of great diversity in the demographic front in the states of India; the success story in the demographic front of all the southern parts of India is different from the northern/central parts of India being represented in good measure by a good number of EAG (erstwhile BIMARU) states. As a matter of fact, demographic character in India is not unique plagued as it is by traits of dualism and that the decadal growth rate, CBR, TFR etc. vary sharply among the states and also within rural and urban areas in a particular state. Since IMR is a big factor for number of CBR, this component has been shown alongside. Let us have a summary view of CBR, IMR and TFR in the states of India at Table 19.5 as shown below:

The Crude Birth Rate (CBR), as earlier mentioned, is defined as the number of live births per thousand population. It is usually the dominant factor in determining the rate of population growth. On the other hand, Total Fertility Rate (TFR) is the number of children which a woman would bear during her life time and the TFR of level 2.1 is generally taken as replacement level fertility. It is usually considered to be a precondition for population stabilization. However, it does not mean the size of population get stabilized immediately as it takes some time to take effect because acceleration in population growth persists due to larger number of couples already existing in the reproductive ages in that population. This is very relevant in the context of population stabilization in India where a very larger number of couples already exists in the reproductive ages. As per the SRS estimates of 2013, eight States have reached TFR of 2.1 and nine States—all of them in the north and east, except for Gujarat—haven't yet reached replacements levels of 2.1. India as whole might be in a position to achieve TFR by 2020, as demographers optimistically believe.

The Table 19.5, as below, demonstrates that for effecting replacement level of population, India has to depend on success on performances in family planning area like CBR (Crude Birth Rate) and TFR (Total Fertility Rate) of most of the EAG states in the country. While the projected time for achieving TFR of 2.1 is very long and optimistic, the soft voluntary approach on family planning in the country may not even ensure the projected date of its realisation. The journey for replacement level population is indeed very long in India and it is very difficult to make any exercise on roadmap for sustainable population on this count. Further, there arises a big question whether replacement level of population can be equated with sustainable population and whether such a replacement level of population of India, say in 2020, would qualify it to be sustainable. Generally speaking, while replacement level of population is a satisfying indicator for any country, there is, however, no one to one correspondence with it with sustainable population. As a matter of fact, sustainable population is a much bigger concept connected with carrying capacity of a country endowed with natural resources, economic resources and technological level of achievement, whereas the attainment of replacement level of population may be looked upon to serve as a necessary condition for its population stability. Therefore, population size of India in 2020, following TFR attainment of 2.1, with backlog of huge population, may not qualify to be called sustainable. It could, at best, be labelled as a stage of stable population after a phase of an upwardly swing of population numbers. The position is indeed very dismal in the population front in India today. The burgeoning number is no longer our asset, notwithstanding a section of our demographers' claim of demographic dividend on ground of age-mix considerations. The huge size of poverty, the little scope of full

Table 19.5 Summary view of CBR and TFR in the states of India

India/States	CBR in 2012	IMR (recent status)**	IMR (12th Plan target)**	TFR (2013)
All India	21.6	44	25	2.3
Andhra Pradesh	17.5	43	25	1.8
Arunachal Pradesh	19.4	32	19	
Assam	22.5	55	32	2.3
Bihar	27.7	44	26	3.4
Chhattisgarh	24.5	48	28	2.6
Goa	13.1	11	6	
Gujarat	21.1	41	24	2.3
Haryana	21.6	44	26	2.2
Himachal Pradesh	16.2	38	22	1.7
Jammu & Kashmir	17.6	41	24	1.9
Jharkhand	24.7	39	23	2.7
Karnataka	18.5	35	15	1.9
Kerala	14.9	12	6	1.8
Madhya Pradesh	26.6	59	34	2.9
Maharashtra	16.6	11	15	1.8
Manipur	14.6	23	6	
Meghalaya	24.1	52	30	
Mizoram	16.3	34	20	
Nagaland	15.6	21	12	
Odisha	19.9	57	33	1.8
Punjab	15.9	30	16	1.7
Rajasthan	25.9	52	30	2.8
Sikkim	17.2	26	15	
Tamil Nadu	15.7	22	13	1.7
Tripura	13.9	29	17	
Uttar Pradesh	27.4	57	32	3.1
Uttarakhand	18.5	36	21	
West Bengal	16.1	32	11	1.6
A&N Islands#	15	23	12	
NCT of Delhi	17.3	28	15	1.7
Chandigarh#	14.8	20	12	
D&N Haveli#	25.6	35	20	
Daman & Diu#	18.1	22	13	
Lakshadweep#	14.8	24	14	
puducherry	15.8	19	11	

Source ** Planning Commission-12th Plan document, SRS 2013; # stands for Union Territories

employment even of this younger segment and their joining the reproductive phase sooner, could even signal another spell of population boom.

The formal take off to SDGs journey is yet to begin. For understanding the very difficult but future shaping journey towards SDGs, it is required to look at the relevant status of MDG performances as mentioned at Chap. 13. India is yet to eradicate extreme poverty and hunger and could not halve the same between 1990 and 2015; the proportion of people who suffer from hunger and under-weight children remains substantially high. The status of IMR and MMR, though improved considerably, could not reach the MDG targets by 2015. While dedicated efforts appeared to have been made to reach MDG targets, the population load stood in the way to achieve desired results.

In the given context let us look if there is any policy in place in our country at this material point of time or any National goal or any National declaration for containing the runaway population situation by a Target Year, and whether intervening initiatives are good enough and positive enough to effectively put a brake on alarming addition of population and ensure stable/sustainable population. As discussed earlier, the National Population Policy (NPP), 2000 happened to be the first structured population policy in India, lifespan of which expired at 2000. Since then, no Population policy worth the name has been in place in India. The NPP, 2000 had demographic goals and identified a host of issues which needed to be addressed to bring the TFR to replacement levels by 2010 and stable population by 2045. However, the basic postulate of TFR to reach at replacement level did not take place and the question of reaching stable population remains as a wish-dream in view of other hard data on the field. Further, the concept of stable population, as perceived, is different from sustainable population which the country is required to put in place to reach SDGs. The draft National Health policy, 2015 with a health agenda on population stabilisation has still to be adopted formally and therefore, the National Health policy (NHP), 2002, to all intents and purposes, remains as the official policy of the country for the issues relevant to the family welfare aspect of it. Incidentally, NHP, 2002 addresses items on better and safer contraceptive choice, age of Marriage, birth spacing, family planning services on any fixed day at non-camp site, increased male sterilisation and non-coercive approach to family planning with improved access etc. There was no separate goal for population stabilization nor of any projected population size of the country to be put in place by a target date. In the given context, whatever has been incorporated in the Twelfth Five Year Plan document on related demographic goal or projection would, if so facto, be taken as the current population policy of the day. Unfortunately, the 12th Plan document on Health and Family Welfare portrays only the vision on Health; it has not incorporated separate vision on Family Welfare or for that matter population stabilisation. Sadly enough, the country does not have any robust view on population stabilisation or policy to address the critical runaway population scenario of the country.

Sustainable population concept is much bigger and more complex than the traditional concept of stable population, as discussed earlier in Chap. 13. For a country to achieve sustainable population, it is necessary to have the postulates of a separate and dedicated policy goal, a holistic approach, a defined strategy and a work platform to execute them firmly and effectively. While the country would be addressing way forward for SDGs in the near future with all the stakeholders and partners, on the lines of the SDG model of sustainable summit at the UN, the missing component of sustainable population on the SDG platform need to be duly addressed while working on blue-print for operation of SDGs in a country situation.

In the face of grim scenario and looming threat of numbers, India is pushed back to a situation of no choice and has to adopt an unfamiliar approach. The fractured political bodies will raise a huge hue and cry; a group of social activists will take up right to procreation at free will and the religious lobby would create hysteria on ground of being on attack on personal matter and a social group will come out openly for fear of loss of its demographic agenda and vote bank politics. The country has to take such risk as it is poised to an impending civilization crisis triggered by demographic disaster and human wellbeing assault. In every year, thousands of acres of farmland and forests are being destroyed killing off hundreds of species, fresh water tables are falling; large size of consumption of many "renewable" resources are depleting stocks of non-renewable resources—fossil fuels, metals and minerals—at ever-increasing rates. As per Global Footprint Network and other research organizations, humans are currently consuming renewable resources over 50 % faster than its truly sustainable rate. While there are no firm estimates for India's consumption overshoot, it would occupy a very good percentage of global over-consumption. For a sustainable quality of life, as per estimates of the Global Footprint Network, the total world population needs to range in between one and three billion people. Given the world's current population in the range of 7 billion, India's current estimated population of 1.3 billion need to be reduced in the range of 0.5 billion, a size of population India can maintain to have quality of life on a sustainable basis. Since it is absurd to think of any reduction of such size of population, the option before the government is to take hard decision to adopt the much maligned Chinese prescription to have one child one family norm even in this democratic country like India. The unusual situation calls for unusual step.

The country has urgency to firm-up its actionable role on Sustainable population in the context of SDGs as adopted at the UN's Sustainable Summit, 2015. India is in need of framing a Sustainable Population Policy and also its Strategy. Accordingly, India's Sustainable Population Policy and Strategy is needed to be adopted by the parliament of India. A Parliamentary Committee is needed to address all mundane issues connected with sustainable population and draw up a fully baked Report for intensive discussions in the Parliament on sustainable population policy and strategy. It would be worthwhile that before preparing its draft, the Sub-committee may undertake wider discussions with civil society organisations for a cohesive, pragmatic and future shaping policy and strategy on India's perceived role on sustainable population in the country in general and SDGs in particular. The uppermost consideration should always be the realisation that there are physical limits to the earth-space of India and its life support systems. Technology may give temporary respite in some areas with perceptible fortune for short duration; the limits to population growth are permanent. India is a case of limitless population growth. It is a classic case of over population. India has now no choice but to pursue a regime of hard options for a period of time as there is no quick fix for sustainable population.

Appendix A
FAQs on Sustainable Development Summit, 2015

(Press Kit for the Sustainable Development Summit 2015: Time for Global Action for People and Planet)

Frequently Asked Questions

What is sustainable development?

- Sustainable development has been defined as development that meets the needs of the present without compromising the ability of future generations to meet their own needs.
- Sustainable development calls for concerted efforts towards building an inclusive, sustainable and resilient future for people and planet.
- For sustainable development to be achieved, it is crucial to harmonize three core elements: economic growth, social Inclusion and environmental protection. These elements are interconnected and all are crucial for the wellbeing of individuals and societies.
- Eradicating poverty in all its forms and dimensions is an indispensable requirement for sustainable development. To this end, there must be promotion of sustainable inclusive and equitable economic growth, creating greater opportunities for all, reducing inequalities, raising basic standards of living, fostering equitable social development and inclusion, and promoting integrated and sustainable management of natural resources and ecosystems.

What are the Sustainable Development Goals?

- The 193 Member States of the United Nations reached consensus on the outcome document of a new sustainable development agenda entitled, "Transforming Our World: The 2030 Agenda for Sustainable Development". This agenda contains 17 goals and 169 targets. The complete list of goals and targets are available at: http://www.un.org/sustainabledevelopment/sustainable–developmentgoals/. The outcome document is available at: https://sustainabledevelopment.un.org/post2015/transformingourworld

- World leaders will officially adopt this universal, integrated and transformative agenda in September to spur actions that will end poverty and build a more sustainable world over the next 15 years.
- This agenda builds on the achievements of the Millennium Development Goals (MDGs), which were adopted in 2000 and guided development action for the last 15 years. The MDGs have proven that global goals can lift millions out of poverty.
- The new goals are part of an ambitious, bold sustainable development agenda that will focus on the three interconnected elements of sustainable development: economic growth, social inclusion and environmental protection.
- The Sustainable Development Goals (SDGs) and targets are global in nature and universally applicable, taking into account different national realities, capacities and levels of development and respecting national policies and priorities. They are not independent from each other—they need to be implemented in an integrated manner.
- The SDGs are the result of a three year long transparent, participatory process inclusive of all stakeholders and people's voices. They represent an unprecedented agreement around sustainable development priorities among 193 Member States. They have received world wide support from civil society, business, Parliamentarians and other actors. The decision to launch a process to develop a set of SDGs was made by UN Member States at the United Nations Conference on Sustainable Development (Rio+20), held in Rio de Janeiro in June 2012.

What are the elements underpinning the Sustainable Development Goals?

- The Goals and targets will stimulate action over the next 15 years in areas of critical importance: people, planet, prosperity, peace and partnership.
  - People, as we are determined to end poverty and hunger, in all their forms and dimensions, and to ensure that all human beings can fulfil their potential in dignity and equality and in a healthy environment.
  - Planet, to protect the planet from degradation, including through sustainable consumption and production, sustainably managing its natural resources and taking urgent action on climate change, so that it can support the needs of the present and future generations.
  - The Goals and targets will stimulate action over the next 15 years in areas of critical importance: people, planet, prosperity, peace and partnership.
  - People, as we are determined to end poverty and hunger, in all their forms and dimensions, and to ensure that all human beings can fulfil their potential in dignity and equality and in a healthy environment.
  - Prosperity, to ensure that all human beings can enjoy prosperous and fulfilling lives and that economic, social and technological progress occurs in harmony with nature.
  - Peace, to foster peaceful, just and inclusive societies free from fear and violence. There can be no sustainable development without peace and no peace without sustainable development.
  - Partnership, to mobilize the means required to implement this agenda through a revitalised global partnership for sustainable development, based

on a spirit of strengthened global solidarity, focussed in particular on the needs of the poorest and most vulnerable and with the participation of all countries, all stakeholders and all people.

Why are new goals being adopted this year?

- The Millennium Development Goals that were launched in 2000 set 2015 as the target year. Recognizing the success of the Goals—and the fact that a new development agenda was needed beyond 2015—countries agreed in 2012 at Rio +20, the UN Conference on Sustainable Development, to establish an open working group to develop a set of sustainable development goals for consideration and appropriate action.
- After more than a year of negotiations, the Open Working Group presented its recommendation for the 17 sustainable development goals.
- In early August 2015, the 193 member states of the United Nations reached consensus on the outcome document of the new agenda "Transforming Our World: The 2030 Agenda for Sustainable Development".
- Member States decided that the UN summit for the adoption of new sustainable development agenda with its 17 goals will be held from 25 to 27 September 2015, in New York and convened as a highlevel plenary meeting of the General Assembly.

Why are the Sustainable Development Goals so broad in comparison to the Millennium Development Goals which were very specific?

- There are 17 sustainable development goals with 169 targets, in contrast to the 8 Millennium Development Goals with 21 targets. The complex challenges that exist in the world today demand that a wide range of issues be covered. It is, also, critical to address the root causes of the problems and not only the symptoms.
- The Sustainable Development Goals are the result of a negotiation process that involved the 193 UN Member States and also saw unprecedented participation of civil society and other stakeholders. This led to the representation of a wide range of interests and perspectives. On the other hand, the MDGs were produced by a group of experts behind closed doors.
- The SDGs are broad in scope because they will address the interconnected elements of sustainable development: economic growth, social inclusion and environmental protection. The MDGs focused primarily on the social agenda.
- The MDGs targeted developing countries, particularly the poorest, while the Sustainable Development Goals will apply to the entire world, developed and developing countries.

Aren't 17 SDGs and 169 targets too many, too ambiguous and unrealistic?

- Poverty eradication, shared prosperity and planetary sustainability cannot be reduced to a simple formula.
- The SDGs represent the shared global goals and targets that will be tailored at the country level, informed by context—based evidence.

How are the SDGs different from the MDGs?

- The 17 Sustainable Development Goals with 169 targets are broader in scope and will go further than the MDGs by addressing the root causes of poverty and the universal need for development that works for all people. These goals will cover the three dimensions of sustainable development: economic growth, social inclusion and environmental protection.
- Building on the success and momentum of the MDGs, the new global goals will cover more ground with ambitions to address inequalities, economic growth, decent jobs, cities and human settlements, industrialization, energy, climate change, sustainable consumption and production, peace and justice.
- The new goals are universal and apply to all countries, whereas the MDGs were intended for action in developing countries only.
- A core feature of the SDGs has been the means of implementation–the mobilization of financial resources as well as capacity building and the transfer of environmentally sound technologies.
- The new goals recognize that tackling climate change is essential for sustainable development and poverty eradication. SDG 13 aims to promote urgent action to combat climate change and its impacts. Civil society has participated in the process of negotiations for the new sustainable development agenda. How can we quantify their contribution to the final document?
- The negotiating process on the sustainable development goals involved the unprecedented participation of civil society and other stakeholders, such as the private sector and mayors.
- During the negotiations, civil society and other stakeholders were able to speak directly to government representatives.
- Many young people were also involved from the beginning on social media platforms and the UN's global My World survey that received more than 7 million votes from around the world, with approximately 75 % of participants under 30 years of age.

How much will the implementation of this new sustainable development agenda cost?

- The means of implementation—how to mobilize the financial resources to achieve the sustainable development agenda is a core feature of the new agenda.
- This ambitious agenda will require the mobilization of significant resources—in the trillions of dollars. But these resources already exist. There are far more than enough savings in the world to finance the new agenda.
- Resources need to be mobilized from domestic and international sources, as well as from the public and private sectors.
- Official development assistance is still necessary to help finance sustainable development to assist the least developed countries.
- The agenda can be met within the framework of a revitalized global partnership for sustainable development, supported by the concrete policies and actions as

outlined in the Addis Ababa Action Agenda, the outcome document of the Third International Conference on Financing for Development held in July.

How will the new development agenda be implemented?

- Implementation and success will rely on countries' own sustainable development policies, plans and programmes, and will be led by countries. The SDGs will be a compass for aligning countries' plans with their global commitments.
- Nationally owned and country–led sustainable development strategies will require equivalent resource mobilization and financing strategies.
- The 17 SDGs and 169 targets of the new agenda will be Monitored and reviewed using a set of global indicators. The global indicator framework, to be developed by the Inter Agency and Expert Group on SDG Indicators, will be agreed on by the UN Statistical Commission by March 2016.
- Governments will also develop their own national indicators to assist in monitoring progress made on the goals and targets.
- The follow–up and review process will be undertaken on an annual basis by the High Level Political Forum on Sustainable development through a SDG Progress Report to be prepared by the Secretary General.
- The means of implementation of the SDGs will be monitored and reviewed as outlined in the Addis Ababa Action Agenda, the outcome document of the Financing for Development Conference, to ensure that financial resources are effectively mobilized to support the new sustainable development agenda.
- A Technology Facilitation Mechanism, to be launched at the September Summit, will address the technology needs of developing countries, the options to address those needs and capacity building. Recognizing the central role of technological cooperation for the achievement of sustainable development, countries agreed on this mechanism at the Financing for Development Conference.

How does climate change relate to sustainable development?

- We are already seeing that climate change is impacting public health, food and water security, migration, peace and security. Investments in sustainable development will help address climate change by reducing emissions and building climate resilience. Action on climate change will drive sustainable development and vice versa.
- Climate change, left unchecked, will roll back the development gains we have made over the last decades and will make further gains impossible.
- Tackling climate change and fostering sustainable development are two mutually reinforcing sides of the same coin; sustainable development cannot be achieved without climate action, as many of the SDGs are actually addressing the core drivers of climate change.

How does the climate component of the SDGs influence the debate on climate change and the upcoming climate change conference in Paris later this year?

- The consensus reached on the outcome document of the new sustainable development agenda does not aim to anticipate or usurp the role of the United Nations Framework Convention on Climate Change, the body responsible for the Climate Change Conference in Paris in December.
- Nevertheless, the agreement will send a strong signal that the world has high expectations that the time has come for decisive positive outcomes on matters of climate change.
- Given that many of the SDGs guide action on the core drivers of climate change, their implementation, beginning on 1 January 2016, will accelerate transition towards the implementation of the climate agreement that will enter into force in 2020.

What have the MDGs accomplished?

- The MDGs have produced the most successful anti–poverty movement in history and will serve as the springboard for the new sustainable development agenda.
- Poverty and hunger: only two short decades ago, nearly half of the developing world lived in extreme poverty. The number of people now living in extreme poverty has declined by more than half, falling from 1.9 billion in 1990 to 836 million in 2015.
- Gender equality: The world has also witnessed dramatic improvement in gender equality in schooling since the MDGs, and gender parity in primary school has been achieved in the majority of countries. More girls are now in school, and women have gained ground in parliamentary representation over the past 20 years in nearly 90 % of the 174 countries with data.
- Child mortality: globally, the under–five mortality rate dropped from 90 to 43 deaths per 1000 live births between 1990 and 2015.
- Maternal health: the maternal mortality ratio shows a decline of 45 % worldwide, with most of the reduction occurring since 2000.
- Fighting diseases: new infection rates from HIV fell approximately by 40 % between 2000 and 2013. Over 6.2 million malaria deaths were averted between 2000 and 2015, while tuberculosis prevention, diagnosis and treatment interventions saved an estimated 37 million lives between 2000 and 2013.
- Sanitation: Worldwide, 2.1 billion have gained access to improved sanitation and the proportion of people practicing open defecation has fallen almost by half since 1990.
- Global partnership: official development assistance from developed countries saw an increase of 66 % in real terms from 2000 and 2014, reaching $135.2 billion.

What are the remaining gaps left by the MDGs?

- About 800 million people still live in extreme poverty and 795 million still suffer from hunger.

Appendix A: FAQs on Sustainable Development Summit, 2015

- Between 2000 and 2015, the number of children out of school declined by almost half. However, there are still 57 million children who are denied the right to primary education.
- Gender inequality persists in spite of more representation of women in parliament and more girls going to school. Women continue to face discrimination in access to work, economic assets and participation in private and public decision–making.
- Economic gaps still exist between the poorest and richest households and rural and urban areas. Children from the poorest 20 % of households are more than twice as likely to be stunted as those from the wealthiest 20 % and are also four times as likely to be out of school. Improved sanitation facilities are only covering half of rural population, as opposed to 82 % in urban areas.
- While the mortality rate for children under five dropped by 53 % between 1990 and 2015, child deaths continue to be increasingly concentrated in the poorest regions and in the first month of life.

How will progress of the SDGs be measured? How many indicators will be developed for the 169 targets of the Sustainable Development Goals?

- The 17 goals and 169 targets will be monitored and reviewed using a set of global indicators. These will be complemented by indicators at the regional and national levels, which will be developed by Member States.
- The Inter Agency and Expert Group on SDG Indicators will develop the global indicator framework which the UN Statistical Commission will subsequently agree on in March 2016. Thereafter, the Economic and Social Council and the General Assembly will adopt these indicators.
- Chief statisticians from Member States are working on the identification of the targets with the aim to have 2 indicators for each target. There will be approximately 300 indicators for all the targets. Where the targets cover cross— cutting issues, however, the number of indicators may be reduced.

When are the SDGs expected to start and end?

- The SDGs are expected to start on 1 January 2016 and to be achieved by 31 December 2030. However, some targets that build on pre–set international agreements are expected to be achieved even earlier than the end of 2030.

Appendix B
The Paris Agreement on Climatic Change Conference at COP 21

Conference of the Parties
Twenty-first session
Paris, 30 November to 11 December 2015

ADOPTION OF THE PARIS AGREEMENT
Welcoming the adoption of United Nations General Assembly resolution A/RES/70/1, "Transforming our world: the 2030 Agenda for Sustainable Development", in particular its goal 13, and the adoption of the Addis Ababa Action Agenda of the third International Conference on Financing for Development and the adoption of the Sendai Framework for Disaster Risk Reduction.

Recognizing that climate change represents an urgent and potentially irreversible threat to human societies and the planet and thus requires the widest possible cooperation by all countries, and their participation in an effective and appropriate international response, with a view to accelerating the reduction of global greenhouse gas emissions.

Also recognizing that deep reductions in global emissions will be required in order to achieve the ultimate objective of the Convention and emphasizing the need for urgency in addressing climate change.

Acknowledging that climate change is a common concern of humankind, Parties should, when taking action to address climate change, respect, promote and consider their respective obligations on human rights, the right to health, the rights of indigenous peoples, local communities, migrants, children, persons with disabilities and people in vulnerable situations and the right to development, as well as gender equality, empowerment of women and intergenerational equity.

Also acknowledging the specific needs and concerns of developing country Parties arising from the impact of the implementation of response measures and, in this regard, decisions 5/CP.7, 1/CP.10, 1/CP.16 and 8/CP.17.

Emphasizing with serious concern the urgent need to address the significant gap between the aggregate effect of Parties' mitigation pledges in terms of global annual emissions of greenhouse gases by 2020 and aggregate emission pathways consistent with holding the increase in the global average temperature to well below 2 °C above pre-industrial levels and pursuing efforts to limit the temperature increase to 1.5 °C above pre-industrial levels.

Also emphasizing that enhanced pre-2020 ambition can lay a solid foundation for enhanced post-2020 ambition.

Stressing the urgency of accelerating the implementation of the Convention and its Kyoto Protocol in order to enhance pre-2020 ambition.

Recognizing the urgent need to enhance the provision of finance, technology and capacity-building support by developed country Parties, in a predictable manner, to enable enhanced pre-2020 action by developing country Parties.

Emphasizing the enduring benefits of ambitious and early action, including major reductions in the cost of future mitigation and adaptation efforts.

Acknowledging the need to promote universal access to sustainable energy in developing countries, in particular in Africa, through the enhanced deployment of renewable energy.

Agreeing to uphold and promote regional and international cooperation in order to mobilize stronger and more ambitious climate action by all Parties and non-Party stakeholders, including civil society, the private sector, financial institutions, cities and other subnational authorities, local communities and indigenous peoples.

Adoption

1. Decides to adopt the Paris Agreement under the United Nations Framework Convention on Climate Change (hereinafter referred to as "the Agreement") as contained in the annex;
2. Requests the Secretary-General of the United Nations to be the Depositary of the Agreement and to have it open for signature in New York, United States of America, from 22 April 2016 to 21 April 2017;
3. Invites the Secretary-General to convene a high-level signature ceremony for the Agreement on 22 April 2016;
4. Also invites all Parties to the Convention to sign the Agreement at the ceremony to be convened by the Secretary-General, or at their earliest opportunity, and to deposit their respective instruments of ratification, acceptance, approval or accession, where appropriate, as soon as possible;
5. Recognizes that Parties to the Convention may provisionally apply all of the provisions of the Agreement pending its entry into force, and requests Parties to provide notification of any such provisional application to the Depositary;
6. Notes that the work of the Ad Hoc Working Group on the Durban Platform for Enhanced Action, in accordance with decision 1/CP.17, paragraph 4, has been completed;
7. Decides to establish the Ad Hoc Working Group on the Paris Agreement under the same arrangement, mutatis mutandis, as those concerning the election of

officers to the Bureau of the Ad Hoc Working Group on the Durban Platform for Enhanced Action;
8. Also decides that the Ad Hoc Working Group on the Paris Agreement shall prepare for the entry into force of the Agreement and for the convening of the first session of the Conference of the Parties serving as the meeting of the Parties to the Paris Agreement;
9. Further decides to oversee the implementation of the work programme resulting from the relevant requests contained in this decision;
10. Requests the Ad Hoc Working Group on the Paris Agreement to report regularly to the Conference of the Parties on the progress of its work and to complete its work by the first session of the Conference of the Parties serving as the meeting of the Parties to the Paris Agreement;
11. Decides that the Ad Hoc Working Group on the Paris Agreement shall hold its sessions starting in 2016 in conjunction with the sessions of the Convention subsidiary bodies and shall prepare draft decisions to be recommended through the Conference of the Parties to the Conference of the Parties serving as the meeting of the Parties to the Paris Agreement for consideration and adoption at its first session;

Intended Nationally Determined Contributions

12. Welcomes the intended nationally determined contributions that have been communicated by Parties in accordance with decision 1/CP.19, paragraph 2(b);
13. Reiterates its invitation to all Parties that have not yet done so to communicate to the secretariat their intended nationally determined contributions towards achieving the objective of the Convention as set out in its Article 2 as soon as possible and well in advance of the twenty-second session of the Conference of the Parties (November 2016) and in a manner that facilitates the clarity, transparency and understanding of the intended nationally determined contributions;
14. Requests the secretariat to continue to publish the intended nationally determined contributions communicated by Parties on the UNFCCC website;
15. Reiterates its call to developed country Parties, the operating entities of the Financial Mechanism and any other organizations in a position to do so to provide support for the preparation and communication of the intended nationally determined contributions of Parties that may need such support;
16. Takes note of the synthesis report on the aggregate effect of intended nationally determined contributions communicated by Parties by 1 October 2015, contained in document FCCC/CP/2015/7;
17. Notes with concern that the estimated aggregate greenhouse gas emission levels in 2025 and 2030 resulting from the intended nationally determined contributions do not fall within least-cost 2 °C scenarios but rather lead to a projected level of 55 Gt in 2030, and also notes that much greater emission reduction

efforts will be required than those associated with the intended nationally determined contributions in order to hold the increase in the global average temperature to below 2 °C above pre-industrial levels by reducing emissions to 40 Gt or to 1.5 °C above pre-industrial levels by reducing to a level to be identified in the special report referred to in paragraph 21 below;
18. Also notes, in this context, the adaptation needs expressed by many developing country Parties in their intended nationally determined contributions;
19. Requests the secretariat to update the synthesis report referred to in paragraph 16 above so as to cover all the information in the intended nationally determined contributions communicated by Parties pursuant to decision 1/CP.20 by 4 April 2016 and to make it available by 2 May 2016;
20. Decides to convene a facilitative dialogue among Parties in 2018 to take stock of the collective efforts of Parties in relation to progress towards the long-term goal referred to in Article 4, paragraph 1, of the Agreement and to inform the preparation of nationally determined contributions pursuant to Article 4, paragraph 8, of the Agreement;
21. Invites the Intergovernmental Panel on Climate Change to provide a special report in 2018 on the impacts of global warming of 1.5 °C above pre-industrial levels and related global greenhouse gas emission pathways;

Decisions To Give Effect To The Agreement

Mitigation

22. Invites Parties to communicate their first nationally determined contribution no later than when the Party submits its respective instrument of ratification, accession, or approval of the Paris Agreement. If a Party has communicated an intended nationally determined contribution prior to joining the Agreement, that Party shall be considered to have satisfied this provision unless that Party decides otherwise;
23. Urges those Parties whose intended nationally determined contribution pursuant to decision 1/CP.20 contains a time frame up to 2025 to communicate by 2020 a new nationally determined contribution and to do so every five years thereafter pursuant to Article 4, paragraph 9, of the Agreement;
24. Requests those Parties whose intended nationally determined contribution pursuant to decision 1/CP.20 contains a time frame up to 2030 to communicate or update by 2020 these contributions and to do so every five years thereafter pursuant to Article 4, paragraph 9, of the Agreement;
25. Decides that Parties shall submit to the secretariat their nationally determined contributions referred to in Article 4 of the Agreement at least 9–12 months in advance of the relevant meeting of the Conference of the Parties serving as the

Appendix B: The Paris Agreement on Climatic Change Conference at COP 21 351

meeting of the Parties to the Paris Agreement with a view to facilitating the clarity, transparency and understanding of these contributions, including through a synthesis report prepared by the secretariat;

26. Requests the Ad Hoc Working Group on the Paris Agreement to develop further guidance on features of the nationally determined contributions for consideration and adoption by the Conference of the Parties serving as the meeting of the Parties to the Paris Agreement at its first session;

27. Agrees that the information to be provided by Parties communicating their nationally determined contributions, in order to facilitate clarity, transparency and understanding, may include, as appropriate, inter alia, quantifiable information on the reference point (including, as appropriate, a base year), time frames and/or periods for implementation, scope and coverage, planning processes, assumptions and methodological approaches including those for estimating and accounting for anthropogenic greenhouse gas emissions and, as appropriate, removals, and how the Party considers that its nationally determined contribution is fair and ambitious, in the light of its national circumstances, and how it contributes towards achieving the objective of the Convention as set out in its Article 2;

28. Requests the Ad Hoc Working Group on the Paris Agreement to develop further guidance for the information to be provided by Parties in order to facilitate clarity, transparency and understanding of nationally determined contributions for consideration and adoption by the Conference of the Parties serving as the meeting of the Parties to the Paris Agreement at its first session;

29. Also requests the Subsidiary Body for Implementation to develop modalities and procedures for the operation and use of the public registry referred to in Article 4, paragraph 12, of the Agreement, for consideration and adoption by the Conference of the Parties serving as the meeting of the Parties to the Paris Agreement at its first session;

30. Further requests the secretariat to make available an interim public registry in the first half of 2016 for the recording of nationally determined contributions submitted in accordance with Article 4 of the Agreement, pending the adoption by the Conference of the Parties serving as the meeting of the Parties to the Paris Agreement of the modalities and procedures referred to in paragraph 29 above;

31. Requests the Ad Hoc Working Group on the Paris Agreement to elaborate, drawing from approaches established under the Convention and its related legal instruments as appropriate, guidance for accounting for Parties' nationally determined contributions, as referred to in Article 4, paragraph 13, of the Agreement, for consideration and adoption by the Conference of the Parties serving as the meeting of the Parties to the Paris Agreement at its first session, which ensures that:

    (a) Parties account for anthropogenic emissions and removals in accordance with methodologies and common metrics assessed by the Intergovernmental Panel on Climate Change and adopted by the Conference of the Parties serving as the meeting of the Parties to the Paris Agreement;

(b) Parties ensure methodological consistency, including on baselines, between the communication and implementation of nationally determined contributions;
(c) Parties strive to include all categories of anthropogenic emissions or removals in their nationally determined contributions and, once a source, sink or activity is included, continue to include it;
(d) Parties shall provide an explanation of why any categories of anthropogenic emissions or removals are excluded;

32. Decides that Parties shall apply the guidance mentioned in paragraph 31 above to the second and subsequent nationally determined contributions and that Parties may elect to apply such guidance to their first nationally determined contribution;
33. Also decides that the Forum on the Impact of the Implementation of response measures, under the subsidiary bodies, shall continue, and shall serve the Agreement;
34. Further decides that the Subsidiary Body for Scientific and Technological Advice and the Subsidiary Body for Implementation shall recommend, for consideration and adoption by the Conference of the Parties serving as the meeting of the Parties to the Paris Agreement at its first session, the modalities, work programme and functions of the Forum on the Impact of the Implementation of response measures to address the effects of the implementation of response measures under the Agreement by enhancing cooperation amongst Parties on understanding the impacts of mitigation actions under the Agreement and the exchange of information, experiences, and best practices amongst Parties to raise their resilience to these impacts;[1]
36. Invites Parties to communicate, by 2020, to the secretariat mid-century, long-term low greenhouse gas emission development strategies in accordance with Article 4, paragraph 19, of the Agreement, and requests the secretariat to publish on the UNFCCC website Parties' low greenhouse gas emission development strategies as communicated;
37. Requests the Subsidiary Body for Scientific and Technological Advice to develop and recommend the guidance referred to under Article 6, paragraph 2, of the Agreement for adoption by the Conference of the Parties serving as the meeting of the Parties to the Paris Agreement at its first session, including guidance to ensure that double counting is avoided on the basis of a corresponding adjustment by Parties for both anthropogenic emissions by sources and removals by sinks covered by their nationally determined contributions under the Agreement;
38. Recommends that the Conference of the Parties serving as the meeting of the Parties to the Paris Agreement adopt rules, modalities and procedures for the mechanism established by Article 6, paragraph 4, of the Agreement on the basis of:

---

[1]Paragraph 35 has been deleted, and subsequent paragraph numbering and cross references to other paragraphs within the document will be amended at a later stage.

Appendix B: The Paris Agreement on Climatic Change Conference at COP 21   353

      (a) Voluntary participation authorized by each Party involved;
      (b) Real, measurable, and long-term benefits related to the mitigation of climate change;
      (c) Specific scopes of activities;
      (d) Reductions in emissions that are additional to any that would otherwise occur;
      (e) Verification and certification of emission reductions resulting from mitigation activities by designated operational entities;
      (f) Experience gained with and lessons learned from existing mechanisms and approaches adopted under the Convention and its related legal instruments;

39. Requests the Subsidiary Body for Scientific and Technological Advice to develop and recommend rules, modalities and procedures for the mechanism referred to in paragraph 38 above for consideration and adoption by the Conference of the Parties serving as the meeting of the Parties to the Paris Agreement at its first session;
40. Also requests the Subsidiary Body for Scientific and Technological Advice to undertake a work programme under the framework for non-market approaches to sustainable development referred to in Article 6, paragraph 8, of the Agreement, with the objective of considering how to enhance linkages and create synergy between, inter alia, mitigation, adaptation, finance, technology transfer and capacity-building, and how to facilitate the implementation and coordination of non-market approaches;
41. Further requests the Subsidiary Body for Scientific and Technological Advice to recommend a draft decision on the work programme referred to in paragraph 40 above, taking into account the views of Parties, for consideration and adoption by the Conference of the Parties serving as the meeting of the Parties to the Paris Agreement at its first session;

Adaptation

42. Requests the Adaptation Committee and the Least Developed Countries Expert Group to jointly develop modalities to recognize the adaptation efforts of developing country Parties, as referred to in Article 7, paragraph 3, of the Agreement, and make recommendations for consideration and adoption by the Conference of the Parties serving as the meeting of the Parties to the Paris Agreement at its first session;
43. Also requests the Adaptation Committee, taking into account its mandate and its second three-year workplan, and with a view to preparing recommendations for consideration and adoption by the Conference of the Parties serving as the meeting of the Parties to the Paris Agreement at its first session:

(a) To review, in 2017, the work of adaptation-related institutional arrangements under the Convention, with a view to identifying ways to enhance the coherence of their work, as appropriate, in order to respond adequately to the needs of Parties;

(b) To consider methodologies for assessing adaptation needs with a view to assisting developing countries, without placing an undue burden on them;

44. Invites all relevant United Nations agencies and international, regional and national financial institutions to provide information to Parties through the secretariat on how their development assistance and climate finance programmes incorporate climate-proofing and climate resilience measures;

45. Requests Parties to strengthen regional cooperation on adaptation where appropriate and, where necessary, establish regional centres and networks, in particular in developing countries, taking into account decision 1/CP.16, paragraph 13;

46. Also requests the Adaptation Committee and the Least Developed Countries Expert Group, in collaboration with the Standing Committee on Finance and other relevant institutions, to develop methodologies, and make recommendations for consideration and adoption by the Conference of the Parties serving as the meeting of the Parties to the Paris Agreement at its first session on:

    (a) Taking the necessary steps to facilitate the mobilization of support for adaptation in developing countries in the context of the limit to global average temperature increase referred to in Article 2 of the Agreement;

    (b) Reviewing the adequacy and effectiveness of adaptation and support referred to in Article 7, paragraph 14(c), of the Agreement;

47. Further requests the Green Climate Fund to expedite support for the least developed countries and other developing country Parties for the formulation of national adaptation plans, consistent with decisions 1/CP.16 and 5/CP.17, and for the subsequent implementation of policies, projects and programmes identified by them;

Loss and Damage

48. Decides on the continuation of the Warsaw International Mechanism for Loss and Damage associated with Climate Change Impacts, following the review in 2016;

49. Requests the Executive Committee of the Warsaw International Mechanism to establish a clearinghouse for risk transfer that serves as a repository for information on insurance and risk transfer, in order to facilitate the efforts of Parties to develop and implement comprehensive risk management strategies;

50. Also requests the Executive Committee of the Warsaw International Mechanism to establish, according to its procedures and mandate, a task force to complement, draw upon the work of and involve, as appropriate, existing bodies and expert groups under the Convention including the Adaptation

Appendix B: The Paris Agreement on Climatic Change Conference at COP 21   355

Committee and the Least Developed Countries Expert Group, as well as relevant organizations and expert bodies outside the Convention, to develop recommendations for integrated approaches to avert, minimize and address displacement related to the adverse impacts of climate change;

51. Further requests the Executive Committee of the Warsaw International Mechanism to initiate its work, at its next meeting, to operationalize the provisions referred to in paragraphs 49 and 50 above, and to report on progress thereon in its annual report;

52. Agrees that Article 8 of the Agreement does not involve or provide a basis for any liability or compensation.

Finance

53. Decides that, in the implementation of the Agreement, financial resources provided to developing countries should enhance the implementation of their policies, strategies, regulations and action plans and their climate change actions with respect to both mitigation and adaptation to contribute to the achievement of the purpose of the Agreement as defined in Article 2;

54. Also decides that, in accordance with Article 9, paragraph 3, of the Agreement, developed countries intend to continue their existing collective mobilization goal through 2025 in the context of meaningful mitigation actions and transparency on implementation; prior to 2025 the Conference of the Parties serving as the meeting of the Parties to the Paris Agreement shall set a new collective quantified goal from a floor of USD 100 billion per year, taking into account the needs and priorities of developing countries;

55. Recognizes the importance of adequate and predictable financial resources, including for results-based payments, as appropriate, for the implementation of policy approaches and positive incentives for reducing emissions from deforestation and forest degradation, and the role of conservation, sustainable management of forests and enhancement of forest carbon stocks; as well as alternative policy approaches, such as joint mitigation and adaptation approaches for the integral and sustainable management of forests; while reaffirming the importance of non-carbon benefits associated with such approaches; encouraging the coordination of support from, inter alia, public and private, bilateral and multilateral sources, such as the Green Climate Fund, and alternative sources in accordance with relevant decisions by the Conference of the Parties;

56. Decides to initiate, at its twenty-second session, a process to identify the information to be provided by Parties, in accordance with Article 9, paragraph 5, of the Agreement with the view to providing a recommendation for

consideration and adoption by the Conference of the Parties serving as the meeting of the Parties to the Paris Agreement at its first session;

57. Also decides to ensure that the provision of information in accordance with Article 9, paragraph 7 of the Agreement shall be undertaken in accordance with modalities, procedures and guidelines referred to in paragraph 96 below;

58. Requests Subsidiary Body for Scientific and Technological Advice to develop modalities for the accounting of financial resources provided and mobilized through public interventions in accordance with Article 9, paragraph 7, of the Agreement for consideration by the Conference of the Parties at its twenty-fourth session (November 2018), with the view to making a recommendation for consideration and adoption by the Conference of the Parties serving as the meeting of the Parties to the Paris Agreement at its first session;

59. Decides that the Green Climate Fund and the Global Environment Facility, the entities entrusted with the operation of the Financial Mechanism of the Convention, as well as the Least Developed Countries Fund and the Special Climate Change Fund, administered by the Global Environment Facility, shall serve the Agreement;

60. Recognizes that the Adaptation Fund may serve the Agreement, subject to relevant decisions by the Conference of the Parties serving as the meeting of the Parties to the Kyoto Protocol and the Conference of the Parties serving as the meeting of the Parties to the Paris Agreement;

61. Invites the Conference of the Parties serving as the meeting of the Parties to the Kyoto Protocol to consider the issue referred to in paragraph 60 above and make a recommendation to the Conference of the Parties serving as the meeting of the Parties to the Paris Agreement at its first session;

62. Recommends that the Conference of the Parties serving as the meeting of the Parties to the Paris Agreement shall provide guidance to the entities entrusted with the operation of the Financial Mechanism of the Convention on the policies, programme priorities and eligibility criteria related to the Agreement for transmission by the Conference of the Parties;

63. Decides that the guidance to the entities entrusted with the operations of the Financial Mechanism of the Convention in relevant decisions of the Conference of the Parties, including those agreed before adoption of the Agreement, shall apply mutatis mutandis;

64. Also decides that the Standing Committee on Finance shall serve the Agreement in line with its functions and responsibilities established under the Conference of the Parties;

65. Urges the institutions serving the Agreement to enhance the coordination and delivery of resources to support country-driven strategies through simplified and efficient application and approval procedures, and through continued readiness support to developing country Parties, including the least developed countries and small island developing States, as appropriate;

Technology Development and Transfer

66. Takes note of the interim report of the Technology Executive Committee on guidance on enhanced implementation of the results of technology needs assessments as referred to in document FCCC/SB/2015/INF.3;
67. Decides to strengthen the Technology Mechanism and requests the Technology Executive Committee and the Climate Technology Centre and Network, in supporting the implementation of the Agreement, to undertake further work relating to, inter alia:

    (a) Technology research, development and demonstration;
    (b) The development and enhancement of endogenous capacities and technologies;

68. Requests the Subsidiary Body for Scientific and Technological Advice to initiate, at its forty-fourth session (May 2016), the elaboration of the technology framework established under Article 10, paragraph 4, of the Agreement and to report on its findings to the Conference of the Parties, with a view to the Conference of the Parties making a recommendation on the framework to the Conference of the Parties serving as the meeting of the Parties to the Paris Agreement for consideration and adoption at its first session, taking into consideration that the framework should facilitate, inter alia:

    (a) The undertaking and updating of technology needs assessments, as well as the enhanced implementation of their results, particularly technology action plans and project ideas, through the preparation of bankable projects;
    (b) The provision of enhanced financial and technical support for the implementation of the results of the technology needs assessments;
    (c) The assessment of technologies that are ready for transfer;
    (d) The enhancement of enabling environments for and the addressing of barriers to the development and transfer of socially and environmentally sound technologies;

69. Decides that the Technology Executive Committee and the Climate Technology Centre and Network shall report to the Conference of the Parties serving as the meeting of the Parties to the Paris Agreement, through the subsidiary bodies, on their activities to support the implementation of the Agreement;
70. Also decides to undertake a periodic assessment of the effectiveness of and the adequacy of the support provided to the Technology Mechanism in supporting the implementation of the Agreement on matters relating to technology development and transfer;
71. Requests the Subsidiary Body for Implementation to initiate, at its forty-fourth session, the elaboration of the scope of and modalities for the periodic assessment referred to in paragraph 70 above, taking into account the review of

the Climate Technology Centre and Network as referred to in decision 2/CP.17, annex VII, paragraph 20 and the modalities for the global stocktake referred to in Article 14 of the Agreement, for consideration and adoption by the Conference of the Parties at its twenty-fifth session (November 2019);

Capacity-Building

72. Decides to establish the Paris Committee on Capacity-building whose aim will be to address gaps and needs, both current and emerging, in implementing capacity-building in developing country Parties and further enhancing capacity-building efforts, including with regard to coherence and coordination in capacity-building activities under the Convention;
73. Also decides that the Paris Committee on Capacity-building will manage and oversee the work plan mentioned in paragraph 74 below;
74. Further decides to launch a work plan for the period 2016–2020 with the following activities:

    (a) Assessing how to increase synergies through cooperation and avoid duplication among existing bodies established under the Convention that implement capacity-building activities, including through collaborating with institutions under and outside the Convention;
    (b) Identifying capacity gaps and needs and recommending ways to address them;
    (c) Promoting the development and dissemination of tools and methodologies for the implementation of capacity-building;
    (d) Fostering global, regional, national and subnational cooperation;
    (e) Identifying and collecting good practices, challenges, experiences, and lessons learned from work on capacity-building by bodies established under the Convention;
    (f) Exploring how developing country Parties can take ownership of building and maintaining capacity over time and space;
    (g) Identifying opportunities to strengthen capacity at the national, regional, and subnational level;
    (h) Fostering dialogue, coordination, collaboration and coherence among relevant processes and initiatives under the Convention, including through exchanging information on capacity-building activities and strategies of bodies established under the Convention;
    (i) Providing guidance to the secretariat on the maintenance and further development of the web-based capacity-building portal;

Appendix B: The Paris Agreement on Climatic Change Conference at COP 21   359

75. Decides that the Paris Committee on Capacity-building will annually focus on an area or theme related to enhanced technical exchange on capacity-building, with the purpose of maintaining up-to-date knowledge on the successes and challenges in building capacity effectively in a particular area;
76. Requests the Subsidiary Body for Implementation to organize annual in-session meetings of the Paris Committee on Capacity-building;
77. Also requests the Subsidiary Body for Implementation to develop the terms of reference for the Paris Committee on Capacity-building, in the context of the third comprehensive review of the implementation of the capacity-building framework, also taking into account paragraphs 75, 76, 77 and 78 above and paragraphs 82 and 83 below, with a view to recommending a draft decision on this matter for consideration and adoption by the Conference of the Parties at its twenty-second session;
78. Invites Parties to submit their views on the membership of the Paris Committee on Capacity-building by 9 March 2016;
79. Requests the secretariat to compile the submissions referred to in paragraph 78 above into a miscellaneous document for consideration by the Subsidiary Body for Implementation at its forty-fourth session;
80. Decides that the inputs to the Paris Committee on Capacity-building will include, inter alia, submissions, the outcome of the third comprehensive review of the implementation of the capacity-building framework, the secretariat's annual synthesis report on the implementation of the framework for capacity-building in developing countries, the secretariat's compilation and synthesis report on capacity-building work of bodies established under the Convention and its Kyoto Protocol, and reports on the Durban Forum and the capacity-building portal;
81. Requests the Paris Committee on Capacity-building to prepare annual technical progress reports on its work, and to make these reports available at the sessions of the Subsidiary Body for Implementation coinciding with the sessions of the Conference of the Parties;
82. Also requests the Conference of the Parties at its twenty-fifth session (November 2019), to review the progress, need for extension, the effectiveness and enhancement of the Paris Committee on Capacity-building and to take any action it considers appropriate, with a view to making recommendations to the Conference of the Parties serving as the meeting of the Parties to the Paris Agreement at its first session on enhancing institutional arrangements for capacity-building consistent with Article 11, paragraph 5, of the Agreement;
83. Calls upon all Parties to ensure that education, training and public awareness, as reflected in Article 6 of the Convention and in Article 12 of the Agreement are adequately considered in their contribution to capacity-building;
84. Invites the Conference of the Parties serving as the meeting of the Parties to the Paris Agreement at its first session to explore ways of enhancing the implementation of training, public awareness, public participation and public access to information so as to enhance actions under the Agreement;

Transparency of Action and Support

85. Decides to establish a Capacity-building Initiative for Transparency in order to build institutional and technical capacity, both pre- and post-2020. This initiative will support developing country Parties, upon request, in meeting enhanced transparency requirements as defined in Article 13 of the Agreement in a timely manner;
86. Also decides that the Capacity-building Initiative for Transparency will aim:

    (a) To strengthen national institutions for transparency-related activities in line with national priorities;
    (b) To provide relevant tools, training and assistance for meeting the provisions stipulated in Article 13 of the Agreement;
    (c) To assist in the improvement of transparency over time;

87. Urges and requests the Global Environment Facility to make arrangements to support the establishment and operation of the Capacity-building Initiative for Transparency as a priority reporting-related need, including through voluntary contributions to support developing countries in the sixth replenishment of the Global Environment Facility and future replenishment cycles, to complement existing support under the Global Environment Facility;
88. Decides to assess the implementation of the Capacity-building Initiative for Transparency in the context of the seventh review of the financial mechanism;
89. Requests that the Global Environment Facility, as an operating entity of the financial mechanism include in its annual report to the Conference of the Parties the progress of work in the design, development and implementation of the Capacity-building Initiative for Transparency referred to in paragraph 85 above starting in 2016;
90. Decides that, in accordance with Article 13, paragraph 2, of the Agreement, developing countries shall be provided flexibility in the implementation of the provisions of that Article, including in the scope, frequency and level of detail of reporting, and in the scope of review, and that the scope of review could provide for in-country reviews to be optional, while such flexibilities shall be reflected in the development of modalities, procedures and guidelines referred to in paragraph 92 below;
91. Also decides that all Parties, except for the least developed country Parties and small island developing States, shall submit the information referred to in Article 13, paragraphs 7, 8, 9 and 10, as appropriate, no less frequently than on a biennial basis, and that the least developed country Parties and small island developing States may submit this information at their discretion;
92. Requests the Ad Hoc Working Group on the Paris Agreement to develop recommendations for modalities, procedures and guidelines in accordance with Article 13, paragraph 13, of the Agreement, and to define the year of their first

Appendix B: The Paris Agreement on Climatic Change Conference at COP 21   361

and subsequent review and update, as appropriate, at regular intervals, for consideration by the Conference of the Parties, at its twenty-fourth session, with a view to forwarding them to the Conference of the Parties serving as the meeting of the Parties to the Paris Agreement for adoption at its first session;

93. Also requests the Ad Hoc Working Group on the Paris Agreement in developing the recommendations for the modalities, procedures and guidelines referred to in paragraph 92 above to take into account, inter alia:

    (a) The importance of facilitating improved reporting and transparency over time;
    (b) The need to provide flexibility to those developing country Parties that need it in the light of their capacities;
    (c) The need to promote transparency, accuracy, completeness, consistency, and comparability;
    (d) The need to avoid duplication as well as undue burden on Parties and the secretariat;
    (e) The need to ensure that Parties maintain at least the frequency and quality of reporting in accordance with their respective obligations under the Convention;
    (f) The need to ensure that double counting is avoided;
    (g) The need to ensure environmental integrity;

94. Further requests the Ad Hoc Working Group on the Paris Agreement, when developing the modalities, procedures and guidelines referred to in paragraph 92 above, to draw on the experiences from and take into account other on-going relevant processes under the Convention;

95. Requests the Ad Hoc Working Group on the Paris Agreement, when developing modalities, procedures and guidelines referred to in paragraph 92 above, to consider, inter alia:

    (a) The types of flexibility available to those developing countries that need it on the basis of their capacities;
    (b) The consistency between the methodology communicated in the nationally determined contribution and the methodology for reporting on progress made towards achieving individual Parties' respective nationally determined contribution;
    (c) That Parties report information on adaptation action and planning including, if appropriate, their national adaptation plans, with a view to collectively exchanging information and sharing lessons learned;
    (d) Support provided, enhancing delivery of support for both adaptation and mitigation through, inter alia, the common tabular formats for reporting support, and taking into account issues considered by the Subsidiary Body for Scientific and Technological Advice on methodologies for reporting on financial information, and enhancing the reporting by developing countries on support received, including the use, impact and estimated results thereof;

(e) Information in the biennial assessments and other reports of the Standing Committee on Finance and other relevant bodies under the Convention;
(f) Information on the social and economic impact of response measures;

96. Also requests the Ad Hoc Working Group on the Paris Agreement, when developing recommendations for modalities, procedures and guidelines referred to in paragraph 92 above, to enhance the transparency of support provided in accordance with Article 9 of the Agreement;
97. Further requests the Ad Hoc Working Group on the Paris Agreement to report on the progress of work on the modalities, procedures and guidelines referred to in paragraph 92 above to future sessions of the Conference of the Parties, and that this work be concluded no later than 2018;
98. Decides that the modalities, procedures and guidelines developed under paragraph 92 above, shall be applied upon the entry into force of the Paris Agreement;
99. Also decides that the modalities, procedures and guidelines of this transparency framework shall build upon and eventually supersede the measurement, reporting and verification system established by decision 1/CP.16, paragraphs 40–47 and 60–64, and decision 2/CP.17, paragraphs 12–62, immediately following the submission of the final biennial reports and biennial update reports;

Global Stocktake

100. Requests the Ad Hoc Working Group on the Paris Agreement to identify the sources of input for the global stocktake referred to in Article 14 of the Agreement and to report to the Conference of the Parties, with a view to the Conference of the Parties making a recommendation to the Conference of the Parties serving as the meeting of the Parties to the Paris Agreement for consideration and adoption at its first session, including, but not limited to:

   (a) Information on:

   (i) The overall effect of the nationally determined contributions communicated by Parties;
   (ii) The state of adaptation efforts, support, experiences and priorities from the communications referred to in Article 7, paragraphs 10 and 11, of the Agreement, and reports referred to in Article 13, paragraph 7, of the Agreement;
   (iii) The mobilization and provision of support;

   (b) The latest reports of the Intergovernmental Panel on Climate Change;
   (c) Reports of the subsidiary bodies;

101. Also requests the Subsidiary Body for Scientific and Technological Advice to provide advice on how the assessments of the Intergovernmental Panel on Climate Change can inform the global stocktake of the implementation of the Agreement pursuant to its Article 14 of the Agreement and to report on this matter to the Ad Hoc Working Group on the Paris Agreement at its second session;
102. Further requests the Ad Hoc Working Group on the Paris Agreement to develop modalities for the global stocktake referred to in Article 14 of the Agreement and to report to the Conference of the Parties, with a view to making a recommendation to the Conference of the Parties serving as the meeting of the Parties to the Paris Agreement for consideration and adoption at its first session;

Facilitating Implementation and Compliance

103. Decides that the committee referred to in Article 15, paragraph 2, of the Agreement shall consist of 12 members with recognized competence in relevant scientific, technical, socio-economic or legal fields, to be elected by the Conference of the Parties serving as the meeting of the Parties to the Paris Agreement on the basis of equitable geographical representation, with two members each from the five regional groups of the United Nations and one member each from the small island developing States and the least developed countries, while taking into account the goal of gender balance;
104. Requests the Ad Hoc Working Group on the Paris Agreement to develop the modalities and procedures for the effective operation of the committee referred to in Article 15, paragraph 2, of the Agreement, with a view to the Ad Hoc Working Group on the Paris Agreement completing its work on such modalities and procedures for consideration and adoption by the Conference of the Parties serving as the meeting of the Parties to the Paris Agreement at its first session;

Final Clauses

105. Also requests the secretariat, solely for the purposes of Article 21 of the Agreement, to make available on its website on the date of adoption of the Agreement as well as in the report of the Conference of the Parties at its twenty-first session, information on the most up-to-date total and per cent of

greenhouse gas emissions communicated by Parties to the Convention in their national communications, greenhouse gas inventory reports, biennial reports or biennial update reports;

Enhanced Action Prior to 2020

106. Resolves to ensure the highest possible mitigation efforts in the pre-2020 period, including by:
    (a) Urging all Parties to the Kyoto Protocol that have not already done so to ratify and implement the Doha Amendment to the Kyoto Protocol;
    (b) Urging all Parties that have not already done so to make and implement a mitigation pledge under the Cancun Agreements;
    (c) Reiterating its resolve, as set out in decision 1/CP.19, paragraphs 3 and 4, to accelerate the full implementation of the decisions constituting the agreed outcome pursuant to decision 1/CP.13 and enhance ambition in the pre-2020 period in order to ensure the highest possible mitigation efforts under the Convention by all Parties;
    (d) Inviting developing country Parties that have not submitted their first biennial update reports to do so as soon as possible;
    (e) Urging all Parties to participate in the existing measurement, reporting and verification processes under the Cancun Agreements, in a timely manner, with a view to demonstrating progress made in the implementation of their mitigation pledges;

107. Encourages Parties to promote the voluntary cancellation by Party and non-Party stakeholders, without double counting of units issued under the Kyoto Protocol, including certified emission reductions that are valid for the second commitment period;

108. Urges host and purchasing Parties to report transparently on internationally transferred mitigation outcomes, including outcomes used to meet international pledges, and emission units issued under the Kyoto Protocol with a view to promoting environmental integrity and avoiding double counting;

109. Recognizes the social, economic and environmental value of voluntary mitigation actions and their co-benefits for adaptation, health and sustainable development;

110. Resolves to strengthen, in the period 2016–2020, the existing technical examination process on mitigation as defined in decision 1/CP.19, paragraph 5 (a), and decision 1/CP.20, paragraph 19, taking into account the latest scientific knowledge, including by:
    (a) Encouraging Parties, Convention bodies and international organizations to engage in this process, including, as appropriate, in cooperation with

Appendix B: The Paris Agreement on Climatic Change Conference at COP 21   365

relevant non-Party stakeholders, to share their experiences and suggestions, including from regional events, and to cooperate in facilitating the implementation of policies, practices and actions identified during this process in accordance with national sustainable development priorities;
(b) Striving to improve, in consultation with Parties, access to and participation in this process by developing country Party and non-Party experts;
(c) Requesting the Technology Executive Committee and the Climate Technology Centre and Network in accordance with their respective mandates:

   (i) To engage in the technical expert meetings and enhance their efforts to facilitate and support Parties in scaling up the implementation of policies, practices and actions identified during this process;
   (ii) To provide regular updates during the technical expert meetings on the progress made in facilitating the implementation of policies, practices and actions previously identified during this process;
   (iii) To include information on their activities under this process in their joint annual report to the Conference of the Parties;

(d) Encouraging Parties to make effective use of the Climate Technology Centre and Network to obtain assistance to develop economically, environmentally and socially viable project proposals in the high mitigation potential areas identified in this process;

111. Encourages the operating entities of the Financial Mechanism of the Convention to engage in the technical expert meetings and to inform participants of their contribution to facilitating progress in the implementation of policies, practices and actions identified during the technical examination process;

112. Requests the secretariat to organize the process referred to in paragraph 110 above and disseminate its results, including by:

   (a) Organizing, in consultation with the Technology Executive Committee and relevant expert organizations, regular technical expert meetings focusing on specific policies, practices and actions representing best practices and with the potential to be scalable and replicable;
   (b) Updating, on an annual basis, following the meetings referred to in paragraph 112(a) above and in time to serve as input to the summary for policymakers referred to in paragraph 112(c) below, a technical paper on the mitigation benefits and co-benefits of policies, practices and actions for enhancing mitigation ambition, as well as on options for supporting their implementation, information on which should be made available in a user-friendly online format;
   (c) Preparing, in consultation with the champions referred to in paragraph 122 below, a summary for policymakers, with information on specific policies, practices and actions representing best practices and with the potential to be scalable and replicable, and on options to support their implementation,

as well as on relevant collaborative initiatives, and publishing the summary at least two months in advance of each session of the Conference of the Parties as input for the high-level event referred to in paragraph 121 below;

113. Decides that the process referred to in paragraph 110 above should be organized jointly by the Subsidiary Body for Implementation and the Subsidiary Body for Scientific and Technological Advice and should take place on an ongoing basis until 2020;
114. Also decides to conduct in 2017 an assessment of the process referred to in paragraph 110 above so as to improve its effectiveness;
115. Resolves to enhance the provision of urgent and adequate finance, technology and capacity-building support by developed country Parties in order to enhance the level of ambition of pre-2020 action by Parties, and in this regard strongly urges developed country Parties to scale up their level of financial support, with a concrete roadmap to achieve the goal of jointly providing USD 100 billion annually by 2020 for mitigation and adaptation while significantly increasing adaptation finance from current levels and to further provide appropriate technology and capacity-building support;
116. Decides to conduct a facilitative dialogue in conjunction with the twenty-second session of the Conference of the Parties to assess the progress in implementing decision 1/CP.19, paragraphs 3 and 4, and identify relevant opportunities to enhance the provision of financial resources, including for technology development and transfer and capacity-building support, with a view to identifying ways to enhance the ambition of mitigation efforts by all Parties, including identifying relevant opportunities to enhance the provision and mobilization of support and enabling environments;
117. Acknowledges with appreciation the results of the Lima-Paris Action Agenda, which build on the climate summit convened on 23 September 2014 by the Secretary-General of the United Nations;
118. Welcomes the efforts of non-Party stakeholders to scale up their climate actions, and encourages the registration of those actions in the Non-State Actor Zone for Climate Action platform;3
119. Encourages Parties to work closely with non-Party stakeholders to catalyse efforts to strengthen mitigation and adaptation action;
120. Also encourages non-Party stakeholders to increase their engagement in the processes referred to in paragraph 110 above and paragraph 125 below;
121. Agrees to convene, pursuant to decision 1/CP.20, paragraph 21, building on the Lima-Paris Action Agenda and in conjunction with each session of the Conference of the Parties during the period 2016–2020, a high-level event that:

    (a) Further strengthens high-level engagement on the implementation of policy options and actions arising from the processes referred to in paragraph 110 above and paragraph 125 below, drawing on the summary for policymakers referred to in paragraph 112(c) above;

Appendix B: The Paris Agreement on Climatic Change Conference at COP 21   367

(b) Provides an opportunity for announcing new or strengthened voluntary efforts, initiatives and coalitions, including the implementation of policies, practices and actions arising from the processes referred to in paragraph 110 above and paragraph 125 below and presented in the summary for policymakers referred to in paragraph 112(c) above;

(c) Takes stock of related progress and recognizes new or strengthened voluntary efforts, initiatives and coalitions;

(d) Provides meaningful and regular opportunities for the effective high-level engagement of dignitaries of Parties, international organizations, international cooperative initiatives and non-Party stakeholders;

122. Decides that two high-level champions shall be appointed to act on behalf of the President of the Conference of the Parties to facilitate through strengthened high-level engagement in the period 2016–2020 the successful execution of existing efforts and the scaling-up and introduction of new or strengthened voluntary efforts, initiatives and coalitions, including by:

(a) Working with the Executive Secretary and the current and incoming Presidents of the Conference of the Parties to coordinate the annual high-level event referred to in paragraph 121 above;

(b) Engaging with interested Parties and non-Party stakeholders, including to further the voluntary initiatives of the Lima-Paris Action Agenda;

(c) Providing guidance to the secretariat on the organization of technical expert meetings referred to in paragraph 112(a) above and paragraph 130 (a) below;

123. Also decides that the high-level champions referred to in paragraph 122 above should normally serve for a term of two years, with their terms overlapping for a full year to ensure continuity, such that:

(a) The President of the Conference of the Parties of the twenty-first session should appoint one champion, who should serve for one year from the date of the appointment until the last day of the Conference of the Parties at its twenty-second session;

(b) The President of the Conference of the Parties of the twenty-second session should appoint one champion who should serve for two years from the date of the appointment until the last day of the Conference of the Parties at its twenty-third session (November 2017);

(c) Thereafter, each subsequent President of the Conference of the Parties should appoint one champion who should serve for two years and succeed the previously appointed champion whose term has ended;

124. Invites all interested Parties and relevant organizations to provide support for the work of the champions referred to in paragraph 122 above;

125. Decides to launch, in the period 2016–2020, a technical examination process on adaptation;

126. Also decides that the technical examination process on adaptation referred to in paragraph 125 above will endeavour to identify concrete opportunities for strengthening resilience, reducing vulnerabilities and increasing the understanding and implementation of adaptation actions;
127. Further decides that the technical examination process referred to in paragraph 125 above should be organized jointly by the Subsidiary Body for Implementation and the Subsidiary Body for Scientific and Technological Advice, and conducted by the Adaptation Committee;
128. Decides that the process referred to in paragraph 125 above will be pursued by:

   (a) Facilitating the sharing of good practices, experiences and lessons learned;
   (b) Identifying actions that could significantly enhance the implementation of adaptation actions, including actions that could enhance economic diversification and have mitigation co-benefits;
   (c) Promoting cooperative action on adaptation;
   (d) Identifying opportunities to strengthen enabling environments and enhance the provision of support for adaptation in the context of specific policies, practices and actions;

129. Also decides that the technical examination process on adaptation referred to in paragraph 125 above will take into account the process, modalities, outputs, outcomes and lessons learned from the technical examination process on mitigation referred to in paragraph 110 above;
130. Requests the secretariat to support the technical examination process referred to in paragraph 125 above by:

   (a) Organizing regular technical expert meetings focusing on specific policies, strategies and actions;
   (b) Preparing annually, on the basis of the meetings referred to in paragraph 130(a) above and in time to serve as an input to the summary for policymakers referred to in paragraph 112(c) above, a technical paper on opportunities to enhance adaptation action, as well as options to support their implementation, information on which should be made available in a user-friendly online format;

131. Decides that in conducting the process referred to in paragraph 125 above, the Adaptation Committee will engage with and explore ways to take into account, synergize with and build on the existing arrangements for adaptation-related work programmes, bodies and institutions under the Convention so as to ensure coherence and maximum value;
132. Also decides to conduct, in conjunction with the assessment referred to in paragraph 120 above, an assessment of the process referred to in paragraph 125 above, so as to improve its effectiveness;
133. Invites Parties and observer organizations to submit information on the opportunities referred to in paragraph 126 above by 3 February 2016;

Non-party Stakeholders

134. Welcomes the efforts of all non-Party stakeholders to address and respond to climate change, including those of civil society, the private sector, financial institutions, cities and other subnational authorities;
135. Invites the non-Party stakeholders referred to in paragraph 134 above to scale up their efforts and support actions to reduce emissions and/or to build resilience and decrease vulnerability to the adverse effects of climate change and demonstrate these efforts via the Non-State Actor Zone for Climate Action platform4 referred to in paragraph 118 above;
136. Recognizes the need to strengthen knowledge, technologies, practices and efforts of local communities and indigenous peoples related to addressing and responding to climate change, and establishes a platform for the exchange of experiences and sharing of best practices on mitigation and adaptation in a holistic and integrated manner;
137. Also recognizes the important role of providing incentives for emission reduction activities, including tools such as domestic policies and carbon pricing;

Administrative and Budgetary Matters

138. Takes note of the estimated budgetary implications of the activities to be undertaken by the secretariat referred to in this decision and requests that the actions of the secretariat called for in this decision be undertaken subject to the availability of financial resources;
139. Emphasizes the urgency of making additional resources available for the implementation of the relevant actions, including actions referred to in this decision, and the implementation of the work programme referred to in paragraph 9 above;
140. Urges Parties to make voluntary contributions for the timely implementation of this decision.

Appendix C
China's Population Policy and Family Planning Law—An Unofficial Version

China: Law of 2001, Population and Family Planning Law

Publisher	National legislative bodies/national authorities
Publication Date	29 December 2001
Cite as	*China: Law of 2001, Population and Family Planning Law* [], 29 December 2001, available at: http://www.refworld.org/docid/42417cb54.html [accessed 22 August 2015]
Comments	This is an unofficial translation
Disclaimer	This is not a UNHCR publication. UNHCR is not responsible for, nor does it necessarily endorse, its content. Any views expressed are solely those of the author or publisher and do not necessarily reflect those of UNHCR, the United Nations or its Member States

Chapter I. General provisions

Article 1. This law is enacted, in accordance with the Constitution, so as to bring population into balance with social economic development, resources, and the environment: to promote family planning; to protect citizens' legitimate rights and interests; to enhance family happiness, and to contribute to the nation's prosperity and social progress.

Article 2. China is a populous country. Family planning is a fundamental state policy.

The State shall adopt a comprehensive approach to controlling population size and improving socio-economical and public health characteristics of population. The State shall rely on publicity and education, advances in science and technology, comprehensive services and the establishment and improvement of the incentive and social security systems to carry out the family planning program.

Article 3. Population and family planning programs shall act in concert with programs that expand women's educational and employment opportunities, enhance their health, and elevate their status.

Article 4. The People's Governments and staff at all levels implementing the family planning program shall act strictly within the law, enforcing it in a civil manner, and must not infringe on citizens' legitimate rights and interest.

The family planning administrative departments and their staff acting within the law are protected by law.

Article 5. The State Council shall exercise authority over the national population and family planning program. Local people's governments at all levels shall exercise authority over the population and family planning programs in their respective jurisdictions.

Article 6. The family planning administrative department of the State Council shall be in charge of the national family planning program and population programs related to family planning.

Family planning administrative departments of people's governments at county level and above shall be in charge of family planning programs and population programs related to family planning in their respective jurisdictions.

Other government administrative departments at county level and above shall be in charge of aspects of the population and family planning programs falling within their mandates.

Article 7. Social organizations such as Trade Unions, Communist Youth Leagues, Women's Federations, and Family Planning Associations; enterprises; institutions; and individual citizens shall assist the people's government in carrying out population and family planning programs.

Article 8. Organizations and individuals making outstanding achievements in the population and family planning programs shall be recognized and rewarded by the State.

Chapter II. Formulation and implementation of population development plans

Article 9. The State Council shall devise population development plans and incorporate them into the national economic and social development plans.

Based on plans at the next highest and national levels, people's governments at country level and above shall devise population development plans in line with local conditions and incorporate them into their economic and social development plans.

Article 10. People's governments at country level and above shall devise population and above shall be responsible for routine implementation of population and family planning action plans.

People's governments of township, ethnic township, and town, and subdistrict offices in urban areas, shall be in charge of population and family planning programs in their respective jurisdictions and shall implement population and family planning action plans.

Article 11. Population and family planning action plans shall stipulate measures to govern population size, strengthen maternal and child health care services, and improve socio-economical and public health characteristics of population.

Article 12. Villager's committees and neighbourhood committees shall follow the law and endeavor to implement family planning programs.

State organs, the armed forces, social organizations, enterprises, and institutions shall endeavor to implement their own family planning programs.

Article 13. Government departments responsible for family planning, education science and technology, culture, public health, civil affairs, press and publication, and broadcasting and television shall organize and carry out public education campaigns on population and family planning.

The mass media are obligated to carry pro bono public service messages on population and family planning.

Schools shall provide human physiology, adolescence, and sexual health education to students in a planned and audience-appropriate manner.

Article 14. Family planning programs among the migrant population shall be jointly managed by their local governments of origin and local governments of residence, with efforts focused where they reside.

Article 15. The State shall gradually increase the overall level of funding for population and family planning programs, based on the national economic and social development. People's governments at all levels shall provide the necessary funding for the implementation of their population and family planning programs.

People's governments at all levels shall give special support to population and family planning programs in impoverished areas and minority ethnic areas.

The State shall encourage social organizations, enterprises, institutions, and individuals to contribute support to population and family planning programs.

No unit or individual shall withhold, reduce or redirect funds appropriated to population and family planning programs.

Article 16. The State shall encourage scientific research and international exchange and cooperation in the fields of population and family planning.

Chapter III. Regulation of fertility

Article 17. Citizens have the right to reproduction as well as the responsibility for practicing family planning according to law. Husbands and wives bear equal responsibility for family planning.

Article 18. The State shall maintain its current fertility policy encouraging late marriage and childbearing and advocating one child per couple; arrangements for a second child, if requested, being subject to law and regulation. Specific measures shall be enacted by the People's Congress or its standing committee in each province, autonomous region, and municipality.

Ethnic minorities shall also practice family planning. Specific measures shall be enacted by the People's Congress or its standing committee in each province, autonomous region, and municipality.

Article 19. In implementing family planning, the primary emphasis shall be on contraception.

The State shall create conditions conducive to individuals being assured of an informed choice of safe, effective, and appropriate contraceptive methods. Safety or recipients of birth control procedures must be ensured.

Article 20. Couples of reproductive age shall be conscientious in adopting contraceptive methods and in accepting the guidance of family planning technical services.

Incidences of unwanted pregnancies shall be prevented and reduced.

Article 21. Couples of reproductive age who practice family planning shall be able to obtain technical services free of charge under the basic items as specified by the State.

The cost of the aforesaid services shall be itemized n public appropriations made in accordance with applicable State regulations or be guaranteed by social insurance plans.

Article 22. Discrimination against and mistreatment of women who give birth to female children or who suffer from infertility are prohibited. Discrimination against, mistreatment, and abandonment of female infants are prohibited.

Chapter IV. Incentives and social security

Article 23. The State shall, in accordance with regulations, recognize and reward couples who practice family planning.

Article 24. To facilitate family planning programs, the State shall establish and improve social security arrangements providing basic old-age insurance, basic medical insurance, childbearing insurance, and welfare benefits.

In rural areas where conditions are favorable, various types of old-age support schemes should be set up following the principle of government guidance and rural people's willingness.

Article 25. Citizens who marry late and delay childbearing are entitled to longer nuptial and maternity leaves or other welfare benefits.

Article 26. In accordance with applicable State regulations, women shall have special job-safety protections and be entitled to assistance and subsidies during pregnancy, delivery, and while breast-feeding.

Citizens who undergo surgical procedures for family planning shall be granted leave as stipulated by the State. Local people's governments may award them incentives.

Article 27. The State shall award a "Certificate of Honor for Single-Child Parents" to couples who volunteer to have one child in their lifetime.

Couples awarded such a certificate shall enjoy the incentives provided for in State regulations and in the regulations of their respective provinces, autonomous regions, or municipalities.

Employers shall obligatorily implement those incentive measures, stipulated by law and regulation, favoring couples in their employ who have one child over a lifetime.

Local people's governments shall provide necessary assistance to couples whose only child is disabled or killed in accidents, and who decide not to bear or adopt another child.

Article 28. Local governments at all levels shall give households that practice family planning preferential access to funding, technology, and training.

Households in poverty that practice family planning shall be given priority for poverty-alleviation loans, work relief, and other social assistance.

Article 29. The People's Congress and their standing committees in provinces, autonomous regions, municipalities and larger cities, or local people's governments, shall devise detailed implementation procedures for the incentives stipulated in this chapter in accordance with the provisions of this law and other applicable laws and regulations and in line with local conditions.

Chapter V. Family planning technical services

Article 30. The State shall establish premarital health care and maternal health care systems to prevent or reduce the incidence of birth defects and improve the health of newborns.

Article 31. People's governments at all levels shall take steps to assure all citizens access to family planning technical services in order to enhance their reproductive health.

Article 32. Local people's governments at all levels shall rationally allocate and coordinate the use of health resources; establish and improve family planning service networks comprising family planning clinics and health and medical institutions providing such services; upgrade facilities and improve the conditions under which care is provided; and raise the level of technical services.

Article 33. Family planning technical service institutions, medical and healthcare institutions providing family planning services shall, within the scope of their respective responsibilities, direct publicity and education about basic population and family planning information at people of reproductive age; provide pregnancy check-ups and follow-up for married women of reproductive age; offer counseling and guidance; and provide technical services in family planning and reproductive health.

Article 34. Family planning technical service providers shall give guidance to individuals in choosing safe, effective, and appropriate contraceptive methods. Couples who have given birth are encouraged to choose long-acting contraceptive methods.

The State shall encourage the research, development, and promotion of new family planning technologies and products.

Article 35. Use of ultrasonography or other techniques to identify fetal gender for non-medical purposes is strictly prohibited. Sex-selective pregnancy termination for non-medical purposes is strictly prohibited.

Chapter VI. Legal liability

Article 36. Anyone who violates this law by one of the following acts shall be ordered to rectify the violation and warned by the family planning or public health agency, and all gins derived from such illegal acts shall be confiscated by the family planning or public health administrative departments.

If the illegal gains exceed RMB 10,000, a fine of no less than two times and no more than six times the amount shall be imposed. If no illegal gains is involved or the amount is less than RMB 10,000, a fine of no less than RMB 10,000 and no more than RMB 30,000 shall be imposed. In serious cases, licenses shall be revoked by the issuing agency. Acts constituting a crime shall be referred for criminal prosecution.

1. Illegal performance of a surgical procedure related to family planning at another's behest.
2. Use of ultrasonogrphy or other techniques to identify fetal gender for non-medical purposes or sex-selective pregnancy termination for non-medical purposes, at another's behest.
3. Faking a birth control procedure related to family planning, falsifying a medical report, or counterfeiting certificates related to family planning.

Article 37. Anyone who forges, alters, buys or sells certificates related to family planning shall have the illegal gains confiscated by the family planning administrative departments. If the illegal gains exceed RMB 5000, a fine of no less than two times and no more than ten times the amount shall be imposed. If no illegal gains is involved or the amount is less than RMB 5000, a fine of no less than RMB 5000 and no more than RMB 20,000 shall be imposed. Acts constituting a crime shall be referred for criminal prosecution.

The family planning administrative departments shall render void improperly-obtained certificates related to family planning. Administrative penalties shall be imposed on both the executive in charge of the agency issuing flawed certificates and the individuals directly responsible.

Article 38. Family Planning service providers who commit malpractice or who delay emergency response, diagnosis or treatment with dire results shall be held liable under the applicable laws and regulations.

Article 39. Staff of state organs who commit one of the following acts in the course of family planning activities shall, if the act constitutes a crime, be referred for criminal prosecution; or, if the act does not constitute a crime, be subject to both administrative penalties and confiscation of any illegal gains.

1. Infringing on a citizen's personal rights, property rights or other legitimate rights and interests.
2. Abuse of power, dereliction of duty or graft.
3. Seeking or accepting a bribe.
4. Withholding, reducing, redirecting or embezzling family planning program funds or social compensation fees.
5. Distorting, under-reporting, fabricating, modifying or refusing to report statistical data on population or family planning.

Article 40. Those who violate provisions of this law or are derelict in family planning program management shall be ordered to rectify the violation and rebuked in a circular by their local government. Administrative penalties shall be imposed on both the executive in charge of the agency and the individuals directly responsible.

Article 41. Citizens who give birth not in accordance with the stipulations in Article 18 shall pay a social compensation fee prescribed by this law. Those failing to pay the full amount before the due date shall be levied a late payment penalty specified in applicable State regulations. Those who persist in non-payment shall be sued for payment in People's Court by the family planning administrative departments that levied the social compensation fee.

Article 42. The state employees levied the social compensation fee described in Article 41 shall be subject to additional administrative penalties, according to law. Others levied such a fee shall be subject to additional disciplinary measures imposed by their employing units.

Article 43. Those who resist or hinder family planning administrative departments and staff in the performance of their legitimate duties shall be subject to criticism and ordered to amend their conduct by the family planning administrative departments involved. Conduct breaching public security regulations shall be subject to public security penalties. Acts constituting a crime shall be referred for criminal prosecution.

Article 44. Citizens, entities treated as legal persons or other organizations deeming that an administrative organ has infringed on their legitimate rights and interests while implementing family planning policy may appeal for review or sue for redress.

Chapter VII. Supplementary provisions

Article 45. The State Council shall devise specific measures for managing family planning program among migrants, specific measures for managing family planning technical services, and measures for the administration of collecting social compensation fees.

Article 46. Detailed measures for implementing this law by the Chinese People's Liberation Army shall be devised by the Central Military Commission in accordance with this law.

Article 47. This law shall enter into effect on 1 September 2002.

Source: Downloaded website of UNHCR dated 02.08.2016—reference: refworld.[2]

Extracts from Documents of Population and Development Review, Volume 42, Number 4, December 2015

China's Abandonment of the One-Child Policy

The Fifth Plenary Session of the 18th Central Committee of the Communist Party of China was held in Beijing over 26–29 October 2015. The main business of the

---

[2] No copyright conditionality

Session was to consider the draft of the Thirteenth Five Year Plan, set to begin in 2016. At the meeting's conclusion, the Committee issued a brief communiqué summarizing the principles and goals of the Plan and, a few days later, a fuller report covering the same ground. A translation captured in Population and Development Review noted the intention to 'universally implement the that a couple can have two children'. This sentence was in effect the first public announcement of the ending after some 35 years, of China's one-child policy.

The reform is presented as a shift from a one-to a two-child limit on births, with the implication that the same family planning bureaucracy as before would continue to oversee birth planning.

Appendix D
Quotations on Population Stabilisation

Quotes:

- Unlike plagues of the dark ages or contemporary diseases we do not understand, the modern plague of overpopulation is soluble by means we have discovered and with resources we posses. What is lacking is not sufficient knowledge of the solution but universal consciousness of the gravity of the problem and education of the billions who are its victim.

—Martin Luther King, Jr., civil rights leader and Nobel laureate

- The point of population stabilization is to reduce or minimize misery.

—Roger Bengston, founding board member, World Population Balance

- Pressures resulting from unrestrained population growth put demands on the natural world that can overwhelm any efforts to achieve a sustainable future. If we are to halt the destruction of our environment, we must accept limits to that growth.

—World Scientists' Warning to Humanity, signed by 1600 senior scientists from 70 countries, including 102 Nobel Prize laureates

- If we don't halt population growth with justice and compassion, it will be done for us by nature, brutally and without pity- and will leave a ravaged world.

—Nobel Laureate Dr. Henry W. Kendall

- When the family is small, whatever little they have they are able to share. There is peace.

—Philip Njuguna, pastor, Nairobi, Kenya

- Once it was necessary that the people should multiply and be fruitful if the race was to survive. But now to preserve the race it is necessary that people hold back the power of propagation.

—Helen Keller, world-renowned deaf and blind author and lecturer

- The key problem facing humanity in the coming century is how to bring a better quality of life – for 8 billion or more people – without wrecking the environment entirely in the attempt.

—Edward O. Wilson, scientist, Pulitzer prize winning author and father of biodiversity

- If the world is to save any part of its resources for the future, it must reduce not only consumption but the number of consumers.

—B.F. Skinner, psychologist and author (Introduction to Walden Two, 1976 edition)

- 'Smart growth' destroys the environment. 'Dumb growth' destroys the environment. The only difference is that 'smart growth' does it with good taste. It's like booking passage on the Titanic. Whether you go first-class or steerage, the result is the same.

—Dr. Albert A. Bartlett, Emeritus Professor of Physics, University of Colorado; World Population Balance Board of Advisors

- We must stabilize population. This will be possible only if all nations recognize that it requires improved social and economic conditions, and the adoption of effective, voluntary family planning.

—Lester Milbrath, professor emeritus and author, Learning to Think Environmentally (While there is Still Time)

- Which is the greater danger - nuclear warfare or the population explosion? The latter absolutely! To bring about nuclear war, someone has to DO something; someone has to press a button. To bring about destruction by overcrowding, mass starvation, anarchy, the destruction of our most cherished values-there is no need to do anything. We need only do nothing except what comes naturally - and breed. And how easy it is to do nothing.

—Dr. Isaac Asimov, biochemist and science writer (in this 1966 interview he predicted that world population would reach 6 billion around 2000. Most leaders dismissed his prediction as outrageous. Population passed 6 billion in 1999.)

- The greatest shortcoming of the human race is our inability to understand the exponential function.

—Dr. Albert A. Bartlett, Emeritus Professor of Physics, University of Colorado; World Population Balance Board of Advisors

Index

A
Absolute employment, 44
Acceptor data, 179
Access to information/counselling, 158
Accountable and inclusive institutions, 228, 240
Accredited Social Health Activists (ASHA), 302, 306, 307
Acquired Immunodeficiency Syndrome (AIDS), 159, 259
Action plan, 269, 271
Action taken report, 213
Activities
  at international level, 166
  at national level, 166
Adaptation, 220, 230, 237, 238, 244
Adolescents, 165
  care for, 106
Age-mix, 333
Agenda for population stabilisation, 261, 268, 271
Age of marriage, 68, 70, 74, 123, 158, 256
Aids to family welfare, 188
Anaemia, 37
Analysis and dissemination, 166
Annual Sentinel Surveillance for HIV Infection, 311
20A of the concurrent list, 198
Article 243G, 194, 195
Article 243W, 194, 195
Audio visual aids, 188
Auxiliary Nursing Midwifery (ANM), 300, 302, 303, 307
Availability of geographical, 330

B
Balika Samridhi Yojana, 270
Basic data collection, 166
Basket of choices, 259

Below poverty line, 27
Better health for all, 248
Bhore committee, 255
BIMARU, 333
Bio capacity, 222, 223
Biodiversity, 222, 239, 240
  loss, 228, 239
Birth
  avertion, 327
  interval, 179, 318, 320
  order, 318
  rate, 6, 7
  spacing, 98, 102, 123, 158
Block Health & Family Welfare Samity, 195
Block panchayat, 196
Block wise and municipality wise data, 184
BMI, 37
BPL, 32
  families, 309
  population, 328
Breast feeding, 102–104, 107, 108, 123, 211, 212
Buddhist, 173
Build resilient infrastructure, 228, 235
Burgeoning number, 333

C
Calorie intake, 34
Carrying capacity, 218, 220, 221, 328, 330, 333
Census, 2011, 10, 15, 21, 24
Census Act, 256
Census Report, 2011, 271
Central assistance, 256
Centrally sponsored scheme, 198, 297, 310
Central plan allocations, 256
Reproductive, Maternal, New born, Child and Adolescent Health Services (RMNCH + A), 301, 305

Child care centres, 270
Child care issues including child nutrition, 106
Child health, 258, 259, 261–265, 269, 271
Child Marriage Restraint Act, 1976, 265, 270
Child mortality, 167–169, 224, 225
Child Mortality Rate (CMR), 178
Child nutrition, 34, 35
Child population by sex and religion, 183, 184
Child sex ratio, 172
Child survival, 165
Christian, 173
Civilization crisis, 336
Civil registration system, 66, 67
Civil societies, 196
Clarion call, 329
Climate change, 52, 218, 220, 227–230, 237, 238, 243, 244
Combat desertification, 228, 239, 240
Command system, 313
Commercial crops, 24
Communicable diseases, 159, 259
Communication, 264
Community incentive scheme, 310
Community-level health care services, 270
Compound annual growth rate, 252
Compulsory birth planning system, 313
Conceptual clarifications, 177
Concurrent evaluation, 213
Concurrent list, 21
42nd Constitutional Amendment, 268, 270
Constitutional provisions, 132
Constitution of India, 21
Consumption expenditure, 26, 27
Contraception, 158, 258–261, 269
  issues, 90, 92
  prevalence rate, 179
  technology, 266
Convergence, 259, 261, 271
  of health related interventions, 308
Coordination cell, 269
Co-partner for population control, 328
Copper-T, 299
Cost of living index, 27
Counselling services, 262, 270
Counselling set-up, 92, 123
Couple currently and effectively protected, 325
Couple Protection Rate (CPR), 179, 184–186, 258, 327
Crèches, 270
Crude Birth Rate (CBR), 58, 64, 177, 184, 253, 258, 316, 328, 332, 333
Crude Death Rate (CDR), 178, 255, 314

Cultural issues, 71
Current status of human development, 331

D

Data structure, 212
Death rate, 6, 7, 184
Deaths, 158, 159, 255, 259–262
Decadal growth
  of muslim population, 21
  incremental size of population, 10
  increment of, 184, 185
  rate, 183–185, 328
  rate of SC population, 173
  of ST population, 173
Decentralisation, 194–196
Decentralised planning, 261
Decentralized governance, 193
Declaration at Alma Ata in 1978, 248
Deconcentration, 194
Delayed marriage, 158, 259
Delegation, 194, 195
Deliveries by trained persons, 158
Delivery Points (DPs), 305
Demographic
  challenges, 171
  data, 58, 59, 172, 183–185
  disaster, 336
  dividend, 333
  indicators, 314, 328
  issues, 58
Demographically weaker states, 210, 211
Density of population, 15, 172, 182–184
Department of Health Research, 189
Departments of Health and Family Welfare, 189
Dependency ratio, 179
Devolution, 194, 195
Devolved functional responsibilities, 195, 196
Directorate of Health Services, 190
Distribution of population, 183, 184
District Family Welfare Bureau, 297
District Health & Family Welfare Samity, 195
District level data, 183–185
Diverse Health Care Providers, 265
DLHS III, 327
Documented migrants, 164, 166
Dowry, 70, 75, 136–139
Dowry Prohibition Act, 70, 136, 137, 140
Draft National Health Policy, The, 254

E

EAG states, 333
Early neonatal mortality rate, 178
Earth-space of India, 337

Index 383

Ecological economics, 222
Ecological footprint, 222, 223
Ecology, 222
Economic and social life, 331
Economic resources, 333
Economy, 222
Education, 188, 255, 258, 261–265, 268, 271
Education and communication issues, 122
Effective couple protection rate, 179
Effect of economic growth, 6
Effect of growing population, 6
Eight five year plan, 286
Eleventh five year plan, 290, 292
Eleventh schedule, 194
Eligible Couple and Children Register, 186, 210, 212
Eligible couples, 184, 185
Emergency contraceptive pills, 299, 300
Emergency method, 299
Employment, 228, 234
Employment scenario, 328
Empowered Action Group (EAG), 188, 301, 309, 310, 333
Empowerment of women, 162, 165, 167, 224, 225
End hunger, 218, 229
Energy requirement, 34
Environmental and climatic scenario, 54
Environmental protection, 259
Environmental sustainability, 167, 222
Equity, 218, 222
Establishment and Maintenance of Rural Family Planning Sub-centres, 297
Establishment and Maintenance of Sterilisation Beds, 298
Establishment and Maintenance of Urban Family Planning, 298
Estimated children, 186
Estimated eligible couples, 186
Estimated utilizable water, 53
Expert group, 257
Extension education, 188
Extent of coverage, 318, 322
External assistance, 188
Extreme poverty and hunger, 167

F

Family planning, 4, 162, 163, 165, 167
   indicators, 314, 318
   for male, 76
   services, 158
   welfare data, 185, 194, 196
   for women, 79

Family Planning Cell at the State Secretariat, 297
Family welfare, 194, 196
   health insurance plan, 270
   matters, 188–190, 192
   schemes, 188
   statistics, 316, 327
   structure, 187, 188, 190, 191
Famine, 4, 5
FAO, 34
Female education, 256
Fertility regulation, 158
Field inspection, 213
Fifth five year plan, 279
Financial monitoring, 210
First five year plan, 274, 275
Food grain production scenario, 22, 26
Food production, 4, 5, 7
Food security, 228, 229
Forest coverage area, 53
Forest resources, 52
Foster innovation, 228, 235
Fourth five year plan, 278
Free diagnostics service initiative, 305
Free drugs and free diagnostic service, 305
Free drug service initiative, 305
Free supply scheme, 299
Freezing the population base, 256
Full employment, 335
Fundamentals, 331
Future generations, 218, 220–223, 243

G

GDP, 24, 34
Gender equality, 162, 165, 167, 224, 225, 228, 232
Gender inequalities, 265
Gender issues, 75
Gender mainstreaming, 254
Geographical area, 51, 53
Girl child, 71, 73, 74, 123, 162, 163, 165
Global CO_2 emission, 54
Global Footprint Network, 222, 336
Global partnership, 162, 167, 224, 228, 229, 242
Global warming, 54
Globe's land area, 257
Goal, 228
Gram panchayat, 195
Green house gases, 54
Group incentives, 256
Growth
   of absolute number, 10

Growth (cont.)
    of food grains, 24
    rate, 179

H
HDI, 47
Head count ratio, 27, 30
Health agenda, 335
Health care infrastructure, 259, 261, 262, 265, 269
Health–care sector, 165
Health issues, 90, 99, 149
Health personnel, 259
Health sector, 247, 248, 250, 252, 254
Health Survey and Development Committee, 255
Healthy lives, 228, 230
Height for age, 35
Higher order live births, 320
Higher training, 188
Hill areas, 264
Hindustan Latex Limited, 188
HIV/AIDS, 69, 78, 92, 97, 104, 106, 167, 224, 226
Homeopathy, 266
Hospital, 321
Housing & Urban Poverty Alleviation, 307
Huge size of poverty, 333
Human development, 4, 6, 328
    indicators, 269
    profile, 47
    report, 47
Human resource development, 166
Human Resource Women & Child Development, 307
Human settlements, 228, 236
Human sexuality, 165

I
Immediate supervisory authority, 212
Immunisation, 211, 212
Immunisation programme, 264
Immunisation coverage, 186
Improved access, 158
Improved life expectancy, 267
IMR over the years for Rural and Urban years, 184
Incentives versus disincentives, 153
Increased participation of men, 265
Incremental per year population, 331
Indian systems of medicine, 259, 266
Inequality, 235
Inequities in health outcomes, 253

Infant Mortality Rate (IMR), 178, 258, 314, 333
Infiltration, 210, 211
Influx, 210, 211
Information
    education and communication, 267
    in relation to population and family welfare, 188
Informed consent, 270
In-service training, coordinating, 311
Institutional deliveries, 158, 259
Integrated service delivery, 259, 262, 264
Integration of Indian Systems of Medicine (ISM), 159
Internal migration, 166
International conference, 248
International Conference on Population and Development(ICPD), 161
International cooperation, 166
International Institute for Population Sciences (IIPS), 188, 312
International migration, 166
Inter-sectoral coordination, 188
IPAT equation, 222
IUD insertion, 310

J
Jain, 173
Janani Shishu Suraksha Karyakarm (JSSK), 303
Janani Suraksha Yojana(JSY), 303
Jansankhya Sthirata Kosh (JSK), 188, 308
Justice, access to, 240

L
Lakdawala methodology, 29
Land degradation, 228, 239
Lassiaze faire approach, 328
Legal age of marriage, 83, 132
Legal issues, 123, 131
Legislation, 268
Legislature of a State, 194
Life expectancy, 258
Life support systems, 337
List of business, 188
Literacy growth rate, 185
Literates, 183–185
Literates and Literacy Rates by sex/SC/ST/Religion, 183, 184
Local self-governments, 195, 196

M
Maharashtra Family Act, 1976, The, 148
Malaria, 167, 224, 226, 230

Index 385

Male and female population, 183–185
Male sterilisation, 158
Malthu, 4, 5
Malthusian theory of population, 4, 5
Marginalized communities, 253
Marriage and pregnancy, 159, 259, 265
Maternal, 301, 303–305
Maternal health, 167, 169, 224, 225
Maternal mortality rate, 178
Maternal mortality ratio, 178, 253, 259, 262, 314
Maternity benefit scheme, 270
Media issues, 121
Medical attention, 318, 321
Medical Termination Of Pregnancy Act, 1971, 146, 150
Medium term objective, 259
Methods of family planning, 258, 318, 322
Migrant population, 264
Migration, 184
Migration related data, 183
Millennium development, 167
Millennium Development Goals (MDGs), 167, 224, 227
Ministry of Health and Family Welfare, 187, 189, 191, 192, 327, 328
Ministry of Home Affairs, The, 328
Minoritism, 83
Miscellaneous business, 188
Mobility of ANMs, 270
Monetary compensation, 256
Monitoring, 209–213
  authority, 211, 213
  indicators, 211
  items, 210
  mechanism, 211
  meeting, 213
  of birth spacing, 212
  report, 213
  responsibility, 211
  review, 212
  technique, 213
Mother and Child Health Wings (MCH), 304
Mother and Child Tracking Facilitation Centre (MCTFC), 306
Mother and Child Tracking System (MCTS), 304, 306
Mudaliar Committee of 1962, The, 248
Municipalities, 194, 196
Muslim population, 173
Muslim population in India, 19
Muslim rate of growth more than Hindu rate of growth, 15

N
National AIDS control organisation, 159
National average, 172, 173
National Commission on Population, 188, 268
National Development Committee, 257
National Family Welfare programme, 299, 310
National Health Mission (NHM), 301, 305–307, 328
National Health Policy, 158
National Institute of Health and Family Welfare (NIHFW), 188, 306, 311
National Iron + Initiative, 305
National Mobile Medical Units (NMMU), 303
National policies and plans of action, 166
National Population Commission (NPC), 311, 327
National Population Policy, 157, 219, 247, 250, 255–258, 268, 271
National Population Stabilization Fund (NPSF), 308
National Rural Health Mission (NRHM), 188, 189, 301–303, 305, 308, 311
National socio-demographic goals, 259, 261
National Urban Health Mission (NUHM), 301, 307, 308
Natural growth rate, 184
Natural resources, 331, 333
Natural sex ratio, 172, 173
Natural stock of resources, 328
NDC Committee on Population, 199
NDC report, 198, 199
Necessary condition, 330, 333
Neo-natal care, 263
Neonatal mortality rate, 178
New born, 305
New framework, 329
New multi-media national strategy, 256
NFHS-1, 35, 327
NFHS-2, 35, 327
NFHS-3, 35
NGO guidelines, 306, 307
NHP, 1983, 248
Ninth five year plan, 287, 288
Non-agriculture, 24
Non-coercive approach, 158, 335
Non-Government Organisations (NGOs), 196, 265
Non-renewable resources, 336
NSS, 27, 30, 35, 41
Number of births, 322
Number of married couples, 184
Nutritional services, 265
Nutritional status, 34, 35, 37, 41
Nutrition profile, 328

O

Older population, 267
Optimum population, 5, 218, 219, 243
Optimum theory of population, 5
Oral contraceptive pills, 298–300
Organisation, 188–190, 192
Organisational issues, 149
Organisational set-up, 150, 151
Output food grains, 24
Outreach, 254
Over population, 218

P

Panchayats, 194, 195, 261, 262, 269
Parliamentary Committee, 336
Partnership with the Non-Government Sector, 166
Per capita, 26–28, 32, 34, 47, 53
Percentage distribution, 319–321
Percentage growth of food grains production, 26
Percentage growth of yield, 24
Percentage of FW expenditure, 206
Perinatal mortality rate, 178
Permanent methods, 298, 299
Physical limits, 337
Plan and non-plan expenditure, 199
Plan expenditure, 198
Planned parenthood, 265
Planning commission, 26, 28, 29, 32, 34, 41
12th Plan document, SRS 2013, 199
Plan-wise expenditure, 199
Policy and organisation of Family Welfare, 188
Political issues, 119
Population, 3–7
    and development, 330
    clock, 330
    control, 3, 4, 254
    distribution, 166
    education, 119, 120, 126
    growth and family planning, 21
    growth and structure, 165
    load, 330, 331, 335
    matters, 221, 222
    of Muslims and other minorities, 183, 184
    of SC and ST, 183, 184
    policy, in India, 255
    projection, 59, 258
    scenario, 206
    stabilization, 254, 333, 335
    status, 9
    watch, 328
Post neonatal mortality rate, 178
Post partum care/post natal care, 101

Poverty, 26–30, 32, 34, 35
    end to, 224, 227, 243
    estimation, 27, 32
    line, 26–30, 32
    ratio, 32
Pre-conception and Pre-natal Diagnostic Technique Act, 188
Preconception and Pre-natal Diagnostic Test Act, 140
Pregnancy care/prenatal care, 99
Pregnancy test kits, 299
Pre-natal Diagnostic Techniques Act, 1994, 270
Prerna strategy, 308
Prevention of human immunodeficiency virus (HIV), 165
Primary health care, 165
Private sector, 265, 266
Production patterns, 228, 237
Programme implementation, 261
Programme Implementation Plans (PIP), 308
Programme management, 166
Projected population, 183, 184
    and actual population in India, 315
    size, 335
Promote sustained, 228, 234
Promotional and motivational measures, 269
Proportion
    of muslim population, 19
    to total population, 183, 184
Protect, 228, 231, 233, 234, 236, 238, 239, 241
Public expenditure, 200, 201
Public private partnership, 188
Public support and new structures, 268

Q

Qualified professional, 321
Quality Assurance (QA), 306
Quality education, 228, 231
Quality of life, 9, 41, 218, 219, 221–223, 257
Quick fix, 337

R

Rashtriya Bal Swasthya Karyakram (RBSK), 304
Rashtriya Kishor Swasthya Karyakram (RKSK), 304
Reduction of drop outs, 158
Reduction of Infant Mortality Rate (IMR), 158, 308, 310
Reduction of Maternal Mortality Rate (MMR), 158, 308
Refugees
    asylum-seekers, 166

displaced persons, 166
Regional Family Planning Training Centres, 297
Registration
    of births, 159, 259, 261
    of marriage, 133, 134, 136
Reliable, 228, 233, 235
Religious agenda, 328
Religious issues, 83
Renewable resources, 336
Replacement level population, 333
Replacement levels of TFR, 159
Report card of development, 228
Reported coverage of Eligible Couples and its Percentage, 184, 185
Report of working group, 327
Reproduction, 298, 301, 304, 305
    and child health care, 188, 259, 261–264, 266
    health, 162, 163, 165
    health research, 166
    phase, 335
    rights, 165
    tract illness, 254
Reproductive Tract Infection (RTI), 103, 159
Research on Reproductive and Child health, 266
Resilient and sustainable, 228, 236
Resource mobilization and allocation, 166
Resources scenario, 52
Revisits the policy area, 272
Revolving fund, 270
Rogi Kalyan Samiti, 302
Runaway population, 328
Rural, 26, 27, 30, 32, 34, 45
Rural and Urban Composition of Population, 183, 184

S
Safe, 228, 229, 231–234, 236, 237
Safe abortion services, 254
Safe motherhood, 165
Sample registration system, 66
Santushti strategy, 308
School education, 158, 259
SDG platform, 336
SDGs journey, 335
Secondary and tertiary facilities, 264
Second five year plan, 275, 276
Service delivery, 254
Services for fertility regulation, 259
Seventh five year plan, 282, 284
Seventy Third and the Seventy Forth Amendments, 194

Sex ratio, 172, 173, 179, 183, 184, 253
    at Birth, 179
    of SC/ST population, 173
Sexually Transmitted Infection (STI), 103, 159, 165
Shrivastav Committee of 1975, The, 248
Sikh, 173
Sixth five year plan, 280, 281
Size of population, 33, 41, 52, 54, 171, 182–185, 330, 331, 333, 336
Small family norm, 159, 256, 257, 259–261, 265, 268–270
Social and economic research, 166
Social development, 259
Social issues, 68, 154
Social marketing schemes, 264, 270, 299
Socio-religious Issues, 82
Soft voluntary approach, 333
Sons preference, 75
Spacing methods, 298, 300
Spell of population boom, 335
Stable population, 179, 219, 243, 259, 333, 335
State average, 173
State Family Welfare Bureau, 297
State Health & Family Welfare Samity, 195
State Health Society, 301, 308
State level data, 182, 183, 185
Stationary population, 179
Still birth, 178
Structure of monitoring format, 213
Sub-centre, 196
Sub-district level data, 185
Sub regional and regional activities, 166
Supplementary nutrition, 211, 212
Sustainability, 221, 222, 237, 241
    equation, 222
    agenda, 328
    agriculture, 229
    consumption, 228, 237
    development, 162–164, 166, 220–223, 227, 228, 232, 237, 238, 240–244, 329, 330
    development summit, 272
    economic growth, 162, 164, 228, 234, 243, 259
    forest management, 228, 240
    global partnership for development, 241
    industrialization, 228, 235
    management, 228, 233
    and modern energy, 228, 233
    population, 218–220, 244, 272, 330, 333, 335–337
Sustainable development goals, 218, 227, 229, 242, 243, 331
Sustainable use

of oceans, 228, 238
of terrestrial ecosystems, 228, 239
Swaminathan, M.S. Dr., 257

T
Take off, 335
Target-free approach in family planning, 211
Technological level of achievement, 333
Technology mission, 269
Tendulkar committee, 29, 30
Tenth five year plan, 288
TFR over the years, 184
Theory of demographic transition, 6, 7
Third five year plan, 276
Total expenditure, 206
Total Fertility Rate (TFR), 65, 177, 253, 254, 256–260, 271, 314, 317, 332, 335
Total Marital Fertility Rates (TMFR), 319
Total population, 15, 183, 184
Training and research, 188
Transforming our world
  the 2030 Agenda for Sustainable Development, 329
Trends
  in census population, 183, 184, 314
  of proportion of muslim population, 19
Tribal communities, 264
Tubectomy, 308, 310
Twelfth five year plan, 292, 293
Twelfth schedule, 194

U
UN's Sustainable Summit, 2015, 336
UN, 218, 229
Under-5 child mortality rate, 184
Under population, 218
Underweight children, 35
Undocumented migrants, 164, 166
Un-employment rate, 41, 45
Unemployment scenario, 41
UNICEF, 248
Union business, 188
Universal access to information/counselling, 259
Universal access to reproductive health care, 308
Universal awareness, 258
Universal Health Coverage (UHC), 305
Universal immunization, 158
Universal immunization of children, 259

Universal primary education, 167, 224
Unmet needs
  of monitoring, 210
  for contraception, 259, 319, 327
  for family welfare services, 264
UN Millennium Summit, 223
UN population division, 330
UN projection, 331
Unprotected couples, 327
Untrained functionary, 321
Urban, 26, 27, 32, 34, 45
Urban Community Health Centre (U-CHC), 307
Urbanization, 166
Urban Primary Centre (U-PHC), 307
Urban-rural inequities, 253
Urban slums, 264, 270
User data, 179

V
Vasectomy, 299, 300, 308, 310
Vertical, 194
Village Health Sanitation and Nutrition Committee (VHSNC), 302
Vital statistics, 184, 185
Vocational training schemes for girls, 270
Voluntary efforts, 257
Voluntary family planning, 313
Vulnerable groups, 253

W
Water demand utilimation, 54
Water resource, 52, 53
Weekly Iron and Folic Acid Supplementation (WIFS), 305
Weight for age, 35
Weight for height, 35
Women's health, 165, 254
Women nutrition, 34, 37
Working population, 26
World's surface area, 330
World Health Organization, 248
World population, 330, 336
World summit, 220

Z
ZilaParishads, 269
Zilla Panchayat, 196

CPSIA information can be obtained
at www.ICGtesting.com
Printed in the USA
BVOW06*0839311016
466489BV00003B/12/P